Conceptual Modeling for Discrete-Event Simulation

Edited by

Stewart Robinson
University of Warwick
Coventry, UK

Roger Brooks
University of Lancaster
UK

Kathy Kotiadis
University of Warwick
Coventry, UK

Durk-Jouke van der Zee
University of Groningen
The Netherlands

CRC Press
Taylor & Francis Group
Boca Raton London New York

CRC Press is an imprint of the
Taylor & Francis Group, an **informa** business
A CHAPMAN & HALL BOOK

First published in paperback 2024

First published 2011
by CRC Press
2385 NW Executive Center Drive, Suite 320, Boca Raton FL 33431

and by CRC Press
4 Park Square, Milton Park, Abingdon, Oxon, OX14 4RN

CRC Press is an imprint of Taylor & Francis Group, LLC

Publisher's Note
The publisher has gone to great lengths to ensure the quality of this reprint but points out that some imperfections in the original copies may be apparent.

Library of Congress Cataloging-in-Publication Data

Conceptual modeling for discrete-event simulation / editors, Stewart Robinson ... [et al.].
p. cm.
Includes bibliographical references (p.) and index.
ISBN 978-1-4398-1037-8 (hardcover : alk. paper)
1. Computer simulation. 2. Simulation methods. I. Robinson, Stewart, 1964- II. Title.

QA76.9.C65C6566 2011
003.3--dc22 2010021877

ISBN: 978-1-4398-1037-8 (hbk)
ISBN: 978-1-03-291906-5 (pbk)
ISBN: 978-0-429-14886-6 (ebk)

DOI: 10.1201/9781439810385

Visit the Taylor & Francis Web site at
http://www.taylorandfrancis.com

and the CRC Press Web site at
http://www.crcpress.com

Contents

Part III Soft Systems Methodology for Conceptual Modeling

Part IV Software Engineering for Conceptual Modeling

Part V Domain-Specific Conceptual Modeling

Part VI Conclusion

Preface

Simulation models are abstractions of a real-world, or proposed real-world, system. In other words, the models do not contain every aspect and detail of the system that they are attempting to imitate. A model of military combat might range from a high-level representation to a very detailed model. In the former, much detail will be either omitted from the model or subsumed into high-level concepts. Meanwhile, a detailed model might focus on a specific aspect of combat, and even then not all details will be included, either because there is insufficient knowledge about them or because it is not deemed necessary to model them. The same applies to models in the business (e.g., manufacturing, service, supply chain, and transport) and health sectors. A model of a complete supply chain might include every partner in that chain but contain only scant detail about the individual partners. A more detailed model might focus on just a few partners, representing in detail the processes and decisions involved. A model of a hospital ward might include much detail about the day-to-day running of the ward, but when modeling the regional health service, that same ward may be subsumed into a high-level model of the hospital.

What we are observing is that during the development of simulation models, a set of decisions is made concerning what to include and what to exclude from the model. The modeler must decide which model to develop out of a nearly infinite set of potential models that could be chosen for the system that is being studied. This process of abstraction is what is referred to as *conceptual modeling.*

So, conceptual modeling is not about how to implement, or code, a model on a computer, but it is about how to decide what to include in a model and what to exclude from that model. Unfortunately, this aspect of simulation modeling is not well understood. There is much written on how to code simulation models and on the analysis of simulation output, but there is very little written on conceptual modeling. This is despite the recognition that conceptual modeling is a vital element in performing simulation studies. A good conceptual model lays a strong foundation for successful simulation modeling and analysis.

There have been sporadic bursts of interest in the topic of conceptual modeling over the last four or five decades, but never a concerted effort. However, more recently there has been an increased concentration on the topic, particularly through workshops (e.g., the biennial meeting of the Conceptual Modeling Group, UK), special-interest groups (e.g., the NATO Conceptual Modeling Group), conference sessions (e.g., the Winter Simulation Conference, Operational Research Society Simulation Workshop, and Simulation Interoperability Workshop), and journal special issues (e.g., *Journal of Simulation* 1(3), 2007).

The purpose of this book is to build upon these efforts and to provide a comprehensive view of the current state-of-the-art in conceptual modeling for simulation. It achieves this by bringing together the work of an international group of researchers from different areas of simulation: military, business, and health modeling. In doing this, we look at a range of issues in conceptual modeling:

- What are conceptual models and conceptual modeling?
- How can conceptual modeling be performed?
- What is the role of established approaches in conceptual modeling?
- How is conceptual modeling performed in specific modeling domains?

We cannot claim to fully answer any of these questions, but we are able to present the latest thinking on these topics. The book is aimed at students, researchers, and practitioners with an interest in conceptual modeling for simulation. Indeed, we would argue that all simulation modelers have an interest in conceptual modeling, because all simulation modelers are involved, either consciously or subconsciously, in conceptual modeling.

The focus of the book is on discrete-event simulation (Pidd 2005; Law 2007), which for reasons of simplicity is described as just "simulation." In this approach, the dynamics of a system are modeled as a series of discrete events at which the state of the system changes. It is primarily used for modeling queuing systems that are prevalent in a vast array of applications in the military, business, and health sectors. Despite this focus on discrete-event simulation, many of the ideas will have wider applicability to other forms of simulation (e.g., continuous simulation, system dynamics) and modeling more generally.

In reading this book, it will become clear that there is no single agreed definition of a conceptual model or conceptual modeling. The chapters also present some quite different perspectives on what conceptual modeling entails. These differences are perhaps most stark between those working with military models and those working in business and health. This is largely a function of the scale and complexity of the models that the two groups work with, military models generally being much larger in scale (Robinson 2002). As editors of the book, we have made no attempt to reconcile these differences. The state-of-the-art is such that we are not yet in a position to propose a unified definition of a conceptual model or a unified approach to conceptual modeling. Indeed, it seems unlikely that such a goal is achievable either in the short term, given our limited understanding of conceptual modeling, or even in the long term, given the range of domains over which simulation is used and the complexity of the conceptual modeling task. What this book does provide is a single source in which different perspectives on conceptual modeling are presented and a basis upon which they can be compared.

The book is split into six parts, each focusing on a different aspect of conceptual modeling for simulation. Part I explores the foundations of conceptual modeling. In Chapter 1, Robinson discusses the definition of a conceptual model and conceptual modeling, the purpose and requirements of a conceptual model, and the guidance that is given in the literature on conceptual modeling. The ideas that are presented provide a backdrop for the rest of the book. Brooks (Chapter 2) explores the relationship between the level of detail and complexity of a model, and the performance of that model. He identifies eleven "elements" that can be used for measuring model performance. In an experiment, he investigates the relationship between complexity and model performance. In Chapter 3, Wang and Brooks follow an expert modeler and a number of novice modelers through the conceptual modeling process. As a result, they are able to identify the process followed and differences in concentration on the various elements of conceptual modeling.

Part II includes five chapters on frameworks for conceptual modeling. A framework provides a set of steps and tools that aim to help a modeler through the process of deciding what model to build. Robinson (Chapter 4) presents a framework for modeling operations systems, such as manufacturing and service systems. Meanwhile, van der Zee (Chapter 5) concentrates on conceptual modeling for manufacturing systems using an object-oriented approach. The ABCmod conceptual modeling framework, devised by Arbez and Birta (Chapter 6), provides a detailed procedure that is useful for modeling discrete-event dynamic systems. In Chapter 7, Karagöz and Demirörs describe and compare a series of conceptual modeling frameworks (FEDEP, CMMS (FDMS), DCMF, Robinson's framework, and KAMA) most of which derive from simulation modeling in the military domain. In the final chapter of Part II, Haydon reflects upon his many years of experience in simulation modeling (Chapter 8). He describes how he approaches conceptual modeling from a practical perspective by outlining a series of steps that can be followed. This provides a valuable practice-based reflection on the topic.

Some authors have identified a correspondence between soft systems methodology (SSM) (Checkland 1981) and conceptual modeling. Pidd and Kotiadis discuss this connection in Part III. It is important to correctly identify the problem to be tackled, otherwise a simulation study is set for failure from the outset; Balci (1994) identifies this as a type 0 error. As a result, Pidd (Chapter 9) discusses problem structuring and how SSM can help a simulation modeler ensure that the right problem is tackled. Kotiadis specifically focuses on using SSM to help identify the objectives of a simulation study and describes the approach through a case study on modeling community health care (Chapter 10).

Part IV investigates the links between software engineering and conceptual modeling; this might be described as "conceptual engineering." In Chapter 11, Liston et al. describe and illustrate the use of SysML as an aid to conceptual modeling. Following a review of process modeling methods,

Ryan and Heavey devise simulation activity diagrams to support conceptual modeling (Chapter 12). These aim to aid the communication between the modeler and system users and also to help the modeler gather the data needed for the creation of the model. In Chapter 13, Onggo compares a range of conceptual model representation methods and proposes a multifaceted approach. Tolk et al. look at the issue from the perspective of building large-scale models (Chapter 14). In this case, the composability of the different model components becomes an important issue. By *composability,* they refer to the alignment of the model conceptualizations (or conceptual models). Tolk et al. go on to explore how composability might be achieved. The final chapter in Part IV discusses how conceptual models might be verified and validated. Tanrıöver and Bilgen discuss this in the context of UML-based conceptual models (Chapter 15).

In Part V, conceptual modeling is discussed with respect to two specific domains: military and semiconductor manufacturing. Pace (Chapter 16) discusses conceptual modeling in the military domain, with a specific focus on the discussions that have taken place at the Simulation Interoperability Workshops over the last decade or so. In Chapter 17, Sprenger and Rose demonstrate how a semiconductor wafer factory simulation model can be simplified. Model simplification is identified as an important issue in conceptual modeling in Chapters 1 and 2. This chapter provides a useful illustration of the concept.

In the final part of the book, Chapter 18 provides a review of the previous chapters with a view to identifying the current state-of-the-art in conceptual modeling and directions for future research. It is hoped that this will provide an agenda for researchers working in the field of conceptual modeling and that it will be a means for moving this underrepresented topic forward.

We would like to thank all those who have contributed to this book, particularly the authors and reviewers. We believe it provides an invaluable resource for those working in discrete-event simulation and particularly for those with an interest in conceptual modeling.

Roger J. Brooks
Kathy Kotiadis
Stewart Robinson
Durk-Jouke van der Zee

References

Balci, O. 1994. Validation, verification, and testing techniques throughout the life cycle of a simulation study. *Annals of Operations Research* 53: 121–173.

Checkland, P.B. 1981. *Systems Thinking, Systems Practice.* Chichester: Wiley.
Law, A.M. 2007. *Simulation Modeling and Analysis,* 4th ed. New York: McGraw-Hill.
Pidd, M. 2005. *Computer Simulation in Management Science,* 5th ed. Chichester: Wiley.
Robinson, S. 2002. Modes of simulation practice: Approaches to business and military simulation. *Simulation Practice and Theory* 10: 513–523.

Editors

Roger J. Brooks is a lecturer in the management science department at Lancaster University. He has a BA (Hons) in mathematics from Oxford University, and a MSc (Eng) and a PhD in operational research from Birmingham University. He is coauthor with Stewart Robinson of a textbook on simulation. His research interests include conceptual modeling, Boolean networks, and agent-based simulation.

Kathy Kotiadis is an assistant professor at Warwick Business School at the University of Warwick (UK) and cochair of the UK Simulation Special Interest Group. She holds a BSc and PhD from the University of Kent. Her research interests include conceptual modeling, participative modeling, health service modeling, and combining problem structuring approaches with discrete-event simulation modeling.

Stewart Robinson is a professor of operational research and associate dean for specialist masters programmes at Warwick Business School. He holds a BSc and PhD in management science from Lancaster University. Previously employed in simulation consultancy, he supported the use of simulation in companies throughout Europe and the rest of the world. He is author/coauthor of three books on simulation. His research focuses on the practice of simulation model development and use. His key areas of interest are conceptual modeling, model validation, output analysis, and alternative simulation methods (discrete-event, system dynamics, and agent-based).

Durk-Jouke van der Zee is an associate professor of operations at the Faculty of Economics and Business, University of Groningen, the Netherlands. He holds an MSc and PhD in industrial engineering from the University of Twente, the Netherlands. He teaches in the areas of operations management and industrial engineering. His research interests include simulation methodology and applications, simulation and serious gaming, shop floor control, and flexible manufacturing systems. Publications of his work can be found in leading journals, such as *Decision Sciences, International Journal of Production Research, International Journal of Production Economics, Journal of Simulation, IIE Transactions,* and *Transportation Research B.*

Contributors

Gilbert Arbez
School of Information Technology and Engineering (SITE)
University of Ottawa
Ottawa, Ontario, Canada

Semih Bilgen
Electrical and Electronics Engineering Department
Middle East Technical University
Ankara, Turkey

Louis G. Birta
School of Information Technology and Engineering (SITE)
University of Ottawa
Ottawa, Ontario, Canada

Roger J. Brooks
Department of Management Science
Lancaster University Management School
Lancaster, United Kingdom

James Byrne
Department of Manufacturing and Operations Engineering
University of Limerick
Limerick, Ireland

Onur Demirörs
Informatics Institute
Middle East Technical University
Ankara, Turkey

Saikou Y. Diallo
Virginia Modeling Analysis & Simulation Center
Old Dominion University
Suffolk, Virginia

Peter Dungan
Department of Manufacturing and Operations Engineering
University of Limerick
Limerick, Ireland

David Haydon
Kings Somborne
Hants, United Kingdom

Cathal Heavey
Department of Manufacturing and Operations Engineering
University of Limerick
Limerick, Ireland

Kamil Erkan Kabak
Department of Manufacturing and Operations Engineering
University of Limerick
Limerick, Ireland

N. Alpay Karagöz
Bilgi Grubu Ltd.
Ankara, Turkey

Robert D. King
Virginia Modeling Analysis & Simulation Center
Old Dominion University
Suffolk, Virginia

Kathy Kotiadis
Warwick Business School
University of Warwick
Coventry, United Kingdom

Paul Liston
Department of Manufacturing and Operations Engineering
University of Limerick
Limerick, Ireland

Stephan Onggo
Department of Management Science
Lancaster University Management School
Lancaster, United Kingdom

Dale K. Pace
Formerly of the Applied Physics Laboratory
Johns Hopkins University
Laurel, Maryland

Jose J. Padilla
Engineering Management and Systems Engineering Department
Old Dominion University
Norfolk, Virginia

Michael Pidd
Department of Management Science
Lancaster University Management School
Lancaster, United Kingdom

Stewart Robinson
Warwick Business School
University of Warwick
Coventry, United Kingdom

Oliver Rose
Institute of Applied Computer Science
Dresden University of Technology
Dresden, Germany

John Ryan
Faculty of Tourism and Food
School of Hospitality Management and Tourism
Dublin Institute of Technology
Dublin, Ireland

Ralf Sprenger
Department of Mathematics and Computer Science
University of Hagen
Hagen, Germany

Ö. Özgür Tanriöver
Information Management Department
Banking Regulation and Supervision Agency
Ankara, Turkey

Andreas Tolk
Engineering Management and Systems Engineering Department
Old Dominion University
Norfolk, Virginia

Charles D. Turnitsa
Virginia Modeling Analysis & Simulation Center
Old Dominion University
Suffolk, Virginia

Durk-Jouke van der Zee
Faculty of Economics and Business
University of Groningen
Groningen, The Netherlands

Wang Wang
Lancaster University Management School
Lancaster, United Kingdom

Paul Young
Mechanical and Manufacturing Engineering
Dublin City University
Dublin, Ireland

Part I

Foundations of Conceptual Modeling

1

Conceptual Modeling for Simulation: Definition and Requirements

Stewart Robinson

CONTENTS

1.1 Introduction

Conceptual modeling is the process of abstracting a model from a real or proposed system. It is almost certainly the most important aspect of a simulation project. The design of the model impacts all aspects of the study, in particular the data requirements, the speed with which the model can be developed, the validity of the model, the speed of experimentation and the confidence that is placed in the model results. A well designed model significantly enhances the possibility that a simulation study will be a success.

Although effective conceptual modeling is a vital aspect of a simulation study, it is probably the most difficult and least understood (Law 1991). There is surprisingly little written on the subject. It is difficult to find a book that devotes more than a handful of pages to the design of the conceptual model.

Neither are there a plethora of research papers, with only a handful of well regarded papers over the last four decades. A search through the academic tracks at major simulation conferences on discrete-event simulation reveals a host of papers on other aspects of simulation modeling. There are, however, only a few papers that give any space to the subject of conceptual modeling.

The main reason for this lack of attention is probably due to the fact that conceptual modeling is more of an 'art' than a 'science' and therefore it is difficult to define methods and procedures. Whatever the reason, the result is that the art of conceptual modeling is largely learnt by experience. This somewhat ad hoc approach does not seem satisfactory for such an important part of the simulation modeling process.

The purpose of this chapter is to bring more clarity to the area of conceptual modeling for simulation. The issue is addressed first by defining the meaning of conceptual modeling and then by establishing the requirements of a conceptual model. The meaning of the term conceptual model is discussed in relation to existing definitions in the literature. A refined definition of a conceptual model is then given and the scope of conceptual modeling is defined. There is a pause for thought concerning the purpose of a conceptual model before a discussion on the requirements of a conceptual model. The chapter finishes with a brief review of the guidance that is available for conceptual modeling.

The domain of interest for this discussion is primarily in the use of discrete-event simulation for modeling *operations systems* or *operating systems*. "An operating system is a configuration of resources combined for the provision of goods or services" (Wild 2002). Wild identifies four specific functions of operations systems: manufacture, transport, supply, and service. This is one of the prime domains for simulation in operational research. We might refer to it as "business-oriented" simulation while interpreting *business* in its widest sense to include, for instance, the public sector and health. Models in this domain tend to be of a relatively small scale, with a project life cycle of normally less than 6 months (Cochran et al. 1995). The models are generally developed by a lone modeler acting as an external or internal consultant. Sometimes the models are developed on a "do-it-yourself" basis with a subject matter expert carrying out the development. This is somewhat different to the nature of zsimulation modeling in the military domain, another major application of simulation in operational research, where models tend to be of a much larger scale and where they are developed by teams of people (Robinson 2002). Although the focus is on discrete-event simulation for modeling operations systems, this is not to say that the concepts do not have wider applicability.

Throughout the chapter, three roles in a simulation study are assumed:

- *The Clients*: the problem owners and recipients of the results
- *The Modeler*: the developer of the model
- *Domain Experts*: experts in the domain being modeled who provide data and information for the project

These roles do not necessarily imply individual or separate people. There are often many clients and domain experts involved in a simulation study. In some situations one of the clients or subject matter experts may also act as the modeler.

Before exploring the meaning of conceptual modeling, let us begin with an example that highlights how more than one (conceptual) model can be developed of the same system.

1.2 Example: Modeling the Ford Motor Company's South Wales Engine Assembly Plant

I had been called in to carry out some simulation modeling of the new engine assembly plant that Ford Motor Company (Ford) was planning to build in South Wales. Faced with a meeting room full of engineers I started, as normally I would, by asking what was the problem that they wished to address. There was a unanimous response: "Scheduling! We are not sure that there is enough space by the line to hold sufficient stocks of the key components. Obviously the schedules we run on the key component production lines and on the main engine assembly line will affect the inventory we need to hold." After further questioning it was clear that they saw this as the key issue. In their view, there was no problem with achieving the required throughput, especially because they had designed a number of similar lines previously.

The engine assembly line was planned to consist of three main assembly lines (with well over a hundred operations), a Hot Test facility, and a Final Dress area. Figure 1.1 provides a schematic of the line. On the first line (Line A), engine blocks are loaded onto platens (metal pallets on which engines move around the conveyor system) and then pass through a series of operations. On the Head Line various components are assembled to the

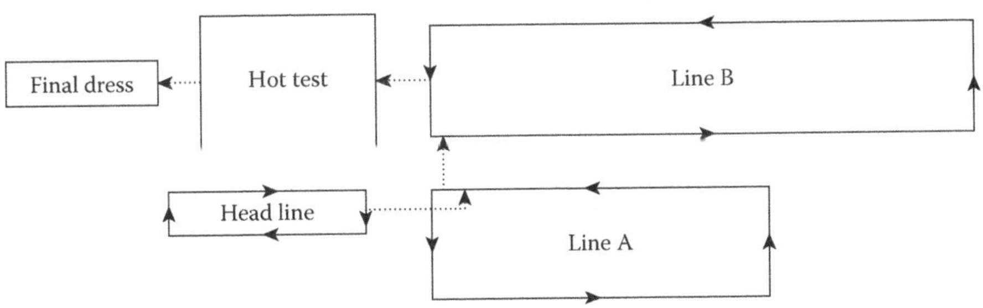

FIGURE 1.1
Schematic showing the layout of the South Wales Engine Assembly Plant.

head before the complete subassembly is joined with the engine block on Line A. On leaving Line A, the engine is loaded to a Line B platen to continue the assembly process. The empty Line A platen is washed and returned so a new engine block can be loaded. At the end of Line B, completed engines are off-loaded and move to the Hot Test facility. In Hot Test, engines are rigged to test machines, run for a few minutes and monitored. Engines that pass Hot Test move to the Final Dress area for completion. Engines that fail Hot Test are rectified and then completed.

The majority of the operations on the three main assembly lines consist of a single automatic machine. Some operations require two parallel machines due to the length of the machine cycle, while a few other operations are performed manually. At various points along the line there are automatic test stations. When an engine fails the test, it is sent to an adjoining rework station, before returning to be tested again. All the operations are connected by a powered roller conveyor system.

The key components are the engine block, head, crankshaft, cam shaft, and connecting rods. These are produced at nearby production facilities, delivered to the main assembly plant and stored line-side ready for assembly. Because various engine derivatives are made on the assembly line, a range of component derivatives need to be produced and stored for assembly. The result was the concern over scheduling the production and the storage of these key components.

As with all such projects, time for developing and using the model was limited. It was important, therefore, to devise a model that could answer the questions about scheduling key components as quickly as possible while maintaining a satisfactory level of accuracy.

In consideration the nature of the problem, it was clear that the key issue was not so much the rate at which engines progressed through the assembly line, but their sequence. The initial sequence of engines was determined by the production schedule, but this sequence was then disturbed by engines being taken out for rework and by the presence of parallel machines for some operations. Under normal operation the parallel machines would not cause a change in the sequence of engines on the line, but if one of the machines breaks down for a period, then the engines queuing for that machine would be delayed and their sequence altered.

It was recommended that the simulation model should represent in detail those elements that determined the sequence of engines on the main assembly line, that is, the schedule, the test and rework areas, and the parallel machines. All other operations could be simplified by grouping sections of the line that consisted of individual machines and representing them as a queue with a delay. The queue capacity needed to equate to the capacity of that section of the line. The delay needed to be equal to the time it took for an engine to pass through the section of the line, allowing for breakdowns. This would give a reasonable approximation to the rate at which engines would progress through the facility. Of course, the operations where the key components are assembled to the engine need to be modeled in detail, along with the line-side storage areas for those components.

Further to this, it was noted that detailed models of the key component production lines already existed. Alternative production schedules for each line could be modeled separately from the engine assembly line model and the output from these models stored. The outputs could then be read into the engine assembly line model as an input trace stating the component derivatives and their time of arrival at the assembly line. Some suitable delay needed to be added to allow for the transportation time between the key component lines and the main assembly line. It was also unnecessary to model the Hot Test and Final Dress, as all of the key components have been assembled prior to reaching these areas.

As a result of these simplifications, the model could be developed much more quickly and the final model ran much faster, enabling a greater amount of experimentation in the time available. The model fulfilled its objectives, sizing the line side storage areas and showing that shortages of key components were unlikely. What the model did suggest, however, was that there may be a problem with throughput.

Although the scheduling model indicated a potential problem with throughput, it did not contain enough detail to give accurate predictions of the throughput of the engine assembly line. As a result, a second model was developed with the objective of predicting and helping to improve the throughput of the facility. This model represented each operation in detail, but on this occasion did not represent the arrival and assembly of key components. It was assumed that the key components would always be available, as had been suggested by the scheduling model.

The second (throughput) model indeed confirmed that the throughput was likely to fall significantly short of that required by Ford and identified a number of issues that needed to be addressed. Over a period of time, by making changes to the facility and performing further simulation experiments, improvements were made such that the required throughput could be achieved.

This example demonstrates how two very different simulation models can be developed of the same system. But which model was the right one? The answer is both, since both answered the specific questions that were being asked of them. Underlying the differences between the models was the difference in the modeling objectives. Neither simulation model would have been useful for meeting the objectives of the other model. Of course, a single all encompassing model could have been developed, which could have answered both sets of questions. This, however, would have taken much longer to develop and it would certainly have run much slower, restricting the extent of the experimentation possible. Anyway, the need for the second model was only identified as a result of indications about throughput from the first model. Up to that point, a throughput model seemed unnecessary.

A more fundamental question that should be asked is if very different models can be developed of the same system, how can a modeler determine which model to use? Indeed, how can a modeler develop a model design, or a set of model designs from which to select? The only clue that comes from the example above is the importance of the modeling objectives in determining the nature of the model. Beyond this, modelers need some means for determining what to model. This process of taking a real-world situation and from it designing a model is what we refer to as conceptual modeling.

1.3 What is Conceptual Modeling?

Conceptual modeling is about abstracting a model from a real or proposed system. All simulation models are simplifications of reality (Zeigler 1976). The issue in conceptual modeling is to abstract an appropriate simplification of reality (Pidd 2003). This provides some sense of what conceptual modeling is, but only in the most general of terms. How can the terms conceptual model and conceptual modeling be more precisely defined? Existing literature may shed some light on this topic.

In general, the notion of conceptual modeling, as expressed in the simulation and modeling literature, is vague and ill-defined, with varying interpretations as to its meaning. What seems to be agreed is that it refers to the early stages of a simulation study. This implies a sense of moving from the recognition of a problem situation to be addressed with a simulation model to a determination of what is going to be modeled and how. Balci (1994) breaks the early parts of a simulation study down into a number of processes: problem formulation, feasibility assessment of simulation, system and objectives definition, model formulation, model representation, and programming. Which of these is specifically included in conceptual modeling is not identified. What is clear from Balci and other authors, for instance Willemain (1995), is that these early stages of a modeling study are not just visited once, but that they are continually returned to through a series of iterations in the life cycle of a project. As such, conceptual modeling is not a one-off process, but one that is repeated and refined a number of times during a simulation study.

Zeigler (1976) sheds some light on the subject by identifying five elements in modeling and simulation from the "real system" through to the "computer" (the computer-based simulation model). In between is the "experimental frame," "base model," and "lumped model." The experimental frame is the limited set of circumstances under which the real system is observed, that is, specific input–output behaviors. The base model is a hypothetical complete explanation of the real system, which is capable of producing all possible input–output behaviors (experimental frames). The base model cannot be fully known since full knowledge of the real system cannot be attained. For instance, almost all systems involve some level of human interaction that will affect its performance. This interaction cannot be fully understood since it will vary from person to person and time to time.

In the lumped model the components of a model are lumped together and simplified. The aim is to generate a model that is valid within the experimental frame, that is, reproduces the input–output behaviors with sufficient fidelity. The structure of the lumped model is fully known. Returning to the example of human interaction with a system, in a lumped model specific rules for interaction are devised, e.g., a customer will not join a waiting line of more than 10 people.

Nance (1994) separates the ideas of conceptual model and communicative model. The conceptual model exists in the mind of a modeler, the communicative model is an explicit representation of the conceptual model. He also specifies that the conceptual model is separate from model execution. In other words, the conceptual model is not concerned with how the computer-based model is coded. Fishwick (1995) takes a similar view, stating that a conceptual model is vague and ambiguous. It is then refined into a more concrete executable model. The process of model design is about developing and refining this vague and ambiguous model and creating the model code. In these terms, conceptual modeling is a subset of model design, which also includes the design of the model code.

The main debate about conceptual modeling and its definition has been held among military simulation modelers. Pace has lead the way in this debate and defines a conceptual model as "a simulation developer's way of translating modeling requirements … into a detailed design framework …, from which the software that will make up the simulation can be built" (Pace 1999). In short, the conceptual model defines what is to be represented and how it is to be represented in the simulation. Pace sees conceptual modeling as being quite narrow in scope viewing objectives and requirements definition as precursors to the process of conceptual modeling. The conceptual model is largely independent of software design and implementation decisions. Pace (2000a) identifies the information provided by a conceptual model as consisting of assumptions, algorithms, characteristics, relationships, and data.

Lacy et al. (2001) further this discussion, reporting on a meeting of the Defense Modelling and Simulation Office (DMSO) to try and reach a consensus on the definition of a conceptual model. The paper describes a plethora of views, but concludes by identifying two types of conceptual model. A *domain-oriented* model that provides a detailed representation of the problem domain and a *design-oriented* model that describes in detail the requirements of the model. The latter is used to design the model code. Meanwhile, Haddix (2001) points out that there is some confusion over whether the conceptual model is an artifact of the user or the designer. This may, to some extent, be clarified by adopting the two definitions above.

The approach of military simulation modelers can be quite different to that of those working in business-oriented simulation (Robinson 2002). Military simulations often entail large scale models developed by teams of software developers. There is much interest in model reuse and distributed simulation, typified by the High Level Architecture (DMSO 2005). Business-oriented simulations tend to be smaller in scale, involve lone modelers normally using a visual interactive modeling system (Pidd 2004), and the models are often thrown away on completion of a project. Interest in distributed simulation is moderate, mostly because the scale and lifetime of the models does not warrant it (Robinson 2005). As a result, although the definition and requirements for conceptual modeling may be similar in both these domains, some account must be made of the differences that exist.

In summary, the discussion above identifies some key facets of conceptual modeling and the definition of a conceptual model:

- Conceptual modeling is about moving from a problem situation, through model requirements to a definition of what is going to be modeled and how.
- Conceptual modeling is iterative and repetitive, with the model being continually revised throughout a modeling study.
- The conceptual model is a simplified representation of the real system.
- The conceptual model is independent of the model code or software (while model design includes both the conceptual model and the design of the code [Fishwick 1995]).
- The perspective of the client and the modeler are both important in conceptual modeling.

It is clear, however, that complete agreement does not exist over these facets.

1.3.1 A Definition of a Conceptual Model

Following the discussion above, Figure 1.2 defines a conceptual model as shown by the area within the dashed ellipse. It also places it within the wider context of a simulation study as defined in Robinson (2004). Figure 1.2 shows four key processes in the development and use of a simulation model: conceptual modeling, model coding, experimentation, and implementation. The outcome of each process is, respectively, a conceptual model, a computer model, solutions to the problem situation and/or a better understanding of the real world, and improvements to the real world. The double arrows illustrate the iterative nature of the process and the circular diagram illustrates the potential to repeat the process of improvement through simulation a number of times. Missing from this diagram are the verification and validation activities involved in a simulation study. These are carried out in parallel with each

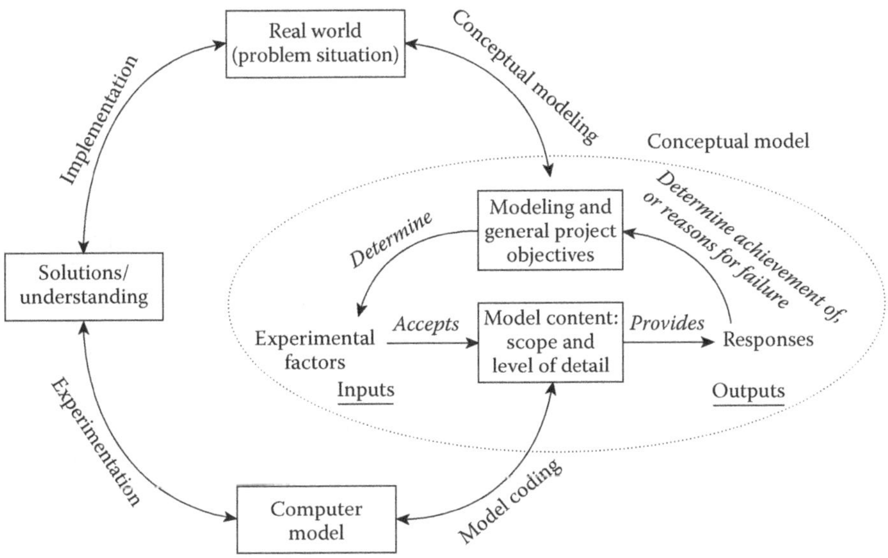

FIGURE 1.2
The conceptual model in the simulation project life cycle. (Adapted from Robinson, S., *Simulation: The Practice of Model Development and Use,* Wiley, Chichester, UK, 2004. With permission.)

of the four processes outlined in Figure 1.2. For a more detailed description of this life cycle and model verification and validation see Robinson (2004).

Based upon an understanding of the problem situation the conceptual model is derived. This model is only a partial description of the real world, but it is sufficient to address the problem situation. The double arrow between the problem situation and objectives signifies the interplay between problem understanding and modeling. While the conceptual model reflects the understanding of the problem situation, the process of developing the conceptual model also changes the understanding of the problem situation. In particular, the nature of the questions that the modeler asks during conceptual modeling can lead to new insights on behalf of the clients and domain experts. At a greater extreme, ideas derived purely from conceptual modeling may be implemented in the real system, changing the actual nature of the problem situation.

The conceptual model itself consists of four main components: objectives, inputs (experimental factors), outputs (responses), and model content. Two types of *objective* inform a modeling project. First, there are the modeling objectives, which describe the purpose of the model and modeling project. Second, there are general project objectives, which include the timescales for the project and the nature of the model and its use (e.g., requirements for the flexibility of the model, run-speed, visual display, ease-of-use, and model/component reuse). The definition of objectives is seen as intrinsic to decisions about the conceptual model. The Ford example above highlighted how different modeling objectives led to different models. Similarly, the general project objectives can affect the nature of the model. A shorter timescale, for instance, may require a simpler conceptual model than would have been devised had more time been available. For this reason, the objectives are included in the definition of the conceptual model.

Including the modeling objectives as part of the definition of a conceptual model is at odds with Pace (1999). He sees the objectives and requirements definition as separate from the conceptual model. The author's view is that while understanding the problem situation and the aims of the organization lies within the domain of the real world (problem situation), the modeling objectives are specific to a particular model and modeling exercise. Different modeling objectives lead to different models within the same problem situation, as in the Ford example. As a result, the modeling objectives are intrinsic to the description of a conceptual model. Without the modeling objectives, the description of a conceptual model is incomplete.

The *inputs* (or experimental factors) are those elements of the model that can be altered to effect an improvement in, or better understanding of, the problem situation. They are determined by the objectives. Meanwhile, the *outputs* (or responses) report the results from a run of the simulation model. These have two purposes: first, to determine whether the modeling objectives have been achieved; second, to point to reasons why the objectives are not being achieved, if they are not.

Finally, the *model content* consists of the components that are represented in the model and their interconnections. The content can be split into two dimensions (Robinson 1994):

- *The scope of the model:* the model boundary or the breadth of the real system that is to be included in the model.
- *The level of detail:* the detail to be included for each component in the model's scope.

The model content is determined, in part, by the inputs and outputs, in that the model must be able to accept and interpret the inputs and to provide the required outputs. The model content is also determined by the level of accuracy required. More accuracy generally requires a greater scope and level of detail.

While making decisions about the content of the model, various assumptions and simplifications are normally introduced. These are defined as follows:

- *Assumptions* are made either when there are uncertainties or beliefs about the real world being modeled.
- *Simplifications* are incorporated in the model to enable more rapid model development and use, to reduce data requirements, and to improve transparency (understanding) (Section 1.5.1).

Assumptions and simplifications are identified as separate facets. Assumptions are ways of incorporating uncertainties and beliefs about the real world into the model; they relate to the problem domain (i.e., the real world). Simplifications are ways of reducing the complexity of the model; they relate to the model domain. As such, assumptions are a facet of limited knowledge or presumptions, while simplifications are a facet of the desire to create simple models.

Based on these ideas a conceptual model is defined as follows:

> The conceptual model is a non-software-specific description of the computer simulation model (that will be, is or has been developed), describing the objectives, inputs, outputs, content, assumptions, and simplifications of the model.

This definition adds the point that the conceptual model is non-software-specific in line with the views of the other authors described above. Considerations as to how the model code will be developed (whether it be a spreadsheet, specialist software, or a programming language) should not dominate debate around the nature of the model that is required to address the problem situation. Conceptual modeling is about determining the right model, not how the software will be implemented.

In saying this, it must be recognized that many simulation modelers only have access to one or possibly two simulation tools. As a result, considerations of software implementation will naturally enter the debate about the nature of the conceptual model. This is recognized by the double arrow, signifying iteration, for the model coding process in Figure 1.2. What this definition for a conceptual model aims to highlight is the importance of separating as far as possible detailed model code considerations from decisions about the conceptual design.

The definition does not place the conceptual model at a specific point in time during a simulation study. This reflects the level of iteration that may exist in simulation work. A conceptual model may reflect a model that is to be developed, is being developed or has been developed in some software. The model is continually changing as the simulation study progresses. Whatever stage has been reached in a simulation study, the conceptual model is a non-software-specific description of the model as it is understood at that point in time.

It should also be noted that this definition does not imply that a formal and explicit conceptual model is developed. Indeed, the conceptual model may not be formally expressed (not fully documented) and/or it may be implicit (in the mind of the modeler). Whether the conceptual model is explicit or not, it still exists, in that the modeler has made decisions about what to model and what not to model. For the rest of this chapter we shall assume that the conceptual model is made explicit, in a more or less formal fashion, with a view to reaching a joint agreement between the modeler, clients, and domain experts concerning its content.

1.3.2 Conceptual Modeling Defined

Put simply, conceptual modeling is the process of creating the conceptual model. Based on the definition given above this requires the following activities:

- Understanding the problem situation (a precursor to conceptual modeling)
- Determining the modeling and general project objectives
- Identifying the model outputs (responses)
- Identify the model inputs (experimental factors)
- Determining the model content (scope and level of detail), identifying any assumptions and simplifications

These activities are explored in more detail in chapter 4. This list suggests a general order in which the elements of a conceptual model might be determined. There is likely to be a lot of iteration forwards and backwards between these activities. Further to this, there is iteration between conceptual modeling and the rest of the process of model development and use (Robinson 2004). Although the conceptual model should be independent of the modeling software, it must be recognized that there is an interplay between the two. Since many modelers use the software that they are familiar with, it is possible (although not necessarily desirable) that methods of representation and limitations in the software will cause a revision to the conceptual model. Continued learning during model coding and experimentation may cause adjustments to the conceptual model as the understanding of the problem situation and modeling objectives change. Model validation activities may result in alterations to the conceptual model in order to improve the accuracy of the model. Availability, or otherwise, of data may require adjustments to the conceptual model. All this implies a great deal of iteration in the process of modeling and the requirement to continually revise the conceptual model. This iteration is illustrated by the double arrows between the stages in Figure 1.2.

1.4 The Purpose of a Conceptual Model

In reflecting on the purpose of a conceptual model, one might question whether it is necessary to have one at all. Indeed, some might argue that the power of modern simulation software negates the need for conceptual modeling. Such software enables a modeler to move straight from developing an understanding of the problem situation to creating a computer model.

Albeit that this argument appears to have some credence, it ignores the fact that whatever practice a modeler might employ for developing the model code, decisions still have to be taken concerning the content and assumptions of the model. Modern simulation software does not reduce this level of decision-making. What the software can provide is an environment for the more rapid development of the model code, enhancing the opportunities for iteration between conceptual modeling and model coding, and facilitating rapid prototyping. This does not negate the need for conceptual modeling, but simply aids the process of model design. It also highlights the point that conceptual modeling is not a one-off step, but part of a highly iterative process, particularly in relation to model coding.

Indeed, the power of modern software (and hardware) and the wider use of distributed processing may actually have increased the need for effective conceptual modeling. Salt (1993) and Chwif et al. (2000) both identify the problem of the increasing complexity of simulation models; a result of the "possibility" factor. People build more complex models because the hardware and software enable them to. While this may have extended the utility of simulation to problems that previously could not have been tackled, it also breads a tendency to develop overly complex models. There are various problems associated with such models including extended development times and onerous data requirements. This trend to develop ever more complex models has been particularly prevalent in the military domain (Lucas and McGunnigle 2003). Indeed, it could be argued that there are some advantages in only having limited computing capacity; it forces the modeler to carefully design the model! As a result of the possibility factor it would seem that careful design of the conceptual model is more important than ever.

Beyond the general sense that careful model design is important, there are a number of reasons why a conceptual model is important to the development and use of simulation models. Pace (2003) puts this succinctly by stating that the conceptual model provides a roadmap from the problem situation and objectives to model design and software implementation. He also recognizes that the conceptual model forms an important part of the documentation for a model. More specifically, a well-documented conceptual model does the following:

- Minimizes the likelihood of incomplete, unclear, inconsistent, and wrong requirements (Borah 2002, Pace 2002)
- Helps build the credibility of the model
- Guides the development of the computer model
- Forms the basis for model verification and guides model validation
- Guides experimentation by expressing the objectives, experimental factors, and responses

- Provides the basis of the model documentation
- Can act as an aid to independent verification and validation when it is required
- Helps determine the appropriateness of the model or its parts for model reuse and distributed simulation (Pace 2000b)

Overall the conceptual model, if made explicit and clearly expressed, provides a means of communication between all parties in a simulation study: the modeler, clients, and domain experts (Pace 2002). In so doing it helps to build a consensus, or least an accommodation, about the nature of the model and its use.

1.5 Requirements of a Conceptual Model

In designing a conceptual model it would be useful to have a set of requirements in mind. These could provide a basis against which to determine whether a conceptual model is appropriate. Indeed, Pritsker (1987) says that "modelling is a difficult process because we do not have measurable criteria for evaluating the worth of a model." In conceptual modeling it may be difficult to identify a complete set of *measurable* criteria, since the model is purely descriptive at this stage. That said, a sense of requirements, even if they are more qualitative, would be helpful.

So what are the requirements for an effective conceptual model? This question is first answered by describing four main requirements after which the overarching need to keep the model as simple as possible is discussed.

Assessment criteria for models have been discussed by a number of authors, for instance, Gass and Joel (1981), Ören (1981, 1984), Robinson and Pidd (1998), and Balci (2001). The majority of this work is in the domain of large scale military and public policy models; Robinson and Pidd are an exception. Furthermore, the criteria focus on assessing models that have been developed rather than on the assessment of conceptual models.

In terms of criteria for conceptual models in operational research there has been little reported. Willemain (1994), who investigates the preliminary stages of operational research interventions, briefly lists five qualities of an effective model: validity, usability, value to the clients, feasibility, and aptness for the clients' problem. Meanwhile, Brooks and Tobias (1996a) identify 11 performance criteria for a good model. Requirements are also briefly discussed by Pritsker (1986), Henriksen (1988), Nance (1994), and van der Zee and Van der Vorst (2005). Outside of operational research there are some discussions, for instance, Teeuw and van den Berg (1997) who discuss the quality of conceptual models for business process reengineering.

Based on the discussions by simulation modelers and operational researchers, here it is proposed that there are four main requirements of a conceptual model: validity, credibility, utility, and feasibility. Table 1.1 shows how the requirements discussed in the literature relate to these.

It is generally agreed that a valid model is one that is sufficiently accurate for the purpose at hand (Carson 1986). However, since the notion of accuracy is of little meaning for a model that has no numeric output, conceptual model validity might be defined as:

> A perception, on behalf of the modeler, that the conceptual model can be developed into a computer model that is sufficiently accurate for the purpose at hand.

The phrase "can be developed into a computer model" is included in recognition that the conceptual model is a description of a model, not the computer model itself. Depending on the status of the simulation project, the conceptual model may be describing a computer model that will be developed, is being developed, or has been developed.

Underlying the notion of validity is the question of whether the model is "right." Note that this definition places conceptual model validity as a perception of the modeler. It also maintains the notion that a model is built for a specific purpose, which is common to most definitions of validity.

Credibility is similar to validity, but is taken from the perspective of the clients rather than the modeler. The credibility of the conceptual model is therefore defined as:

> A perception, on behalf of the clients, that the conceptual model can be developed into a computer model that is sufficiently accurate for the purpose at hand.

The clients must believe that the model is sufficiently accurate. Included in this concept is the need for the clients to be convinced that all the important components and relationships are in the model. Credibility also requires that the model and its results are understood by the clients. Would a model that could not be understood have credibility? An important factor in this respect is the transparency of the model, which is discussed below.

Validity and credibility are seen as separate requirements because the modeler and clients may have very different perceptions of the same model. Although a modeler may be satisfied with a conceptual model, the clients may not be. It is not unusual for additional scope and detail to be added to a model, not because it improves its validity, but because it improves its credibility. Not that adding scope and detail to gain credibility is necessarily a bad thing, but the modeler must ensure that this does not progress so far that the model becomes over complex. Simulation is particularly prone to such a drift through, for instance, the addition of nonvital graphics and the logic required to drive them.

TABLE 1.1

Requirements of a Conceptual Model Related to those Documented in the Literature

	Documented Requirements					
Proposed Requirements	**Pritsker (1986)**	**Henriksen (1988)**	**Nance (1994)**	**Willemain (1994)**	**Brooks and Tobias (1996a)**	**van der Zee and Van der Vorst (2005)**
Validity	Valid	Fidelity	Model correctness Testability	Validity Aptness for client's problem	Model describes behavior of interest Accuracy of the model's results Probability of containing errors Validity Strength of theoretical basis of model	Completeness
Credibility	Understandable				Ease of understanding	Transparency
Utility	Extendible	Execution speed Ease of modification	Adaptability Reusability Maintainability	Value to client Usability	Portability and ease with which model can be combined with others	
Feasibility	Timely	Elegance		Feasibility	Time and cost to build model Time and cost to run model Time and cost to analyze results Hardware requirements	

The third concept, *utility*, is defined as:

> A perception, on behalf of the modeler and the clients, that the conceptual model can be developed into a computer model that is useful as an aid to decision-making within the specified context.

Utility is seen as a joint agreement between the modeler and the clients about the usefulness of the model. This notion moves beyond the question of whether the model is sufficiently accurate, to the question of whether the model is useful for the context of the simulation study. Utility includes issues such as ease-of-use, flexibility (i.e., ease with which model changes can be made), run-speed and visual display. Where the model, or a component of the model, might be used again on the same or another study, reusability would also be subsumed within the concept of utility. The requirements for utility are expressed through the general project objectives.

Within any context a range of conceptual models could be derived. The accuracy of these models would vary, but some or all might be seen as sufficiently accurate and, hence, under the definitions given above, they would be described as valid and credible. This does not necessarily mean that the models are useful. For instance, if a proposed model is large and cumbersome, it may have limited utility due to reduced ease-of-use and flexibility. Indeed, a less accurate (but still sufficiently accurate), more flexible model that runs faster may have greater utility by enabling a wider range of experimentation within a timeframe.

Hodges (1991) provides an interesting discussion around model utility and suggests that a "bad" model (one that is not sufficiently accurate) can still be useful. He goes on to identify specific uses for such models. Bankes (1993) continues with this theme, discussing the idea of inaccurate models for exploratory use, while Robinson (2001) sees a role for such models in facilitating learning about a problem situation.

The final requirement, *feasibility*, is defined as follows:

> A perception, on behalf of the modeler and the clients, that the conceptual model can be developed into a computer model with the time, resource and data available.

A range of factors could make a model infeasible: it might not be possible to build the proposed model in the time available, the data requirements may be too onerous, there may be insufficient knowledge of the real system, and the modeler may have insufficient skill to code the model. Feasibility implies that the time, resource, and data are available to enable development of the computer model.

The four requirements described above are not mutually exclusive. For instance, the modeler's and clients' perspectives on model accuracy are likely to be closely aligned, although not always. An infeasible model could not generally be described as a useful model, although a conceptual model that is infeasible could be useful for aiding problem understanding. Albeit that these concepts are related, it is still useful to identify them as four separate requirements so a modeler can be cognizant of them when designing the conceptual model.

1.5.1 The Overarching Requirement: Keep the Model Simple

The overarching requirement is the need to avoid the development of an overly complex model. In general the aim should be this: *to keep the model as simple as possible to meet the objectives of the simulation study* (Robinson 2004).

There are a number of advantages with simple models (Innis and Rexstad 1983, Ward 1989, Salt 1993, Chwif et al. 2000, Lucas and McGunnigle 2003, Thomas and Charpentier 2005):

- Simple models can be developed faster.
- Simple models are more flexible.
- Simple models require less data.
- Simple models run faster.
- The results are easier to interpret since the structure of the model is better understood.

With more complex models these advantages are generally lost. Indeed, at the center of good modeling practice is the idea of resorting to the simplest explanation possible. Occam's razor puts this succinctly, "plurality should not be posited without necessity" (William of Occam; quoted from Pidd 2003), as does Antoine de Saint-Exupery, who reputedly said that "perfection is achieved, not when there is nothing more to add, but when there is nothing left to take away."

The requirement for simple models does not negate the need to build complex models on some occasions. Indeed, complex models are sometimes required to achieve the modeling objectives. The requirement is to build the simplest model possible, not simple models per se. What should be avoided, however, is the tendency to try and model every aspect of a system when a far simpler more focused model would suffice.

The graph in Figure 1.3 illustrates the notional relationship between model accuracy and complexity (Robinson 1994). Increasing levels of complexity (scope and level of detail) improve the accuracy of the model, but with diminishing returns. Beyond point x there is little to be gained by adding to the complexity of the model. A 100% accurate model will never be achieved because it is impossible to know everything about the real system. The graph illustrates a further point. Increasing the complexity of the

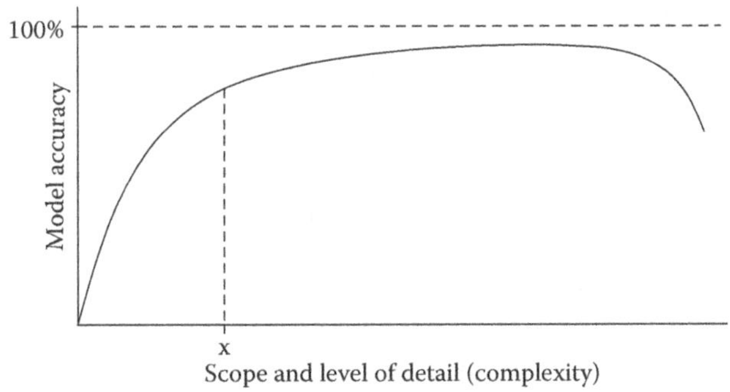

FIGURE 1.3
Simulation model complexity and accuracy. (Adapted from Robinson, S., *Industrial Engineering*, 26 (9), 34–36, 1994. With permission.)

model too far, may lead to a less accurate model. This is because the data and information are not available to support such a detailed model. For instance, it is unlikely that we could accurately model the exact behavior of individuals in a queue, and attempts to do so, beyond very simple rules, may lead to a less accurate result.

Ward (1989) provides a lucid account on the simplicity of models. In doing so, he makes a useful distinction between constructive simplicity and transparency. Transparency is an attribute of the client (how well he/she understands the model), while constructive simplicity is an attribute of the model itself (the simplicity of the model). Because transparency is an attribute of the client, it depends on his/her level of knowledge and skill. A model that is transparent to one client may not be transparent to another. In developing a conceptual model, the modeler must consider transparency as well as simplicity, designing the model with the particular needs of the client in mind. The need for transparency is, of course, confounded by the presence of multiple clients (as is the case in many simulation studies), all of whom must be satisfied with the model. These ideas closely link to the requirement for credibility, as discussed above, since a model that is not transparent is unlikely to have credibility.

Having emphasized the importance of simplicity, there are those that warn against taking this to an extreme. Pritsker (1986) reflects on his experience of developing models of differing complexity of the same system. He concludes that the simplest model is not always best because models need to be able to evolve as the requirements change. The simplest model is not always the easiest to embellish. Schruben and Yücesan (1993) make a similar point, stating that simpler models are not always as easy to understand, code and debug. Davies et al. (2003) point out that simpler models require more extensive assumptions about how a system works and that there is a danger in setting the system boundary (scope) too narrow in case an important facet is missed.

1.6 Guidance on Conceptual Modeling

Exhortations to develop simple models highlight an important consideration in designing a conceptual model. Modeling requirements provide a guide as to whether a conceptual model is appropriate. Neither, however, describes how a modeler might go about determining what the conceptual model should be in a simulation study. So what help is offered in the simulation and modeling literature to guide modelers in designing the conceptual model?

First, it is worth recognizing that conceptual modeling requires creativity (Henriksen 1989). Simulation modeling is both art and science (Shannon 1975) with conceptual modeling lying more at the artistic end! As Schmeiser (2001) points out: "While abstracting a model from the real world is very much an art, with many ways to err as well as to be correct, analysis of the model is more of a science, and therefore easier, both to teach and to do." The need for creativity does not, however, excuse the need for guidelines on how to model (Evans 1992). Ferguson et al. (1997), writing about software development, point out that in "most professions, competent work requires the disciplined use of established practices. It is not a matter of creativity versus discipline, but one of bringing discipline to the work so creativity can happen."

In searching the modeling literature for advice from simulation modelers and operational researchers on how to develop models, three basic approaches can be found: principles of modeling, methods of simplification, and modeling frameworks.

1.6.1 Principles of Modeling

Providing a set of guiding principles for modeling is one approach to advising simulation modelers on how to develop (conceptual) models. For instance, Pidd (1999) describes six principles of modeling:

- Model simple; think complicated
- Be parsimonious; start small and add
- Divide and conquer; avoid megamodels
- Use metaphors, analogies, and similarities
- Do not fall in love with data
- Modeling may feel like muddling through

The central theme is one of aiming for simple models through evolutionary development. Others have produced similar sets of principles (or guidelines), for instance, Morris (1967), Musselman (1992), Powell (1995), Pritsker (1998), and Law (2007). The specific idea of evolutionary model development is further explored by Nydick et al. (2002).

These principles provide some useful guidance for those developing conceptual models. It is useful to encourage modelers to start with small models and to gradually add scope and detail. What such principles do not do, however, is to guide a modeler through the conceptual modeling process. When should more detail be added? When should elaboration stop? There is a difference between giving some general principles and guiding someone through a process.

1.6.2 Methods of Simplification

Simplification entails removing scope and detail from a model or representing components more simply while maintaining a sufficient level of accuracy. In Zeigler's (1976) terms this could be described as further lumping of the lumped model. This is the opposite of the start small and add principle.

There are quite a number of discussions on simplification, both in the simulation and the wider modeling context. Morris (1967) identifies some methods for simplifying models: making variables into constants, eliminating variables, using linear relations, strengthening the assumptions and restrictions, and reducing randomness. Ward (1989) provides a similar list of ideas for simplification. Meanwhile, Courtois (1985) identifies criteria for the successful decomposition of models in engineering and science.

For simulation modeling, Zeigler (1976) suggests four methods of simplification: dropping unimportant components of the model, using random variables to depict parts of the model, coarsening the range of variables in the model, and grouping components of the model. There is an apparent contradiction between Morris's and Zeigler's advice in that the former suggests reducing randomness, while the latter suggests increasing it by representing sections of the model with random variables. This difference in opinion can be reconciled by recognizing that simplification methods are sensitive to the modeling approach that is being applied. Morris is concentrating more on mathematical algorithms where the inclusion of randomness is less convenient. Zeigler is writing about simulation specifically, where complex behaviors can sometimes be reduced to a single random variable.

Yin and Zhou (1989) build upon Zeigler's ideas, discussing six simplification techniques and presenting a case study. Sevinc (1990) provides a semi-automatic procedure based on Zeigler's ideas. Innis and Rexstad (1983) enter into a detailed discussion about how an existing model might be simplified. They provide a list of 17 such methods, although they do not claim that these are exhaustive. They conclude by suggesting that managers should be provided with both a full and a simplified simulation model. There is a sense in which the Ford example followed this approach, with one model being more detailed than the other, although neither could be described as a "full" model. Robinson (1994) also lists some methods for simplifying simulation models. Finally, Webster et al. (1984) describe how they selected an appropriate level of detail for generating samples in a timber harvesting simulation model.

Such ideas are useful for simplifying an existing (conceptual) model, but they do not guide the modeler over how to bring a model into existence. Model simplification acts primarily as a redesign tool and not a design tool.

1.6.3 Modeling Frameworks

A modeling framework goes beyond the idea of guiding principles and methods of model simplification by providing a specific set of steps that guide a modeler through development of a conceptual model. There have been some attempts to provide such frameworks going back to Shannon (1975) who describes four steps: specification of the model's purpose; specification of the model's components; specification of the parameters and variables associated with the components; and specification of the relationships between the components, parameters, and variables.

Both Nance and Pace have devised frameworks that relate primarily to the development of large scale models in the military domain. Nance (1994) outlines the conical methodology. This is an object-oriented, hierarchical specification language that develops the model definition (scope) top-down and the model specification (level of detail) bottom-up. A series of modeling steps are outlined. Balci and Nance (1985) focus specifically on a procedure for problem formulation. Meanwhile, Arthur and Nance (2007) identify the potential to adopt software requirements engineering (SRE) approaches for simulation model development. They also note that there is little evidence of SRE actually being adopted by simulation modelers.

Pace (1999, 2000a) explores a four-stage approach to conceptual model development, similar to that of Shannon: collect authoritative information on the problem domain; identify entities and processes that need to be represented; identify simulation elements; and identify relationships between the simulation elements. He also identifies six criteria for determining which elements to include in the conceptual model. These criteria focus on the correspondence between real-world items and simulation objects (Pace 2000a).

Within our domain of interest, simulation for modeling operations systems, there is quite limited work on conceptual modeling frameworks. Brooks and Tobias (1996b) briefly propose a framework for conceptual modeling, but go no further in expanding upon the idea. Recent papers by Guru and Savory (2004) and van der Zee and Van der Vorst (2005) propose conceptual modeling frameworks in some more detail. Guru and Savory propose a set of modeling templates (tables) useful for modeling physical security systems. Meanwhile, van der Zee and Van der Vorst propose a framework for supply chain simulation. Both are aimed at an object-oriented implementation of the computer-based simulation model. Meanwhile, Kotiadis (2007) looks to the ideas of Soft Operational Research, and specifically soft systems methodology (SSM) (Checkland 1981), for aiding the conceptual modeling process. She uses SSM to help understand a complex health care system and then derives the simulation conceptual model from the SSM "purposeful activity model."

In this book, Robinson proposes a conceptual modeling framework that guides a modeler from identification of the modeling objectives through to determining the scope and level of detail of a model (chapter 4). Arbez and Birta describe the ABCmod conceptual modeling framework that provides a procedure for identifying the components and relationships for a discrete-event simulation model (chapter 6). Meanwhile, van der Zee describes a domain-specific framework for developing conceptual models of manufacturing systems (chapter 5). Karagöz and Demirörs describe and compare a number of frameworks that have largely been developed for the military domain (chapter 7), and Haydon explains how he approaches conceptual modeling from a practice-based perspective (chapter 8).

Such frameworks appear to have potential for aiding the development of conceptual models, but they are not yet fully developed and tested, nor are they in common use. An interesting issue is whether frameworks should be aimed at a specific domain (e.g., supply chain), or whether it is feasible to devise more generic frameworks.

1.7 Conclusion

There is, in large measure, a vacuum of research in the area of conceptual modeling for discrete-event simulation. Albeit that many simulation researchers consider effective conceptual modeling to be vital to the success of a simulation study, there have been few attempts to develop definitions and approaches that are helpful to the development of conceptual models. The discussion above attempts to redress this balance by offering a definition of a conceptual model and outlining the requirements for a conceptual model. The conceptual model definition is useful for providing a sense of direction to simulation modelers during a simulation study. If they do not know what they are heading for, how can they head for it? The requirements provide a means for determining the appropriateness of a conceptual model both during and after development. For researchers, the definition and requirements provide a common foundation for further research in conceptual modeling.

Acknowledgments

This chapter is reproduced, with minor editing, from: Robinson, S. 2008. Conceptual modeling for simulation part I: Definition and requirements. *Journal of the Operational Research Society* 59 (3): 278–290. © 2008 Operational Research Society Ltd. Reproduced with permission of Palgrave Macmillan.

Some sections of this chapter are based on the following:

- Robinson, S. 2004. *Simulation: The Practice of Model Development and Use*. Chichester, UK: Wiley.
- Robinson, S. 2004. Designing the conceptual model in simulation studies. In *Proceedings of the 2004 Operational Research Society Simulation Workshop (SW04)*, ed. S.C. Brailsford, L. Oakshott, S. Robinson, and S.J.E. Taylor, 259–266. Birmingham: Operational Research Society.
- Robinson, S. 2006. Issues in conceptual modelling for simulation: Setting a research agenda. In *Proceedings of the Operational Research Society Simulation Workshop (SW06)*, ed. J. Garnett, S.C. Brailsford, S. Robinson, and S.J.E. Taylor, 165–174. Birmingham: Operational Research Society.
- Robinson, S. 2006. Conceptual modeling for simulation: Issues and research requirements. In *Proceedings of the 2006 Winter Simulation Conference*, ed. L.F. Perrone, F.P. Wieland, J. Liu, B.G. Lawson, D.M. Nicol, and R.M. Fujimoto, 792–800. Piscataway, NJ: IEEE.

The Ford engine plant example is used with the permission of John Ladbrook, Ford Motor Company.

References

Arthur, J.D., and R.E. Nance. 2007. Investigating the use of software requirements engineering techniques in simulation modelling. *Journal of simulation* 1 (3): 159–174.

Balci, O. 1994. Validation, verification, and testing techniques throughout the life cycle of a simulation study. *Annals of operations research* 53: 121–173.

Balci, O. 2001. A methodology for certification of modeling and simulation applications. *ACM transactions on modeling and computer simulation* 11 (4): 352–377.

Balci, O., and R.E. Nance. 1985. Formulated problem verification as an explicit requirement of model credibility. *Simulation* 45 (2): 76–86.

Bankes, S. 1993. Exploratory modeling for policy analysis. *Operations research* 41 (3): 435–449.

Borah, J.J. 2002. Conceptual modeling: The missing link of simulation development. In *Proceedings of the 2002 Spring Simulation Interoperability Workshop*. www.sisostds.org (accessed February 10, 2009).

Brooks, R.J., and A.M. Tobias. 1996a. Choosing the best model: Level of detail, complexity and model performance. *Mathematical and computer modeling* 24 (4): 1–14.

Brooks, R.J., and A.M. Tobias. 1996b. A framework for choosing the best model structure in mathematical and computer modeling. In *Proceedings of the 6th Annual Conference AI, Simulation, and Planning in High Autonomy Systems*, 53–60. University of Arizona.

Carson, J.S. 1986. Convincing users of model's validity is challenging aspect of modeler's job. *Industrial engineering* 18 (6): 74–85.

Checkland, P.B. 1981. *Systems Thinking, Systems Practice.* Chichester, UK: Wiley.

Chwif, L., M.R.P. Barretto, and R.J. Paul. 2000. On simulation model complexity. In *Proceedings of the 2000 Winter Simulation Conference*, ed. J.A. Joines, R.R. Barton, K. Kang, and P.A. Fishwick, 449–455. Piscataway, NJ: IEEE.

Cochran, J.K., G.T. Mackulak, and P.A. Savory. 1995. Simulation project characteristics in industrial settings. *Interfaces* 25 (4): 104–113.

Courtois, P.J. 1985. On time and space decomposition of complex structures. *Communications of the ACM* 28 (6): 590–603.

Davies, R., P. Roderick, and J. Raftery, J. 2003. The evaluation of disease prevention and treatment using simulation models. *European journal of operational research* 150: 53–66.

DMSO. 2005. Defense Modeling and Simulation Office, HLA. www.sisostds.org (accessed February 10, 2009).

Evans, J.R. 1992. Creativity in MS/OR: improving problem solving through creative thinking. *Interfaces* 22 (2): 87–91.

Ferguson, P., W.S. Humphrey, S., Khajenoori, et al. 1997. Results of applying the personal software process. *Computer* 30 (5): 24–31.

Fishwick, P.A. 1995. *Simulation Model Design and Execution: Building Digital Worlds.* Upper Saddle River, NJ: Prentice-Hall.

Gass, S.I., and L.S. Joel. 1981. Concepts of model confidence. *Computers and operations research* 8 (4): 341–346.

Guru, A., and P. Savory. 2004. A template-based conceptual modeling infrastructure for simulation of physical security systems. In *Proceedings of the 2004 Winter Simulation Conference*, ed. R.G. Ingalls, M.D. Rossetti, J.S. Smith, and B.A. Peters, 866–873. Piscataway, NJ: IEEE.

Haddix, F. 2001. Conceptual modeling revisited: a developmental model approach for modeling and simulation. In *Proceedings of the 2001 Fall Simulation Interoperability Workshop*. www.sisostds.org (accessed February 10, 2009).

Henriksen, J.O. 1988. One system, several perspectives, many models. In *Proceedings of the 1988 Winter Simulation Conference*, ed. M. Abrams, P. Haigh, and J. Comfort: 352–356. Piscataway, NJ: IEEE.

Henriksen, J.O. 1989. Alternative modeling perspectives: Finding the creative spark. In *Proceedings of the 1989 Winter Simulation Conference*, ed. E.A. MacNair, K.J. Musselman, and P. Heidelberger, 648–652. Piscataway, NJ: IEEE.

Hodges, J.S. 1991. Six (or so) things you can do with a bad model. *Operations research* 39 (3): 355–365.

Innis, G., and E. Rexstad. 1983. Simulation model simplification techniques. *Simulation* 41 (1): 7–15.

Kotiadis, K. 2007. Using soft systems methodology to determine the simulation study objectives. *Journal of simulation* 1 (3): 215–222.

Lacy, L.W., W. Randolph, B. Harris, et al. 2001. Developing a consensus perspective on conceptual models for simulation systems. In *Proceedings of the 2001 Spring Simulation Interoperability Workshop*. www.sisostds.org (accessed February 10, 2009).

Law, A.M. 1991. Simulation model's level of detail determines effectiveness. *Industrial engineering* 23 (10): 16–18.

Law, A.M. 2007. *Simulation Modeling and Analysis*, 4th ed. New York: McGraw-Hill.

Lucas, T.W., and J.E. McGunnigle. 2003. When is model complexity too much? Illustrating the benefits of simple models with Hughes' salvo equations. *Naval research logistics* 50: 197–217.

Morris, W.T. 1967. On the art of modeling. *Management science* 13 (12): B707–717.

Musselman, K.J. 1992. Conducting a successful simulation project. In *Proceedings of the 1992 Winter Simulation Conference*, ed. J.J. Swain, D. Goldsman, R.C. Crain, and J.R. Wilson, 115–121. Piscataway, NJ: IEEE.

Nance, R.E. 1994. The conical methodology and the evolution of simulation model development. *Annals of operations research* 53: 1–45.

Nydick, R.L., M.J. Liberatore, and Q.B. Chung. 2002. Modeling by elaboration: An application to visual process simulation. *INFOR* 40 (4): 347–361.

Ören, T.I. 1981. Concepts and criteria to assess acceptability of simulation studies: A frame of reference. *Communications of the ACM* 28 (2): 190–201.

Ören, T.I. 1984. Quality assurance in modeling and simulation: a taxonomy. In *Simulation and model-based methodologies: An integrative approach*, ed. T.I. Ören, B.P. Zeigler, and M.S. Elzas, 477–517. Heidelberg, Germany: Springer-Verlag.

Pace, D.K. 1999. Development and documentation of a simulation conceptual model. In *Proceedings of the 1999 Fall Simulation Interoperability Workshop*. www.sisostds.org (accessed February 10, 2009).

Pace, D.K. 2000a. Simulation conceptual model development. In *Proceedings of the 2000 Spring Simulation Interoperability Workshop*. www.sisostds.org (accessed February 10, 2009).

Pace, D.K. 2000b. Ideas about simulation conceptual model development. *Johns Hopkins APL technical digest* 21 (3): 327–336.

Pace, D.K. 2002. The value of a quality simulation conceptual model. *Modeling and simulation magazine* 1 (1): 9–10.

Pace, D.K. 2003. Thoughts about the simulation conceptual model. In *Proceedings of the 2003 spring simulation interoperability workshop*. www.sisostds.org (accessed February 10, 2009).

Pidd, M. 1999. Just modeling through: a rough guide to modeling. *Interfaces* 29 (2): 118–132.

Pidd, M. 2003. *Tools for Thinking: Modeling in Management Science*, 2th ed. Chichester, UK: Wiley.

Pidd, M. 2004. *Computer Simulation in Management Science*, 5th ed. Chichester, UK: Wiley.

Powell. S.G. 1995. Six key modeling heuristics. *Interfaces* 25 (4): 114–125.

Pritsker, A.A.B. 1986. Model evolution: a rotary table case history. In *Proceedings of the 1986 Winter Simulation Conference*, ed. J. Wilson, J. Henriksen, and S. Roberts, 703–707. Piscataway, NJ: IEEE.

Pritsker, A.A.B. 1987. Model evolution II: An FMS design problem. In *Proceedings of the 1987 Winter Simulation Conference*, ed. A. Thesen, H. Grant, and W.D. Kelton, 567–574., Piscataway, NJ: IEEE.

Pritsker, A.A.B. 1998. Principles of simulation modeling. In *Handbook of simulation*, ed. J. Banks, 31–51. New York: Wiley.

Robinson, S. 1994. Simulation projects: Building the right conceptual model. *Industrial Engineering* 26 (9): 34–36.

Robinson, S. 2001. Soft with a hard centre: Discrete-event simulation in facilitation. *Journal of the operational research society* 52 (8): 905–915.

Robinson, S. 2002. Modes of simulation practice: Approaches to business and military simulation. *Simulation practice and theory* 10: 513–523.

Robinson, S. 2004. *Simulation: The Practice of Model Development and Use.* Chichester, UK: Wiley.

Robinson, S. 2005. Distributed simulation and simulation practice. *Simulation: Transactions of the society for modeling and computer simulation* 81 (1): 5–13.

Robinson, S., and M. Pidd. 1998. Provider and customer expectations of successful simulation projects. *Journal of the operational research society* 49 (3): 200–209.

Salt, J. 1993. Simulation should be easy and fun. In *Proceedings of the 1993 Winter Simulation Conference*, ed. G.W. Evans, M. Mollaghasemi, E.C. Russell, and W.E. Biles, 1–5. Piscataway, NJ: IEEE.

Schmeiser, B.W. 2001. Some myths and common errors in simulation experiments. In *Proceedings of the 2001 Winter Simulation Conference*, ed. B.A. Peters, J.S. Smith, D.J. Medeiros, and M.W. Rohrer, 39–46. Piscataway, NJ: IEEE.

Schruben, L., and Yücesan, E. 1993. Complexity of simulation models: A graph theoretic approach. In *Proceedings of the 1993 Winter Simulation Conference*, ed. G.W. Evans, M. Mollaghasemi, E.C. Russell, and W.E. Biles, 641–649. Piscataway, NJ: IEEE.

Sevinc, S. 1990. Automation of simplification in discrete event modeling and simulation. *International journal of general systems* 18: 125–142.

Shannon, R.E. 1975. *Systems Simulation: The Art and Science.* Englewood Cliffs, NJ: Prentice-Hall.

Teeuw, W.B., and H. van den Berg. 1997. On the quality of conceptual models. In *Proceedings of the ER '97 Workshop on Behavioral Models and Design Transformations: Issues and Opportunities in Conceptual Modeling*, ed. S.W. Liddle. osm7.cs.byu.edu/ER97/workshop4/ (accessed February 10, 2009).

Thomas, A., and P. Charpentier. 2005. Reducing simulation models for scheduling manufacturing facilities. *European journal of operational research* 161 (1): 111–125.

van der Zee, D.J., and J.G.A.J. van der Vorst. 2005. A modeling framework for supply chain simulation: Opportunities for improved decision making. *Decision sciences* 36 (1): 65–95.

Ward, S.C. 1989. Arguments for constructively simple models. *Journal of the operational research society* 40 (2): 141–153.

Webster, D.B., M.L. Padgett, G.S. Hines, et al. 1984. Determining the level of detail in a simulation model: A case study. *Computers and industrial engineering* 8 (3/4): 215–225.

Wild, R. 2002. *Operations Management*, 6th ed. London: Continuum.

Willemain, T.R. 1994. Insights on modeling from a dozen experts. *Operations research* 42 (2): 213–222.

Willemain, T.R. 1995. Model formulation: What experts think about and when. *Operations research* 43 (6): 916–932.

Yin, H.Y., and Z.N. Zhou. 1989. Simplification techniques of simulation models. In *Proceedings of Beijing International Conference on System Simulation and Scientific Computing*, 782–786. Piscataway, NJ: IEEE

Zeigler, B.P. 1976. *Theory of Modeling and Simulation*. New York: Wiley.

2

Complexity, Level of Detail, and Model Performance

Roger J. Brooks

CONTENTS

2.1 Introduction

Mathematical and simulation models are used extensively in many areas of science and industry from population genetics to climate modeling and from simulating a factory production line to theories of cosmology. Modeling may be undertaken for a number of reasons but the most common aim is to predict the behavior of a system under future circumstances. A model may be purely predictive or it may be part of a decision making process by predicting the system behavior under alternative decision scenarios. There are other occasions when a model is just descriptive, simply summarizing the modeler's understanding of the system (Jeffers 1991). The understanding of the system gained by the modeler and the user can also be an important benefit of the project (Fripp 1985), particularly in scientific research when it can be the principal objective. Equally, a modeling project may have other objectives such as helping to design experiments or identifying research requirements. Despite the great variation in the types of model and their usage, the modeling process itself will take a similar form for most projects and can typically be split into the steps of problem formulation, conceptual modeling, collection and analysis of data, model construction, verification and validation, experimentation, analysis of results, conclusions, and implementation. The modeling steps do not form a linear process but one that may involve many loops, and so the choice of model affects each of the steps.

The outcome of the conceptual modeling step is a conceptual model, and this chapter follows the definition of Brooks and Robinson (2001) that a conceptual model is "a software independent description of the model that is to be constructed." The conceptual model specifies how the system (virtual world) being simulated should work—the entities that it contains and all the interactions, rules, and equations that determine their behavior. The specification includes the type of model (e.g., whether the model state changes continuously or discretely), and the scope or boundary of the model. Therefore, conceptual modeling consists of choosing the model to use in the project. The actual building of the model is the model construction step.

The step of problem formulation consists of understanding the problem and setting the scope and objectives for the project. There is no agreed definition of conceptual modeling within simulation and operational research (OR) and, for example, setting the project objectives is sometimes included as part of conceptual modeling (Robinson 2008).

This chapter examines how the choice of model and the comparison of alternative models has been discussed and investigated in the literature. It includes both mathematical and simulation models from a range of applications in different areas of science, on the basis that in each case the underlying modeling problem is similar—deciding what to include in the model to produce the best outcome for the project. The characteristics most often used when comparing different models are complexity and level of detail, but without these terms being defined clearly. Three more specific characteristics, namely size, connectedness, and calculational complexity are proposed for comparing models. A better understanding of the relationship between model characteristics and the outcome of the modeling project would help the conceptual modeling process since choosing the model implicitly involves predicting the impact of the model on the project. Eleven model performance elements are set out that cover different aspects of how the model affects the project outcome. An experiment to investigate the effect of size, connectedness, and calculational complexity on some of the performance elements is described.

2.2 Choice of the Best Model

The aim of the conceptual modeling step is to select the conceptual model that will lead to the best overall project outcome. The choice of models is usually very broad because a great deal is known about the system relationships on many different spatial and time scales (Courtois 1985). Finding the best model is often viewed to a large extent as the problem of choosing the appropriate level of detail and this is considered one of the most difficult aspects of the modeling process (Law 1991) and one that has a major effect on the successfulness of the project (Tilanus 1985, Ward 1989, Salt 1993).

By viewing the selection of the conceptual model in this way, the alternative models are effectively being ordered by the characteristic of level of detail, which is the most common characteristic used to compare models. This is done in the hope that there will be similarities with previous studies in the effect of the level of detail on model performance, so that experience from these studies can be applied in the selection of the current model. For example, a model that is too simple will be unrealistic and so its results will be, at best, of little use and, at worst, misleading. On the other hand, considerable resources are usually required to build a complex model and so, if the model is too complex, constraints on resources may prevent the completion of the project (here, it is assumed that a more detailed model will be more complex, although the meaning of level of detail and complexity are discussed further in Section 2.4). It is generally harder to understand the relationships contained in a complex model and this makes the interpretation of the results more difficult, possibly leading to incorrect conclusions being drawn. A complex model is probably more likely to contain errors as it is harder to verify that the model is working as intended.

The advice given on selecting the level of detail seems to consist almost entirely of vague principles and general guidelines. A commonly quoted maxim is Ockham's (or Occam's) razor, attributed to the fourteenth-century philosopher William of Ockham, and translated (from the Latin) as "entities should not be multiplied without necessity," or "it is vain to do by more what can be done by fewer." In other words, choose the simplest model that meets the modeling objectives. Often, the advice given is to start from a simple model and progressively add detail until sufficient accuracy is obtained. It is important to match the level of detail of the model with the modeling objectives and with the available data (Law et al. 1993, Jakeman and Hornberger 1993, Hunt 1994). However, knowledge of these principles is of only limited use to the modeler and the choice of the best model seems to be regarded as more of an art than a science.

2.3 Model Performance

The evaluation of the performance of a model should cover the impact of the model on all aspects of the project. Such an assessment is dependent on the particular project; a model would give different performance if used for two different projects. Based on my own modeling experience, it is proposed that the full evaluation should include the following 11 performance elements:

Results

1. The extent to which the model describes the behavior of interest (i.e., whether it has adequate scope and detail)
2. The accuracy of the model's results
3. The ease with which the model and its results can be understood

Future use of the model

4. The portability of the model and the ease with which it can be combined with other models

Verification and validation

5. The probability of the model containing errors (i.e., the model constructed does not match the conceptual model)
6. The accuracy with which the model fits the known historical data (black box validity)
7. The strength of the theoretical basis of the model including the quality of input data (the credibility of the model or white box validity)

Resources required

8. The time and cost to build the model (including data collection, verification, and validation)
9. The time and cost to run the model
10. The time and cost to analyze the results of the model
11. The hardware requirements (e.g., computer memory) of running the model

An assessment of the modeling project as a whole would compare the benefits of the project with the costs incurred. The performance assessment of a model should consist of the impact of the model on these costs and benefits. The performance elements attempt to focus purely on the effect of the model, but it is not possible to isolate this entirely. For example, the resources required to build the model not only depend upon the model used, but also on a number of other factors such as the ability and experience of the modelers. A single absolute measure of model performance cannot be obtained but a meaningful comparison of alternative models in similar circumstances should be possible.

The quality of the project conclusions depends upon the quality of the results, which is a combination of their accuracy (element 2) and the extent to which they address the modeling objectives (element 1). Clearly, if the results are inaccurate then the decisions taken and conclusions drawn are likely to be incorrect, although the degree of accuracy required depends on the objectives. In certain circumstances, for example, the relative ordering of two values may be all that is important rather than their absolute values. In the terms of Zeigler (1976), element 1 is the extent to which the model results cover the experimental frame. It is also important that the model includes the system elements whose effects are to be investigated, i.e., the decision variables (unless the model specifically excludes them to allow comparison with a previous model that does include them). It is important that the model and the results can be understood (element 3) to facilitate the analysis, and increased understanding of the systems may be a significant benefit in itself. The use of all or part of the model in future projects (element 4) can also be an important benefit.

In most cases the model is predictive and so elements 1 and 2 cannot be assessed until some time after the modeling project has been completed and, indeed, the perceptions of the overall success of the project may change over time (Robinson and Pidd 1998). Acceptance of the conclusions from the modeling and implementation of the recommendations requires that the user has confidence in the model and the user's confidence should be based on elements 5–7. It is therefore important not only that (with the benefit of hindsight) the model produced realistic results (element 2), but also that the model was seen to be sufficiently realistic at the time the project was carried out (elements 5–7). It is possible for a very unrealistic model to produce accurate results (for example, due to compensating errors, particularly if the results are a single value). Even if the results of such a model were accepted and these led to the correct decisions being taken, the understanding gained of the system is likely to be incorrect and this may have serious consequences in the future. Elements 5–7 take this into account by giving an assessment of the underlying quality of the model. Successful reuse of the model also requires the model to have a sound basis, as well as requiring the model to be portable (element 4).

Element 5 relates to the process of verification and so is concerned with errors occurring in the model construction step. It is not possible, for most models, to ensure that the model constructed will operate as intended in all circumstances (i.e., to fully verify it) and so the model may contain errors (Gass 1983, Tobias 1991). The probability of the model containing a particular error is a product of the probability of the initial model containing the error, which partly depends upon the choice of model (as well as the quality of the model building), and the conditional probability of the error not being discovered in the verification process, which also partly depends upon the choice of model (as well as the quality of the verification process). There is a trade-off here between the performance elements; the model is less likely to contain errors if more resources are put into building and verifying the model.

Elements 6 and 7 assess how well the conceptual model matches the real system. It is not sufficient for the model output simply to fit the historical data (black box validation); confidence in the model mechanisms, on a theoretical basis, by comparison with knowledge of the real system or on the basis of successful previous experience (white box validation), is also important. It is possible for a model to fit the historical data well but to give poor predictions if the basis of the model is incorrect, particularly if the conditions for the period being predicted are very different to those in the past. If the system being modeled is one that does not currently exist, validation can consist only of an assessment of model credibility (Gass 1983), i.e., element 7.

The effect of the choice of the model on the costs of the project is addressed by elements 8–11. In some projects, a model already exists and the aim of the project is to simplify it or to make it more complex. In this case the time and cost of building the model becomes the time and cost of modifying the model. Considerable effort can be required in order to simplify an existing model (Rexstad and Innis 1985).

An assessment of model performance requires a measurement to be made of each of the performance elements and this is far from straightforward. It should be possible to evaluate elements 1, 6, 9, and 11 fairly easily in most cases, although care is required in the interpretation of the measures used. However, the remaining elements are hard to quantify and a subjective qualitative assessment may be all that is be possible. The accuracy of the results (element 2) may not be known until a long time after the project was completed and may never be known for decision scenarios that were not implemented. The ease of understanding and the probability of errors both contain a human element, which makes a numerical evaluation difficult (elements 3 and 5). Similarly, the strength of the theory behind the model and the credibility of its structure is subjective (element 7). A comparison of the resources required to build and analyze alternative candidate models should be those required if the model is built from scratch with no prior knowledge of the system (elements 8 and 10). Such an assessment therefore ought ideally to consist of measuring the resources used by independent modeling teams of equal modeling experience and ability but this will not be feasible in most instances. Meaningful measures for the extent to which the scope and detail of the results matches the problem requirements and particularly the model portability are also likely to be difficult to derive (elements 1 and 4). An overall assessment of model performance requires a relative weighting to be given to each of the elements and such a weighting will be subjective and will vary considerably from study to study. It may be possible, for a particular study, to ignore some of the elements as insignificant in terms of the overall performance. However, if a number of studies attempt to measure at least some of the performance elements, the measurement procedure is likely to improve.

2.4 Level of Detail and Complexity

There is no single accepted definition of either level of detail or complexity and, in fact, a formal definition is rarely attempted, but, when applied to a model, the level (or amount) of detail usually means an assessment of the extent to which the observable system elements and the assumed system relationships are included in the model. It is usually assessed qualitatively and is most often used just to rank alternative models. Level of detail tends to refer to the system that the model represents (for example, in the case of a model of a production line, the number of machines, parts, etc. included in the model) rather than to the precise way in which the model is implemented (such as the number of variables used). Models are often described as being detailed, meaning that the model contains most of the elements and interactions thought to exist in the system being modeled. For some examples of the usage of the level or amount of detail of a model, see Shannon (1975), Banks and Carson (1984), Stockle (1992), and Law (2007).

The term complexity is much more common than level of detail although it is used in many different ways (Bunge 1963, Henneman and Rouse 1986, Gell-Mann 1995) such as the difficulty in computing a function (computational complexity), a structural attribute of a piece of software (software complexity), the difficulty experienced by people in perceiving information or solving problems in a particular environment (behavioral complexity) and the complexity of terms, sentences and theories (logical and semantic complexity). Weaver (1948) categorized the problems tackled by science as problems of simplicity, organized complexity (the most difficult), and disorganized complexity, and Chaitin (1975) equated the complexity of a number with its randomness.

More recently, the phrase *science of complexity* has been used to describe a scientific discipline covering the study of complex adaptive systems that give rise to "emergent properties" (Morowitz 1995, Flatau 1995). In complexity science, the system structures are complex in the sense that they usually have many interconnected parts but the term complexity refers, at least to an extent, to the behavior of the system in exhibiting emergent properties. However, there does not appear to be an agreed definitions of precisely what *complexity science* means (Amaral and Uzzi 2007). Some of the focus in this area has been on identifying and understanding different types of behavior (e.g., Langton 1990) often split into three categories: ordered (with very little change in behavior over time), chaos (changing behavior with little or no patterns or regularity), and the category "edge of chaos" at the phase transition between the other two categories (changing behavior but a high degree of regularity and robustness to perturbations). The models studied are often networks and one of the factors determining the type of behavior is the number and type of connections between the network nodes. For example, in Kauffman's work on Boolean networks, as the number of connections between the nodes increases the type of behavior tends to change from the ordered category to the edge of chaos category to the chaos category (Kauffman 1993).

Systems theory has perhaps seen the greatest discussion of the concept of complexity although this has resulted in a number of different meanings. Flood and Carson (1993) in their extensive discussion of complexity considered that complexity meant "anything that we find difficult to understand." They therefore viewed complexity as a combination of the structure of an object (particularly the number of elements and relationships) and of the nature of people and the way in which they interact with the object in the particular context. Golay et al. (1989) also equated complexity with the difficulty in understanding the system, whereas Simon (1964) took a complex system to be "one made up of a large number of parts that interact in a non simple way." Casti (1979) referred to the difficulty in understanding a system's structure as static complexity, which he distinguished from the dynamic complexity of the system output noting that a simple structure can produce complex output (for example, a simple system can be chaotic). Rosen (1977) defined a complex system "as one with which we can interact in many different kinds of ways, each requiring a different mode of system description," which varies according to the point of view of the observer, whereas George (1977) considered a system to be complex when it contained sufficient redundancy to be able to function despite the presence of many defects.

Dictionaries usually define complex as being something consisting of many parts, as well as often defining it as something that is difficult to understanding (see also Ward's [1989] discussion of the meaning of simple). For example, the Collins English dictionary's definition of complex is "1. Made up of various interconnected parts; composite. 2. (of thoughts, writing, etc.) Intricate or involved," and Chamber's English dictionary gives "Composed of more than one, or of many parts: not simple: intricate: difficult." Clearly, objects that have many interacting parts do tend to be difficult to understand and vice versa. The complexity of a model is therefore sometimes used to mean the difficulty in understanding the model or the difficulty (in terms of resources required) in generating model behavior (Zeigler 1976, 1984; Schruben and Yücesan 1993). However, these are performance measures rather than model attributes and so, in the first case, for example, the commonly asserted disadvantage of complex models that they are difficult to understand is just a tautology (Ward 1989).

Complexity is commonly used to refer to a model when comparing the output of alternative models (for example, Blöschl and Kirnbauer [1991], le Roux and Potgieter [1991], Durfee [1993], Palsson and Lee [1993], and Smith and Starkey [1995]). In such cases, complexity is usually not defined. However, it is the structure of the alternative models that is described with the difficulty in understanding being rarely mentioned. It therefore appears that the complexity of a model usually refers to a structural property of the model and this is certainly the appropriate usage when comparing the characteristics and performance of different models.

Complexity is therefore being used in a very similar way to level of detail and, in comparing models, a number of authors appear to equate the terms complex and detailed when referring to the models (for example, Webster et al. [1984], Ward [1989], Stockle [1992], and Durfee [1993]). Certainly a simple model is generally considered to be the opposite of both a detailed model and a complex model. Applying the dictionary definition, complexity would be a measure of the number of constituent parts and relationships in the model and so complexity should differ from level of detail in referring to the actual model elements rather than the system elements and this is assumed here. However, the level of detail largely determines the complexity and, in most cases, the ordering of alternative models by level of detail or complexity will be the same. If the system being modeled is extremely complex then it is possible to build a model that has many parts but which omits many system elements, so that such a model would be complex but not detailed. In comparing two models, occasionally the more detailed model may be less complex if, for example, an approximation to a detailed system relationship requires more model elements and connections than the actual relationship, although such an approximation would be unlikely to have any advantages. There may also be several ways in which a given modeling assumption can be implemented (such as alternative algorithms for generating pseudorandom numbers), so that models of the same level of detail may have different complexity.

In summary, one distinction between the different uses of complexity is that sometimes it refers to the underlying structure of the model or system, and sometimes it refers to the dynamic behavior of the model or system. These are quite separate since a simple structure can sometimes produce complex behavior (e.g., chaos) and a complex structure can sometimes produce simple behavior with a high degree of regularity. This paper focuses on the structural characteristics of conceptual models and how these can be related to the different aspects of model performance on the project. Therefore, in the remainder of the paper the complexity of the model or conceptual model will refer to the complexity of its structure. Assuming that the model can be considered as a number of interconnected parts, or components, the overall complexity of the model is taken here to be a combination of three elements: the number of components, the pattern of the connections, (which components are related) and the nature of the connections (the complexity of the calculations determining the relationships). The aim here is to identify invariant structural attributes of the conceptual model and so the frequency of occurrence of each connection has not been included as an element of complexity as this may depend on the particular model runs carried out. Here the three elements are termed size, connectedness, and calculational complexity, respectively.

2.5 Measuring Model Complexity

If a model can be specified as connected components then it can be represented as a graph, with the nodes of the graph representing the components and the edges representing the connections (i.e., the relationships) and graph theory measures used to measure size and connectedness. However, for any given model there are likely to be several possible graphs and many alternative measures.

Models implemented as computer programs can, for most programming languages, be graphically represented by the program control graph in which the nodes represent blocks of code in which control is sequential and the edges represent branches in the program. The complexity measure proposed by McCabe (1976) was the number of edges less the number of nodes plus twice the number of connected components in the program control graph. For a single program (so the number of connected components is one), this is equal to the cyclomatic number (which is the number of nodes less the number of edges plus the number of connected components) of the program control graph with an additional edge added to join the last component to the first. This additional edge strongly connects the graph so that the cyclomatic number is equal to the maximum number of linearly independent circuits in the graph. It therefore represents the number of basic paths through the program, which McCabe (1976) equated to complexity.

A graph of a discrete-event simulation model is also possible by depicting the events as the nodes and the relationships between the events as the edges (Schruben 1983). Events are activities that alter the state of the model and two events are related if the occurrence of one of the events can cause the other to occur (or can cancel the occurrence of the other). A directed edge from event A to event B indicates that if event A occurs and certain conditions hold then, after a specified time, event B will occur. Schruben and Yücesan (1993) suggested several graph theory measures, including the cyclomatic number, which could be applied to event graphs to measure the complexity of a model. It is usually not necessary to explicitly model all the events occurring in the system and the event graph can be used to identify events not required. This means, however, that several graphs are possible for the same conceptual model. For consistency, the graph with the minimum number of events should be used for the complexity measure (although it is not clear whether there is only one such graph). Activity cycle diagrams, which connect the possible states of each entity in the model are a further way of representing a discrete-event simulation model as a graph (Pidd 2004).

A graph can also be obtained by assigning each possible model state to a node and representing possible transitions between the states by the edges, or by letting the nodes represent the state variables and the edges interactions between the variables (Zeigler 1976). There can be many choices for the state variables. Graphs of the interaction between the state variables were used to measure the complexity of alternative fate models of toxic substances in a lake by Halfon (1983a, 1983b). He used Bosserman's (1982) $\bar{c}$ measure, which is the proportion of paths of length $\leq n$ (where n is the number of nodes) that exist in the graph. The measure can be obtained using the adjacency matrix A, which has $a_{ij} = 1$ if there is a connection from node i to node j and 0 otherwise. The matrix A^k, obtained by k multiplications of matrix A by itself using Boolean algebra, has $a_{ij} = 1$ if and only if there exists a path from node i to node j of length k. The $\bar{c}$ measure is then given by the sum of all the elements in the matrices $A, A^2, \ldots, A^n$ divided by n^3 (the total number of elements in these matrices).

A graph theory measure may not always be static. Neural networks are typically defined in terms of nodes and connections, and many other adaptive systems models can be represented in this way (Farmer 1990), which gives a natural graph of the model structure. In these models, a complexity measure based on such a graph would change as the model runs as a result of the connections changing.

In comparing models, differences in the complexity of the models may be due to differences in the complexity of the calculations and so the graph theory measures may be inappropriate or may need to be combined with other measures.

An alternative approach to graph theory may be to use concepts from information theory. Golay et al. (1989) used the following information entropy measure as a complexity measure:

$$H = -\sum_{i=1}^{n} p_i \log_2(p_i) \tag{2.1}$$

where H = information entropy, n = number of system states, p_i = probability of the ith state.

They justified the use of information entropy as a complexity measure by arguing that entropy measures the amount of uncertainty and that a more complex model is more difficult to understand and therefore more uncertain. This measure can only be used in the very limited cases when it is practical to estimate the probability of each system state. Golay et al. (1989) applied the measure to systems in which each component had only two states. The use of the entropy measure can also be argued on the basis that it measures the complexity of the behavior of the model (Langton 1990), in terms of both the number of systems states in total and relative proportion of time spent at each state. Again this indicates a likely correlation with the difficulty in understanding the model and its results. This measure, as it stands, does not measure the complexity of the conceptual model as it is not a direct measure of a structural property of the model but rather a measure of the complexity of the model behavior for a particular run. However, it may be possible to use similar concepts to measure the amount of information contained in the model, although it is not clear how to do this at present.

In computer science, many measures (usually termed metrics) of the size and complexity of the code have been proposed and these can be used to measure the complexity of a model taking the form of software. Most of the metrics are based on counting the occurrence of particular items in the code, such as the number of decision points or just the number of lines of code. An alternative approach developed by Henry and Kafura (1981) is to identify the flows of information between separate program procedures and to incorporate the number of different flows into the metric. Many of the software metrics, however, are partly dependent on the programming language used, whereas the aim here is to measure the complexity of the conceptual model, which should be independent of the specific implementation of the model.

The purpose of a complexity measure is to characterize the model so that this information can aid the choice of model by predicting model performance. Ideally we would like to have a single, system independent definition and measure of complexity covering all the aspects of the level of detail of a model and applicable to all conceptual models. However no such definition or measure exists and as a result the term complexity itself is a source of confusion due to its usage in many different contexts. The best approach would seem to be to identify more specific model attributes, such as the attributes of size, connectedness, and calculational complexity discussed in Section 2.4 and to devise measures for these. It is important that such measures and the type of measurement scale should match our intuitive notion of the nature of the attribute, which, for example, is not always the case with software metrics (Fenton 1991). In addition, the earlier in the modeling process in which a measure can be obtained, the more useful it is.

2.6 Relationship between Model Performance and the Level of Detail or Complexity of a Model

The level of detail and complexity of a model are widely recognized as having very important effects on model performance and the relationships between either the level of detail or complexity of the model and model performance have been discussed in general terms in a number of places (for example, by Meisel and Collins [1973], Fishwick [1988], Law [1991], and Salt [1993]). A more complex model is expected to have greater validity and to give more detailed and accurate results, but to use more resources, and to be more likely to contain errors, more difficult to understand and less portable (i.e., better for performance elements 1–3, 5, and 6 but worse for the others).

In the fields of management science and operational research, modeling projects are often carried out for a client with little modeling experience. There are a number of additional reasons why the client will prefer a simple model and so is more likely to implement the results. Ward (1989) set out a number of advantages, from a clients point of view, of using a simple model including the quicker generation of results and the production of results that are easier to understand and less specific thus allowing the clients own preferences to be incorporated. A simple model is also more flexible and so can be adapted more easily if the project objectives change (Law et al. 1993).

The precise nature of the relationships between the level of detail or complexity and the aspects of model performance are poorly understood. There are a few studies that have compared alternative models from different points of view. However, the objectives of these studies have not specifically been to compare the level of detail and performance of the models and as a result they have tended not to quantify either of these attributes. These studies are briefly reviewed as they do provide some indication of the possible relationships.

The studies of Stockle (1992) and Rexstad and Innis (1985) took existing ecological models and attempted to simplify them. Stockle simplified a model of the amount of radiation intercepted by plant canopies. The most detailed model had nine leaf inclination classes, nine azimuth angle classes and 20 layers of leaves. The number of classes of each of these three elements could be reduced to simplify the model and Stockle found that by doing this the computation time of the model could be reduced by a factor of 12 with a negligible change in results and by 63 with only a small change in results. This suggests that many models may be more complicated than they need to be and this often seems to be the case in discrete-event simulation modeling. Innis and Rexstad (1983) produced a list of simplification techniques and they subsequently applied some of these techniques to three models (Rexstad and Innis 1985) but this latter paper focused on the applicability of their techniques to the three models rather than on a detailed comparison of the original and simplified models. Consequently, quantitative measures of model performance were not reported apart from fitness measures for one of the models. It is therefore difficult to assess the extent to which the models had been simplified or the effect of the simplifications on model performance.

Costanza and Sklar (1985), by contrast, did carry out a quantitative comparison of different models. They compared 87 freshwater wetland ecosystem models using measures termed articulation and descriptive accuracy. Diminishing returns indices were calculated for the number of components, time steps, and spatial units in the model (with each index having a different scaling factor). The average of the three indices was calculated for the data and for the model and the minimum of these two numbers used for the articulation measure. This was therefore a measure of the scope and complexity of the problem (i.e., of the experimental frame). The descriptive accuracy index was a validity measure of the fit of the model data against the actual historical data. They were able to calculate both of these measures for 26 of the models and they also calculated a combined articulation and accuracy measure called effectiveness. They found that the models with the highest descriptive accuracy had low articulation (although the majority of the 26 models had low articulation), i.e., the models with the greatest validity tended to be those addressing the simpler problems. It is difficult to draw concrete conclusions from this result as the amount of data is relatively small but Costanza and Sklar hypothesized that, ultimately, greater articulation necessitates less accuracy and that there might be a level of articulation that maximizes effectiveness (which they considered to be the best model). This assumes that greater articulation is desirable, in the sense that a model with greater articulation provides more information about the system. Often, however, the modeling objectives are quite specific and only greater information relevant to the problem (i.e., within the experimental frame) is a benefit.

Webster et al. (1984) viewed the selection of the level of detail as part of the validation process and so the only measure they reported was the goodness of fit against actual data for the alternative timber harvesting models that they compared. They considered the appropriate level of detail to be the simplest model of adequate validity that is consistent with the expected system relationships (ignoring the accuracy of results that can only be assessed subsequently). They used three alternative methods to generate sample data for three input variables in a simulation model (giving 27 alternatives in all): mean value, regression, and a histogram of actual data. For one of the variables they found that the histogram method (which they considered the most complex level) gave output of lower validity than the simpler methods. Four of the models gave adequate validity and so they chose the simplest of these as the final model.

Halfon's studies (1983a, 1983b) compared the structure of alternative models of a toxic substance in a lake at six levels of detail. This was done for a model with six state variables and for a model with 10 state variables and repeated in each case with and without internal recycling (giving four sets of results). He compared the structures of the models, mainly using Bosserman's (1982) $\bar{c}$ measure (described earlier), which was applied to the graphs of interactions between state variables. The level of detail of the models was increased by adding the physical processes in stages in a logical order. He found that, in each case, adding the last few levels of detail only caused a small increase in the number of connections. He argued that it was not worth including these processes as they are unlikely to affect model behavior significantly and the additional parameters add to the amount of uncertainty in the model. It is reasonable to expect diminishing returns as complexity is added. However, the actual performance of the models was not assessed to confirm this. Halfon (1983b) also suggested displaying the comparisons of alternative model structures as a Hasse diagram.

The lack of studies that have specifically sought to examine the effect of level of detail or complexity on model performance means that even if the expected relationships described at the beginning of this section are generally true, the nature of the relationships are unclear (linear, increasing returns, decreasing returns, etc.) and the circumstances in which the relationships break down are not understood. The particular elements of model complexity that have the greatest effect on each performance element have also not been identified.

Consider, for example, the accuracy of model results. Generally a more complex model is expected to be more accurate and as the model becomes more complex the increase in accuracy of adding further complexity is likely to reduce (assuming that additional detail is added in order of relevance), i.e., decreasing returns. Certainly, if there is a mapping between the models so that the more complex model can be reduced to the simpler model by a suitable choice of parameters, then the most accurate complex model must be at least as accurate as the most accurate simple model. However, the choice is often between models of different types or between models for which only an approximate relationship exists. In this case, it is possible for the simpler model to be more accurate, although a comparison of the complexity of the models is more difficult. For example, empirical models are sometimes more accurate than quasi-physically based models, which would generally be considered to be more complex (Decoursey 1992). For some modeling (such as physically based distributed parameter models), the input parameters cannot be directly measured but must be inferred by calibrating the model against the historical data (Allison 1979). This is called the inverse problem of parameter identification and its nature means that there may be a wide range of parameter values that give a good fit. In this case, the results of a model should be a range of predictions rather than a single prediction (Brooks et al. 1994) and a more complex model may give a wider range. The range will depend on the number of parameters in the model and the extent to which they are allowed to vary (i.e., the size of the parameter space).

There may also be occasions when a simpler model takes longer to build. Garfinkel (1984) pointed out that in modeling a large system, which is considered to consist of a large number of subsystems, there is a much greater choice of simple models (which just model a few subsystems thought to be important) than complex models and it may take longer to choose between the alternative simple models than it would have taken to build a model of the whole system.

Also, a simple model, by incorporating only some of the system elements, may allow the identification of system relationships that are obscured in a more complex model and so gives a greater understanding of the system. On the other hand, a complex model may extend understanding by allowing the investigation of the effect on the system of many more factors. The process of identifying, building, and comparing models at different levels of detail can greatly increase the understanding of the system. Such a process could be used to link simple strategic models that are difficult to verify with more detailed tactical models that can be tested against available data (Murdoch et al. 1992). If the main purpose of the study is gaining an understanding of the system then the benefits of building models at several levels of detail may be well worth the additional effort involved.

In computer science software metrics have been developed to control and predict the performance of software projects (Demarco 1982). Attempts have been made to predict the resources required for and the likely number of errors in a piece of software from particular software attributes (such as "complexity"). Fairly strong relationships have been found within particular environments (for example, by Boehm [1981]) although none of these appear to be generally applicable. A similar approach in simulation might help in predicting the performance of alternative conceptual models.

The discussion in this section indicates that the relationship between the level of detail or complexity and model performance is more complicated than some of the comments in the literature would suggest and the lack of studies in this area means that the relationship is poorly understood. What is required is a number of studies that measure the elements of the complexity and performance of alternative models and one such experiment is described in Section 2.8. This would provide data from which to develop empirical relationships and may lead to a theoretical basis for the relationships.

2.7 Simplification and Other Related Areas

The selection of the best model requires not just an appreciation of the likely performance of each model but also a knowledge of the possible alternative models. There are very many models that could be built in most cases and so the best model may not even be identified as a possible model. Structuring the models by level of detail or complexity can help in the search for better models and one way of identifying new models is to take an existing model and then attempt to simplify it (Zeigler 1979). Zeigler (1976) set out four categories of simplification methods; dropping unimportant parts of the model, replacing part of the model by a random variable, coarsening the range of values taken by a variable, and grouping parts of the model together. The simplified model using these methods will be of the same type as the original; it is also possible to replace part of a model with a model of a different type such as analyzing the inputs and outputs of the particular part and replacing it with a regression equation, analytical equation or neural network (if the original part was very complex). Sevinc (1990) developed a semiautomatic simplification program for discrete-event simulation models based on Zeigler's (1976) DEVS model formalism and simplification ideas. I have previously simplified models in population genetics (Brooks et al.

1997a, 1997b), wheat simulation (Brooks et al. 2001), and manufacturing systems (Brooks and Tobias 2000) based on sensitivity analysis and detailed analysis of the behavior and workings of the model. In each case the process of simplification provided important insights into the system behavior (Brooks and Tobias 1999). Innis and Rexstad (1983) listed and described 17 simplification techniques. These are specific techniques, some of which fall under Zeigler's (1976) categories, as well as techniques for replacing part of the model with a different type. Innis and Rexstad (1983) also included techniques for identifying which parts of the model might be suitable for simplification, techniques for reducing the number of model runs or run times and techniques for improving the readability of the model code. They stated that their list was not exhaustive, and it would appear that a general simplification methodology does not exist.

Zeigler's (1976) DEVS model formalism provides a framework within which alternative discrete-event simulation models can be compared. Addanki et al. (1991) proposed representing the alternative models as nodes on a graph with the edges representing the changes in assumptions from one model to another. Moving around the graph is an alternative way of searching the space of models to which Addanki et al. (1991) applied artificial intelligence techniques. An approach applied to engineering models has been to generate a database of model fragments and then to automate the process of selecting and combining the fragments to produce the model (Falkenheimer and Forbus 1991, Nayak 1992, Gruber 1993). Developments have also taken place in variable resolution modeling, which allows the level of detail of the model to be changed easily even while the model is running (e.g., Davis and Hillestad 1993), and this may be a suitable environment within which to investigate the effect of level of detail.

2.8 Experiment on Model Characteristics and Performance

In order to try and improve the understanding of the relationship between model characteristics and model performance a small scale experiment was carried out. Among the 11 performance elements set out in Section 2.3 are the time taken to build the model, the likelihood of errors, and the ease of understanding of the model and the results. These were compared for four discrete-event simulation models of production lines by analyzing the performance of the 33 students on the MSc Operational Research course at the University of Birmingham (UK) in answering questions on the models and in building the models. The reason for this approach is that, in order to assess the effects of the differences between the models on these performance elements, the models should be built by different people of roughly equal ability and experience. Otherwise, if more than one model is built by the same person, building, and analyzing the first model helps with the next.

As discussed in the previous sections, the complexity of a model is the most common model characteristic related to performance in the literature, and yet it is not defined clearly. Section 2.4 proposed that the overall complexity of a model can be considered as a combination of its size (the number of nodes or elements), its connectedness (the average number of connections per element), and its calculational complexity (the complexity of the calculations making up the connections). The aim of the experiment was to examine the effects of these characteristics and so the models were devised to differ in these three aspects.

The models used are shown in Figure 2.1. Since they represent production lines, the natural definition for the elements is machines and buffers with the connections being the routes taken by the parts. Models A and B both have eight machines and eight buffers in the same basic layout with model A having more part routes (23 compared to 19) and hence higher connectedness. Model C has five machines and five buffers laid out in the same way as a portion of model A and differs from A mainly in size. Model D has only three machines and three buffers but has the most complex calculations to determine the part routes. Model D has high connectedness and calculational complexity.

The models were assigned at random to the students. The students were quite inexperienced modelers, having received between 14 and 16 hours of tuition, mainly consisting of hands on experience together with some formal teaching and demonstrations. The first stage of the experiment aimed to compare how easy the models were to understand. The students were each asked the same four written questions on aspects of the behavior of the particular model assigned to them, and were provided with the model description and selected model output. The second stage focused on model building and the students were each timed as they built their model using the WITNESS software (Lanner Group Ltd., Redditch, UK). The number of errors in each model was subsequently determined (the students were instructed to build the model as quickly as they could but not to test it). The results are shown in Table 2.1.

Using analysis of variance (ANOVA), the differences between the models are statistically significant at the 5% level for build time ($P = 0.032$), question 2 ($P = 0.014$) and question 3 ($P = 0.022$), but not for question 1, question 4, the average mark for all questions and the number of errors.

For build time calculational complexity appears to have the most effect with model D taking considerably longer to build than the other models. With a package like WITNESS, which is user-friendly and already contains many of the constructs required, thinking time is the most important component of the build time, and so it is the complex and less familiar commands that are the most important. Observations also indicated that the

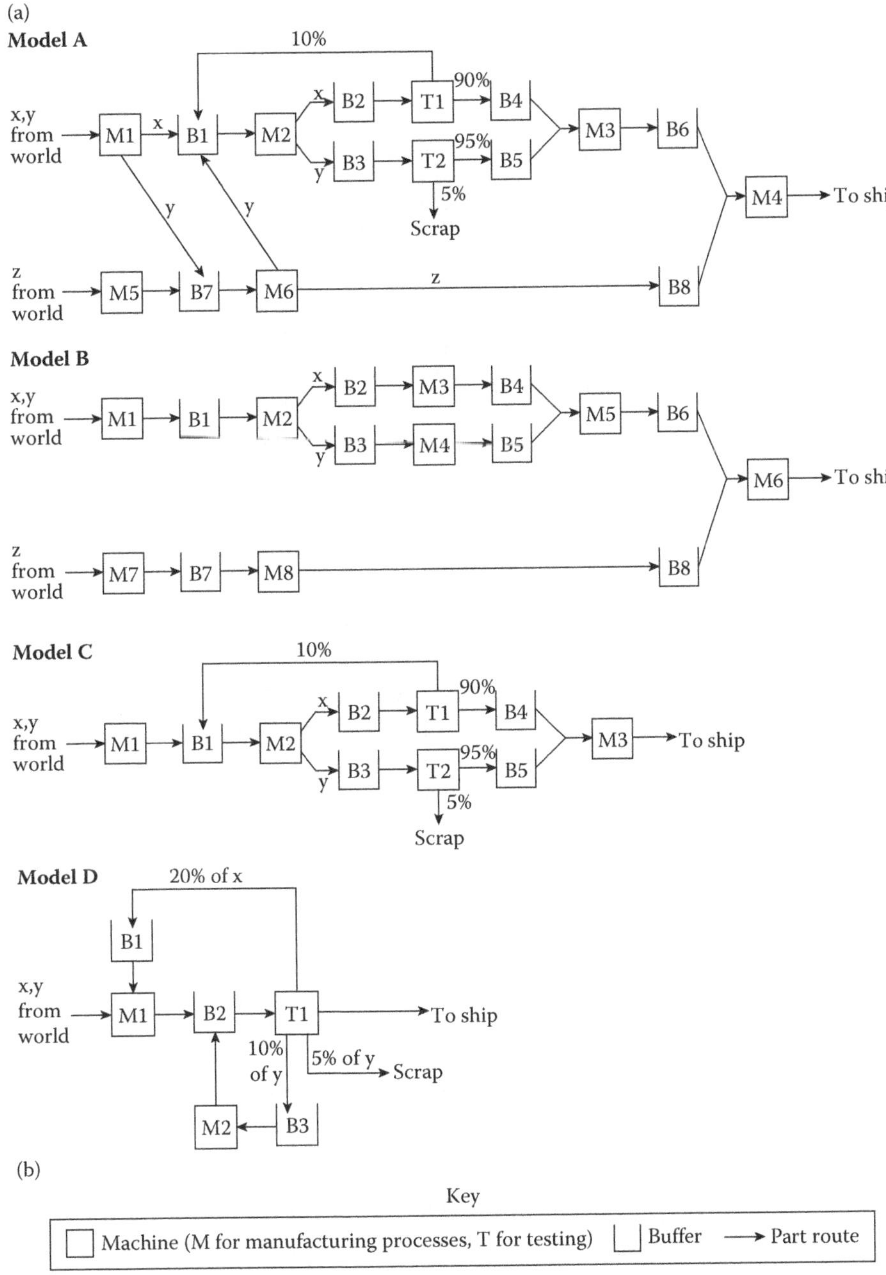

FIGURE 2.1
Process diagrams of the models used in the experiment.

TABLE 2.1

Results from the Experiment

	Aspects of Complexity			Performance		
Model	Size Number of Elements	Connectedness Av. Connections per Element	Calculational Complexity	Understanding Average Marks	Build Time Average Number of Minutes	Errors Average Number of per Model
A	16	1.43	Medium	33%	52.5	2.63
B	16	1.19	Low	45%	44.4	2.25
C	10	1.40	Medium	39%	46.3	1.63
D	6	1.67	High	57%	62.4[a]	2.89[a]

[a] Includes two students who had not quite finished within the maximum time. Six of the 26 errors on this model are omissions in these two models that are probably due to lack of time.

aspects of the model that were easy to code were completed very quickly by the students.

The questions were analyzed both by comparing the marks and by considering the reasoning process required to answer each question, which is discussed in detail in Brooks (1996). Students were asked to give a reason for their answer and considerable importance was given to this since the aim was to assess understanding. Both the correct answer and correct reason were required to score 1 mark. If either the answer or reason was only partially correct then 1/2 mark was awarded. An incorrect answer or the correct answer with an incorrect reason scored 0 marks. As stated above, the significant differences between the models were on questions 2 and 3. Question 2 asked "Which machine(s) is the bottleneck?" and the average mark was much higher for the model D participants (72%) than for the other models (19%, 31%, and 25% for A, B, and C, respectively). The small size of model D made this question easier to answer because there are fewer elements to compare to identify the bottleneck. In fact machine M2 is rarely in operation and so this question only required comparing two machines. This also meant that there were more acceptable reasons for the correct answer than for the other models. Question 3 asked "Which buffer(s) were full at some time during the period?" and could be answered by identifying blocked machines from the output statistics. The average marks were much higher for models B and D (81 and 72%, respectively) than for models A and C (25 and 44%, respectively). Again this reflects the question being inherently easier for models B and D since the blocked machines only sent parts to one buffer, whereas in models A and C they sent parts to several buffers. Therefore, the difference in marks seems to be a result of lower connectedness in the critical section of the models.

The marks were not statistically significant at the 5% level for questions 1 and 4. Question 1 ("How many parts were sent to SHIP in the period?") was expected to be harder for model D since the calculation is more complex but, in fact, the average mark was similar to that for models A and B perhaps again reflecting that the small size means that it is easier to identify the correct part of the model to focus on. Question 4 ("Estimate the % increase in output if [a given machine] cycle time is reduced to 10," where the given machine was chosen not to be the bottleneck) was expected to be easier for model D, but the marks were only slightly higher than for the other models.

Overall the indication is that the difficulty in understanding is mainly affected by size and connectedness with calculational complexity being much less important, although this of course depends on the specific question being considered. This is probably because the fine details can often be ignored in understanding the system with just an appreciation of which elements influence each other being required.

Most of the model building errors for models A, B, and C occurred in the input and output rules for assembly machines, which were relatively complex commands that the students were less familiar with. The number of errors therefore reflects the comparative occurrence of these commands in the models, with models A and B having two assembly machines and model C one (each error was counted separately including repeated errors). Most of the errors for model D were either omissions or occurred in a complex command unique to model D. Generally, the majority of errors are likely to occur in the more complex aspects of the model, and so the number of errors is expected to be most closely related to calculational complexity.

The sample sizes here for each model (8 or 9) are small and the results will depend to some extent on the type of models used and the questions asked. The results can therefore only suggest possible relationships between model attributes and performance and more work is required to investigate this further.

2.9 Conclusions

The lack of research into the process of choosing the best model is surprising given the importance of modeling in science. There are very few studies that have made any quantitative assessment of the effect of different model attributes on the modeling process. This probably stems from the difficulty in measuring either suitable attributes or model performance and also the effort required to build several alternative models. Different models are most often compared by their level of detail or complexity although such a comparison is usually only qualitative and level of detail and complexity are usually not defined clearly. This chapter introduces the more specific model characteristics of size, connectedness, and calculational complexity.

The lack of model comparisons has resulted in only vague guidelines to aid the choice of model. The initial requirement is for a considerable number of studies that compare, preferably quantitatively, some aspects of model performance for alternative models. This chapter describes a small-scale study of this type, which indicated that the difficulty in understanding the model and the results is mainly caused by size and connectedness, whereas build time is mainly related to calculational complexity.

A common piece of advice in conceptual modeling and choosing the level of detail is to use past experience and so, at the very least, the quantitative comparison of alternative models would provide a source of modeling experience from which to draw. Ultimately this approach could lead to the development of general principles and hopefully to a methodology for choosing the best model. A corresponding methodology for simplification is also necessary.

Acknowledgments

Some sections of this chapter are based on Brooks, R. J., and A. M. Tobias. 1996. Choosing the best model: Level of detail, complexity and model performance. *Mathematical and Computer Modelling* 24(4):1–14.

References

Addanki, S., R. Cremonini, and J. S. Penberthy. 1991. Graphs of models. *Artificial Intelligence* 51:145–177.

Allison, H. 1979. Inverse unstable problems and some of their applications. *Mathematical Scientist* 4:9–30.

Amaral, L. A. N., and B. Uzzi. 2007. Complex systems: A new paradigm for the integrative study of management, physical, and technological systems. *Management Science* 53(7):1033–1035.

Banks, J., and J. S. Carson. 1984. *Discrete-Event System Simulation*. Englewood Cliffs, NJ: Prentice-Hall.

Blöschl, G., and R. Kirnbauer. 1991. Point snowmelt models with different degrees of complexity: Internal processes. *Journal of Hydrology* 129:127–147.

Boehm, B. W. 1981. *Software Engineering Economics*. Englewood Cliffs, NJ: Prentice-Hall.

Bosserman, R. W. 1982. Structural comparison for four lake ecosystem models. In *A General Survey of Systems Methodology: Proceedings of the Twenty-sixth Annual Meeting of the Society for General Systems Research*, ed. L. Troncale, 559–568. Washington, DC.

Brooks, R. J. 1996. *A Framework for Choosing the Best Model in Mathematical Modelling and Simulation*. Ph D thesis, University of Birmingham, UK.

Brooks, R. J., D. N. Lerner, and A. M. Tobias. 1994. Determining a range of predictions of a groundwater model which arise from alternative calibrations. *Water Resources Research* 30(11):2993–3000.

Brooks, R. J., and S. Robinson. 2001. *Simulation*, with Inventory Control (author C. Lewis), Operational Research Series. Basingstoke: Palgrave.

Brooks, R. J., M. A. Semenov, and P. D. Jamieson. 2001. Simplifying Sirius: Sensitivity analysis and development of a meta-model for wheat yield prediction. *European Journal of Agronomy* 14(1):43–60.

Brooks, R. J., and A. M. Tobias. 1999. Methods and Benefits of Simplification in Simulation. In *Proceedings of the U.K. Simulation Society (UKSIM 99)*, ed. D. Al-Dabass and R. Cheng, 88–92. U.K. Simulation Society.

Brooks, R. J., and A. M. Tobias. 2000. Simplification in the simulation of manufacturing systems. *International Journal of Production Research* 38(5):1009–1027.

Brooks, R. J., A. M. Tobias, and M. J. Lawrence. 1997a. A time series analysis of the population genetics of the self-incompatibility polymorphism. 1. Allele frequency distribution of a population with overlapping generations and variation in plant size. *Heredity* 79:350–360.

Brooks, R. J., A. M. Tobias, and M. J. Lawrence. 1997b. A time series analysis of the population genetics of the self-incompatibility polymorphism. 2. Frequency equivalent population and the number of alleles that can be maintained in a population. *Heredity* 79:361–364.

Bunge, M. 1963. *The Myth of Simplicity: Problems of Scientific Philosophy*. Englewood Cliffs, NJ: Prentice-Hall.

Casti, J. L. 1979. *Connectivity, Complexity, and Catastrophe in Large Scale Systems*. New York: John Wiley and Sons.

Chaitin, G. J. 1975. Randomness and mathematical proof. *Scientific American* 232(May):47–52.

Costanza, R., and F. H. Sklar. 1985. Articulation, accuracy and effectiveness of mathematical models: A review of freshwater wetland applications. *Ecological Modelling* 27(1–2):45–68.

Courtois, P.-J. 1985. On time and space decomposition of complex structures. *Communications of the ACM* 28(6):590–603.

Davis, P. K., and R. Hillestad. 1993. Families of models that cross levels of resolution: Issues for design, calibration and management. In *Proceedings of the 1993 Winter Simulation Conference*, ed. G. W. Evans, M. Mollaghasemi, E. C. Russell, and W. E. Biles, 1003–1012. New York: IEEE.

Decoursey, D. G. 1992. Developing models with more detail: Do more algorithms give more truth? *Weed Technology* 6(3):709–715.

Demarco, T. 1982. *Controlling Software Projects: Management, Measurement and Estimation*. New York: Yourdon Press.

Durfee, W. K. 1993. Control of standing and gait using electrical stimulation: Influence of muscle model complexity on control strategy. *Progress in Brain Research* 97:369–381.

Falkenheimer, B., and K. D. Forbus. 1991. Compositional modelling: Finding the right model for the job. *Artificial Intelligence* 51:95–143.

Farmer, J. D. 1990. A rosetta stone for connectionism. *Physica D* 42:153–187.

Fenton, N. E. 1991. *Software Metrics: A Rigorous Approach*. London: Chapman and Hall.

Fishwick, P. A. 1988. The role of process abstraction in simulation. *IEEE Transactions on Systems, Man and Cybernetics* 18(1):19–39.

Flatau, M. 1995. Review Article: When order is no longer order—Organising and the new science of complexity. *Organization* 2(3–4):566–575.

Flood, R. L., and E. R. Carson. 1993. *Dealing with Complexity: An Introduction to the Theory and Application of Systems Science*, 2nd edition. New York: Plenum Press.

Fripp, J. 1985. How effective are models? *Omega* 13(1):19–28.

Garfinkel, D. 1984. Modelling of inherently complex biological systems: Problems, strategies, and methods. *Mathematical Biosciences* 72(2):131–139.

Gass, S. I. 1983. What is a computer-based mathematical model? *Mathematical Modelling* 4:467–472.

Gell-Mann, M. 1995. What is complexity? *Complexity* 1(1):16–19.

George, L. 1977. Tests for system complexity. *International Journal of General Systems* 3:253–258.

Golay, M. W., P. H. Seong, and V. P. Manno. 1989. A measure of the difficulty of system diagnosis and its relationship to complexity. *International Journal of General Systems* 16(1):1–23.

Gruber, T. R. 1993. Model formulation as a problem solving task: Computer-assisted engineering modelling. *International Journal of Intelligent Systems* 8(1):105–127.

Halfon, E. 1983a. Is there a best model structure? I. Modelling the fate of a toxic substance in a lake. *Ecological Modelling* 20:135–152.

Halfon, E. 1983b. Is there a best model structure? II. Comparing the model structures of different fate models. *Ecological Modelling* 20:153–163.

Henneman, R. L., and W. B. Rouse. 1986. On measuring the complexity of monitoring and controlling large-scale systems. *IEEE Transactions on systems, man and cybernetics* SMC-16:193–207.

Henry, S., and D. Kafura. 1981. Software quality metrics based on information flow. *IEEE Transactions on Software Engineering* 7(5):510–518.

Hunt, J. C. R. 1994. Presidential address: Contributions of mathematics to the solution of industrial and environmental problems. *IMA Bulletin* 30:35–45.

Innis, G. S., and E. Rexstad. 1983. Simulation model simplification techniques. *Simulation* 41(1):7–15.

Jakeman, A. J., and G. M. Hornberger. 1993. How much complexity is warranted in a rainfall-runoff model? *Water Resources Research* 29:2637–2649.

Jeffers, J. N. R. 1991. From free-hand curves to chaos: Computer modelling in ecology. In *Computer Modelling in the Environmental Sciences*, ed. D. G. Farmer and M. J. Rycroft, The Institute of Mathematics and its Applications Conference Series no. 28:299–308. Oxford: Clarendon Press.

Kauffman, S. A. 1993. *The Origins of Order: Self-Organisation and Selection in Evolution.* New York: Oxford University Press.

Langton, C. G. 1990. Computation at the edge of chaos: Phase transitions and emergent computation. *Physica D* 42:12–37.

Law, A. M. 1991. Simulation model's level of detail determines effectiveness. *Industrial Engineering* 23(10):16–18.

Law, A. M. 2007. *Simulation Modeling and Analysis,* 4th edition. New York: McGraw-Hill.

Law, A. M., J. S. Carson, K. J. Musselman, J. G. Fox, S. K. Halladin, and O. M. Ulgen. 1993. A forum on crucial issues in the simulation of manufacturing systems. In *Proceedings of the 1993 Winter Simulation Conference*, ed. G. W. Evans et al., 916–922. New York: IEEE.

le Roux, J. A., and M.S. Potgieter. 1991. The simulation of Forbush decreases with time-dependent cosmic-ray modulation models of varying complexity. *Astronomy and Astrophysics* 243:531–545.

McCabe, T. J. 1976. A complexity measure. *IEEE Transactions on Software Engineering* 2(4):308–320.

Meisel, W. S., and D. C. Collins. 1973. Repro-modelling: An approach to efficient model utilization and interpretation. *IEEE Transactions on Systems, Man and Cybernetics* SMC-3:349–358.

Morowitz, H. 1995. The emergence of complexity. *Complexity* 1(1):4–5.

Murdoch, W. W., E. McCauley, R. M. Nisbet, W. S. C. Gurney, and A. M. De Roos. 1992. Individual-based models: Combining testability and generality. In *Individual-Based Models and Approaches in Ecology: Populations, Communities and Ecosystems*, ed. D. L. DeAngelis and L. J. Gross, 18–35. New York: Chapman and Hall.

Nayak, P. P. 1992. *Automated modelling of physical systems.* Ph.D. thesis, Computer Science Department, Stanford University, Technical Report STAN-CS-92-1443.

Palsson, B. O., and I. Lee. 1993. Model complexity has a significant effect on the numerical value and interpretation of metabolic sensitivity coefficients. *Journal of Theoretical Biology* 161:299–315.

Pidd, M. 2004. *Computer Simulation in Management Science,* 5th edition. Chichester: John Wiley and Sons.

Rexstad, E., and G. S. Innis. 1985. Model simplification: Three applications. *Ecological Modelling* 27(1–2):1–13.

Robinson, S. 2008. Conceptual modeling for Simulation Part I: Definition and requirements. *Journal of the Operational Research Society* 59:278–290.

Robinson, S., and M. Pidd. 1998. Provider and customer expectations of successful simulation projects. *Journal of the Operational Research Society* 49:200–209.

Rosen, R. 1977. Complexity as a system property. *International Journal of General Systems* 3:227–232.

Salt, J. D. 1993. Keynote address: Simulation should be easy and fun! In *Proceedings of the 1993 Winter Simulation Conference,* ed. G. W. Evans et al., 1–5. New York: IEEE.

Schruben, L. 1983. Simulation modelling with event graphs. *Communications of the ACM* 26(11):957–963.

Schruben, L., and E. Yücesan. 1993. Complexity of simulation models: A graph theoretic approach. In *Proceedings of the 1993 Winter Simulation Conference,* ed. G. W. Evans et al., 641–649. New York: IEEE.

Sevinc, S. 1990. Automation of simplification in discrete event modelling and simulation. *International Journal of General Systems* 18(2):125–142.

Shannon, R. E. 1975. *Systems Simulation: The Art and Science.* Englewood Cliffs, NJ: Prentice-Hall.

Simon, H. A. 1964. The architecture of complexity. *General Systems Yearbook* 10:63–76.

Smith, D. E., and J. M. Starkey. 1995. Effects of model complexity on the performance of automated vehicle steering controllers: Model development, validation and comparison. *Vehicle System Dynamics* 24:163–181.

Stockle, C. O. 1992. Canopy photosynthesis and transpiration estimates using radiation interception models with different levels of detail. *Ecological Modelling* 60(1):31–44.

Tilanus, C. B. 1985. Failures and successes of quantitative methods in management. *European Journal of Operational Research* 19:170–175.

Tobias, A. M. 1991. Verification, validation and experimentation with visual interactive simulation models. *Operational Research Tutorial Papers,* The Operational Research Society.

Ward, S. C. 1989. Arguments for constructively simple models. *Journal of Operational. Research Society* 40(2):141–153.

Webster, D. B., M. L. Padgett, G. S. Hines and D. L. Sirois. 1984. Determining the level of detail in a simulation model: A case study. *Computers and Industrial Engineering* 8(3–4):215–225.

Weaver, W. 1948. Science and complexity. *American Scientist* 36(Autumn):536–544.

Zeigler, B. P. 1976. *Theory of Modelling and Simulation.* New York: John Wiley.

Zeigler, B. P. 1979. Multilevel multiformalism modeling: An ecosystem example. In *Theoretical Systems Ecology: Advances and Case Studies,* ed. E. Halfon, 17–54. New York: Academic Press.

Zeigler, B. P. 1984. *Multifacetted Modelling and Discrete Event Simulation.* London: Academic Press.

3

Improving the Understanding of Conceptual Modeling

Wang Wang and Roger J. Brooks

CONTENTS

3.1 Introduction

Conceptual modeling is a crucial stage of the simulation modeling process, and yet it is poorly understood. Brooks and Robinson (2001) defined a conceptual model is "a software independent description of the model that is to be constructed." Conceptual modeling therefore involves deciding the way in which the virtual world of the simulation model should work (Section 2.1). The conceptual model may be documented fully, such as in an annotated system process flowchart, or it may only be documented partially, or even not documented at all. In the absence of documentation, conceptual modeling still takes places and the conceptual model comprises the combined decisions of the project team in determining the way the model should work. Conceptual modeling is a separate stage to model coding, which consists of writing the computer code for the model (often using a simulation software package). One aspect of conceptual modeling is deciding how much detail to include in the model and Law (1991) considered that for simulation projects "the most difficult aspect of a study is that of determining the appropriate level of model detail." However, little attention is devoted to conceptual modeling in most textbooks.

The advice that is provided often centers on the complexity or level of detail of the model. For example, Robinson (1994) proposed that the basic rule for what to include in a model is to use the minimum components required to achieve the project's objective. In fact, "Model Simple–Think complicated" is one of Pidd's (2003) principles of modeling, and Ward (1989) and Salt (1993) also set out a number of advantages of a simple model. However, definitions of level of detail and complexity are not usually provided in the literature and there are no agreed ways of measuring them.

A particularly interesting study in this area is that of Willemain (1995) who carried out an experiment to investigate the initial stages of a modeling project. The experiment consisted of providing operational research (OR) experts with a description of an OR modeling problem, and asking the expert to speak aloud their thoughts on tackling the problem for a period of an hour, while recording this on tape. Transcripts of the recordings were then analyzed by breaking them into "chunks" (from a phrase to a couple of sentences) and categorizing each one by a topic in the modeling process. The five topics used by Willemain were context, structure, realization, assessment, and implementation.

In Willemain's experiment there were four different problems and four experts tackled all four problems. A further eight experts tackled one problem each, giving a total of 24 sessions. The categorizations were analyzed in various ways including a "topic plot" showing which topic the expert was working on throughout the transcript, the number of transitions between each pair of topics, the proportion of lines of transcript devoted to each topic and a box plot of topic position. One of the main results was that even though the sessions only lasted an hour, the experts spent a considerable proportion of the time on all topics other than implementation, with a lot of alternation between the different topics. In particular, structure (essentially conceptual modeling) was often followed by assessment (essentially verification and validation) and assessment was often followed by structure. In other words, the experts would tend to develop an aspect of the conceptual model, then evaluate it and then often revise the conceptual model based on this evaluation. Recently, Willemain and Powell carried out a similar experiment using novice modelers (Powell and Willemain 2007, Willemain and Powell 2007). They identified five main ways in which the novices fell short of what they considered to be good modeling practice, which were overreliance on data, taking shortcuts, insufficient use of variables and relationships, ineffective self-regulation, and overuse of brainstorming. In his earlier work, Willemain (1994) also carried out a survey of the 12 experts in his experiment, which provided revealing insights on their modeling styles and their views on the ideal qualities of modelers, models, and clients.

Conceptual modeling is often thought of as a skill that improves with experience. One way for all modelers, but particularly novice modelers, to get better at conceptual modeling is therefore to draw on the experience of experts. Knowledge of what both expert and novice modelers actually do in practice is also an essential foundation for conceptual modeling research (Brooks 2007). However, apart from the work just described, there is a lack of empirical studies or data in the literature on how modelers develop conceptual models and on how conceptual modeling relates to the other modeling topics. This chapter describes a study to collect and analyze data on this process for an expert and several novice groups tackling real problems. The results are discussed and the lessons learnt and possible future work outlined.

3.2 Study Objective

The objective of this study was to improve the understanding of the modeling process followed in practice by different modelers, focusing particularly on conceptual modeling. The general approach follows that of Willemain (1995) in collecting data on the topics worked on during the modeling process. However, here data were collected throughout a real project for an expert and nine groups of novices. The study therefore differs from Willemain's in four main ways: first, the projects are all simulation projects; second, they are real projects; third, data were collected for the whole project rather than just the initial stage; and fourth, groups of novices as well as an expert were followed. In fact, moving to real-life projects, looking at novices and looking at groups of modelers were all future experiments suggested by Willemain.

3.3 Data Collection

3.3.1 Expert Project Data Collection

The first project used in the study was conducted by an expert. The expert holds a master's degree in operational research and prior to the study had 4 years of modeling and simulation experience in a variety of application fields, including manufacturing, military, and health care. The project was carried out part-time by the expert over a period of 10 weeks and involved modeling a call center to improve the efficiency of staff usage. The simulation software used was Micro Saint Sharp (Alion MA&D Operation, Colorado, US), which was selected by the client.

The expert was asked to record the total number of hours spent each week on different modeling topics. There was a desire to compare these results with those of Willemain (1995) and so Willemain's paper was used as a basis. The expert preferred to use one of the alternative list of topics (from Hillier and Lieberman 1967) given in Willemain as follows (with the matching topic according to Willemain given in brackets): Formulating the problem (context), constructing a mathematical model (structure), deriving a solution (realization), testing the model and solution (assessment), establishing controls over solution (implementation), and implementing the solution (implementation). Each week, the expert recorded the number of hours spent on each of these topics.

The expert modeler was also interviewed each week and asked whether and how the conceptual model had changed during the week and, if there had been a change, about the process and reasons for changing the model. General issues, for instance, the main task of the week and whether working on one topic influenced the others were also discussed.

3.3.2 Novice Projects Data Collection

Data were obtained for nine Lancaster University (UK) student group projects in two phases. Data for six projects were collected in phase 1 in 2005 and data for a further three projects were collected in phase 2 in 2006. All projects lasted for about 12 weeks. In phase 1, two of the groups were from the simulation module on the master's course in OR and the other four were from the undergraduate simulation course. In phase 2, all three projects were from the undergraduate simulation course. The undergraduate students were in their second or third years in various departments in the Management School. They had little programming and simulation modeling experience prior to the course. The backgrounds for the master's students varied depending on their first degree subjects and previous work experience. Some had programming and modeling experience, but in general, their prior knowledge of simulation was limited. As the main assessment for both courses, the students were required to find a suitable project on a real system (typically from around the university campus) and carry out a complete simulation project. Therefore, although the projects are modeling a real problem there is no external client as such, although the projects are done with the cooperation of the external company if there is one. The master's groups had three students, while the undergraduate groups had five students. The educational version of the simulation software package WITNESS (Lanner Group Ltd., Redditch, UK) was used for all projects. Table 3.1 shows the systems that were modeled.

In the phase 1 study, weekly questionnaires were handed out to each group before the project started. Each group was asked to record the total hours spent on the different topics every day during the week, as well as

TABLE 3.1

Systems Modeled in the Novice Projects

Phase	Course[a]	No. Projects	Systems Modeled
1	UG	4	Food takeaway, post office, coffee shop, Library book loan service points
1	PG	2	Restaurant, traffic crossing
2	UG	3	Convenience store, petrol station, Library photocopiers

[a] UG = undergraduate, PG = postgraduate

whether the conceptual model had changed during the week. In this case, a much more detailed list of topics was provided than the ones used by Willemain (1995) so as to obtain more detailed data and to reduce the amount of interpretation required by the students. In the subsequent analysis, the topics were combined into our own preferred list of simulation tasks. The topics were as follows (with the topic from our list in parentheses): identify alternative potential projects (problem structuring), contact/interview with the client (problem structuring), observe the system (problem structuring), discuss with experts (problem structuring), set project objectives (problem structuring), decide the model structure (conceptual modeling), model coding (model coding), collect data for the model (data collection and analysis), parameter estimation and distribution fitting (data collection and analysis), white box validation (verification and validation), black box validation (verification and validation), verification (verification and validation), experiment with the model and analyze the result (experimentation), and report writing (report writing).

The same data were collected in phase 2 but in an improved way. The limitation of the method used in phase 1 is that the reliability of the data depended on the accuracy of the students in recording the time spent and also on how well they were able to match their tasks against the categories provided. Also data were only recorded on a daily basis. To overcome these drawbacks, in the phase 2 studies, the researcher (Wang Wang) sat in on most of the student group meetings, observed their behavior and recorded the relevant time herself in hourly intervals. Where group members conducted individual work outside the meetings, they reported to the researcher on what task they worked on and the time spent on that task. In addition, the updated computer model was saved at the end of each group meeting so that the changes to the model could be tracked. Collecting data in this way gives more confidence in the reliability of the data. In both studies the hours were not adjusted for the number of people doing each task because of the difficulty in assessing the extra effort this represents. For example, two students working together on coding the model for two hours was recorded as two hours (rather than four).

3.4 Results

3.4.1 Results for Expert

The analysis of the data follows some of Willemain's analysis by calculating the relative weights of the different topics, and showing a graphical representation of the topics over time. Figures 3.1 and 3.2 show these results for the expert project, while Figure 3.3 shows the average weight given to each topic in the 24 sessions in Willemain's experiment measured in number of lines in the transcripts. As Figure 3.1 shows, the expert spent most time on modeling and testing the model. No time was spent by the

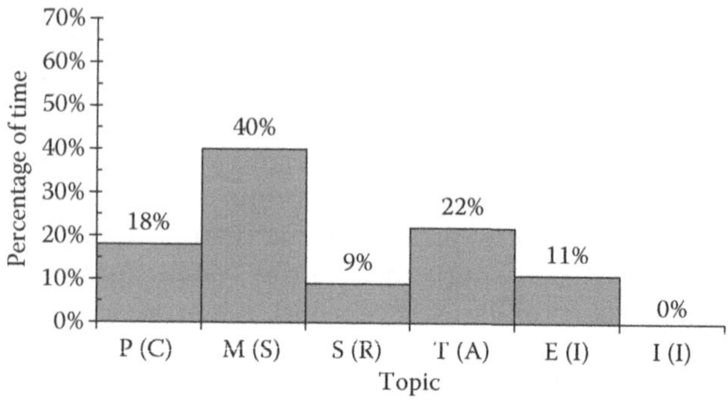

FIGURE 3.1
Proportion of time spent on each topic in the expert project. The topics are (with the matching Willemain topic in parentheses): P (C) = Formulating the problem (context), M (S) = constructing a mathematical model (structure), S (R) = deriving a solution (realization), T (A) = testing the model and solution (assessment), E (I) = establishing controls over solution (implementation), I (I) = implementing the solution (implementation).

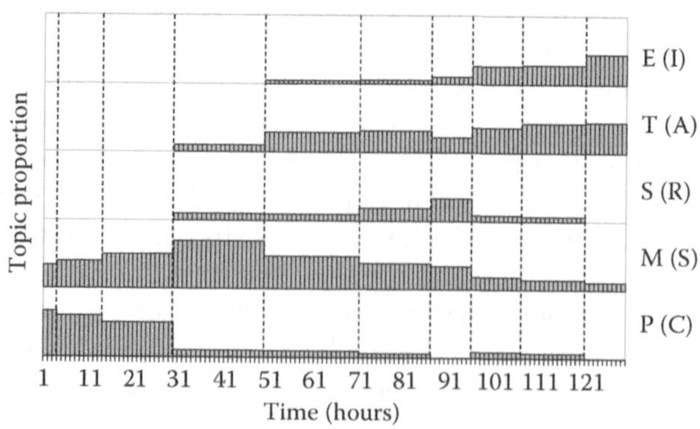

FIGURE 3.2
Timeline plot for expert project. The topics are as in Figure 3.1. The data were collected weekly over 10 weeks, which are shown by the vertical dashed lines.

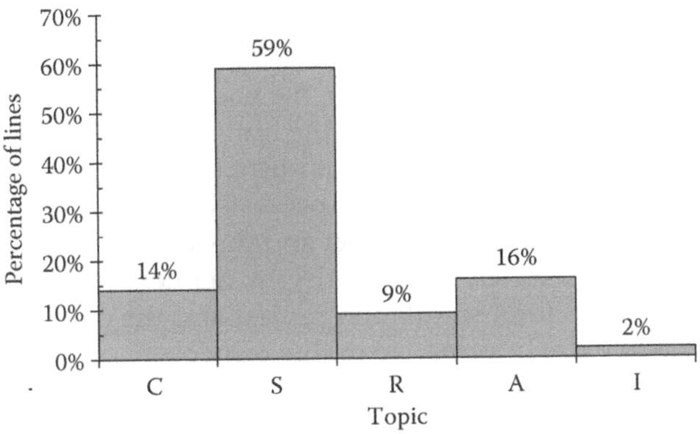

FIGURE 3.3
Percentage of lines devoted to each topic in Willemain's 24 experiments (Redrawn from Willemain, T.R., *Operations Research*, 43(6), 916–932, 1995.)

expert implementing the solution since this was carried out subsequently by the client.

The timeline plot (Figure 3.2) shows the topics worked on during the project. The expert project data were obtained on a weekly basis over the 10-week period of the project. Only the total number of hours spent on the topics in each week were recorded. Since the precise timings during the week are not known the plot spreads the topics evenly during each week. If more than one topic was worked on during the week then this is shown by the bars not being full height in the plot (a full height bar would reach the horizontal line above on the plot). For example, in the second week the expert spent a total of 10 hours working on the project, which consisted of 6 hours on formulating the problem (P) and 4 hours on constructing the model (M). This is shown in the plot by the heights of the bars for P and M being, respectively, 60 and 40% of a full height bar for each hour in a 10-hour period (hours 4–13). This data collection was less detailed than Willemain's data obtained in a laboratory setting, where the protocol recorded what was happening all the time. One consequence is that where more than one topic took place during the week then the order and the interaction between the topics is not known. There could have been a lot of switching between the topics during the week or, on the other hand, the topics could have been worked on completely separately one after the other. This prevented a detailed analysis of the switching between topics as carried out by Willemain. Nevertheless, the topic plots still give useful information about the positions and sequence of the topics throughout the project. In particular, the extensive overlap between the topics does indicate a considerable amount of alternation between the topics rather than a linear process. In general, the topics were in the anticipated order with topics higher up on the y-axis expected to be later.

A comparison of Figures 3.1 and 3.3 show a reasonably similar split between the topics. This perhaps indicates that the relative time spent by the expert on the different topics over the course of the whole project was similar to that spent by the experts in the initial hour of Willemain's experiment. However, this comparison should be treated with caution because it depends on how similar the allocation process was. In particular, an alternative list was used for the expert project and this may not match up perfectly with Willemain's categories. With hindsight, neither list gives a sufficiently detailed list of topics for a simulation project and the data collected for the novice projects is more informative in this respect.

3.4.2 Results for Novices

The proportion of time spent on the topics for the novice projects in phase 1 and phase 2, respectively, are shown in Figures 3.4 and 3.5. In each case, the percentage of time on each topic was calculated for each project and then the project values were averaged. The pattern is reasonably similar for phase 1 and phase 2, which gives some additional confidence that the results for phase 1 are reliable even though they were recorded by the students themselves. A considerable amount of time was spent on data collection and report writing. Conceptual modeling received relatively little attention particularly in the phase 2 projects.

Observation by the researcher of the process for the phase 2 projects gave additional insight into the results. The high proportion of time spent on experimentation was partly due to the technical problems they experienced or mistakes they made. For example, one group did all the experimentation twice as they forgot to consider the warm-up period (the initial transient period before the model reaches the realistic conditions in the system) at

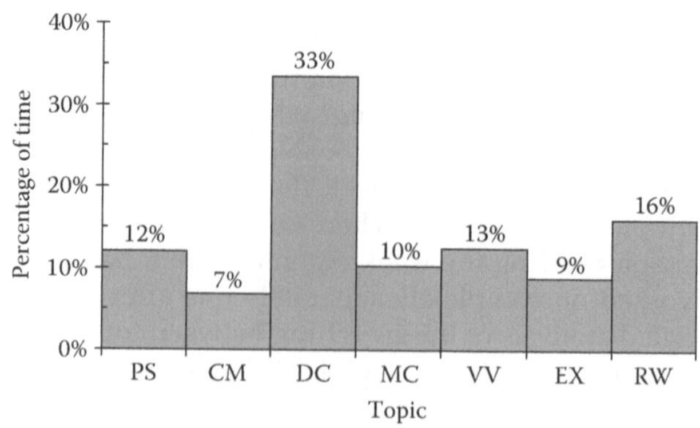

FIGURE 3.4
Proportion of time spent on the topics in the six novice projects in phase 1. PS = problem structuring, CM = conceptual modeling, DC = data collection, MC = model coding, VV = verification and validation, EX = experimentation, RW = report writing.

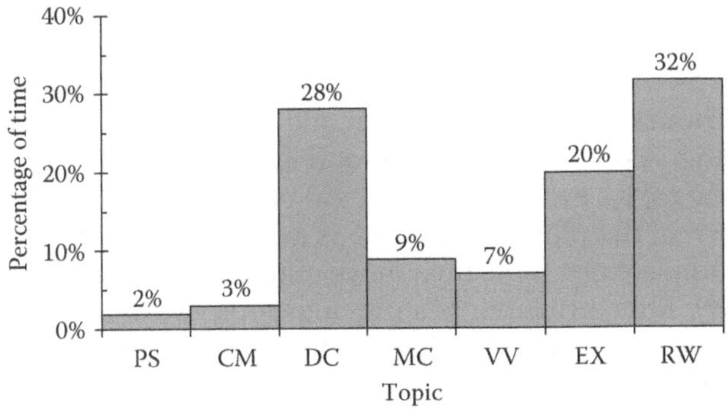

FIGURE 3.5
Proportion of time spent on the topics in the three novice projects in phase 2. The topics are as in Figure 3.4.

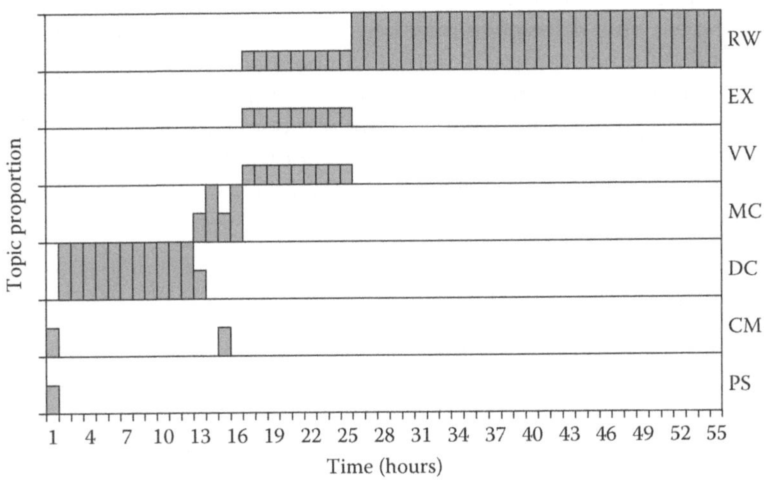

FIGURE 3.6
Timeline plot for one of the phase 2 novice projects. Topics are as in Figure 3.4.

the beginning. Another group had little understanding about warm-up and, as a result, they had to go through the lecture notes first before they could perform this task. Generally with these projects the groups have to collect their own data, which is the reason for the high proportion of time on data collection.

The timeline plot shown in Figure 3.6 is the same general format as Figure 3.2. As already explained, most of the data for the phase 2 novice projects was obtained by observation of the group by the researcher and was recorded on an hourly basis. The overlapping topics in hours 1, 13, and 15 are times when both topics were worked on during the hour. However, some of the work was done individually by the group members and the total time spent was just reported to the researcher. This data are therefore less detailed with the precise interaction between the topics not known. The period from hours 17 to 25 was not observed and instead the group members reported spending 3 hours each on verification and validation, experimentation and report writing. As with the expert plot (Figure 3.2), such data are shown by spreading the topics evenly over the total period.

The pattern of the plot in Figure 3.6 is a fairly linear process. The novices tended to complete one topic then move onto the next with not much overlap and with very little returning to a previous topic. Most of the novice projects were similar in this respect. Although the topic categories are different, this is a quite different pattern to the expert (Figure 3.2) and also to the pattern in Willemain's (1995) experiments. Figure 3.2 shows much more overlap of topics although, as previously explained, the precise pattern within each week for the expert is not known. The overlap of the topics over several weeks shows that there was more switching between topics for the expert than the novices. Since there may also have been several iterations within each week for the expert, this difference may be even more marked than is shown on the graphs. Another comparison that can be made is that the expert started model testing (verification and validation) much earlier than the novice modelers who tended to leave it until after model coding was completed. As for the expert project, the average position of the topics for the novices was in the expected order with topics higher up the *y*-axis on Figure 3.6 expected to be later.

3.4.3 Further Findings and Analysis

Upon the completion of both the expert and novice studies, a further discussion took place with the expert to try and reclassify the topics to enable better comparison with the novice projects. Using our list of topics the expert decomposed each original topic and provided its approximate weighting (for example, establishing controls over solution = 1/4 verification and validation + 3/4 experimentation). This enabled a revised weight breakdown to be produced although not a revised timeline. However, discussion with the expert indicated that the general pattern of task overlapping would be similar. It should be noted that this reallocation took place more than 12 months after the project finished and therefore the values should be regarded as approximate. This revised weighting is shown in Figure 3.7. Again this shows that the data are very different to that of the novice projects (Figures 3.4 and 3.5). With the expert, the topic that received the most attention was conceptual modeling, and much more time was spent on verification and validation than experimentation. In general, it may be that experienced modelers have a greater appreciation than novices of the importance and benefits of both conceptual modeling and verification and validation, perhaps by learning from problems on previous projects. On the other hand, the lower attention on conceptual modeling and validation by the novices could be because the

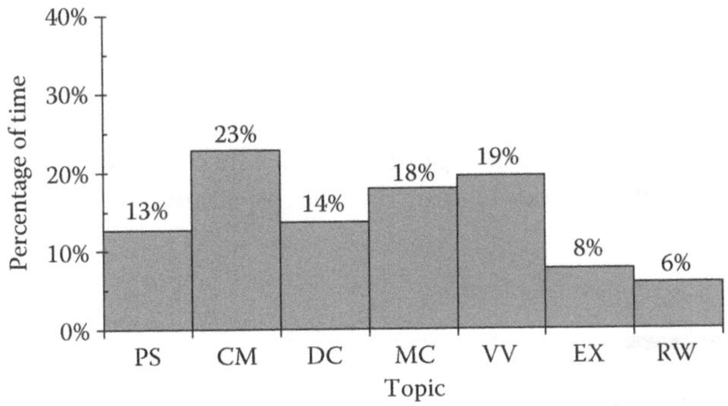

FIGURE 3.7
Proportion of time spent on each topic in the expert project using revised allocation to the same topics as the novices. Topics are as in Figure 3.4.

systems they modeled were so simple that these tasks were straightforward to perform. To investigate this further, the novice project reports were studied. This showed poor performance on validation and verification with this task receiving the lowest marks on average in the assessment of the reports. For example, two groups failed to distinguish between white box validation and verification and out of nine projects investigated, only two groups performed black box validation properly. Sometimes this was due to a lack of planning. For instance, two groups didn't consider collecting data for validation when planning data collection.

The conceptual modeling process can be considered in more detail based on the discussions with the expert and observations of the phase 2 novice groups. The expert developed the conceptual model at the beginning of the project and documented it in a system flow diagram, which guided the construction of the computer model. However, the novice groups devoted little time to understanding how the systems actually worked. They did discuss the process of the system, but rarely drew a diagram. After identifying the project they tended to go straight to collecting data with little prior planning or consideration of the model structure. As a result, some of the data collected proved not to be useful. This is inefficient particularly as data collection is time consuming. Sometimes further discussions on the system process occurred at the model coding stage with conceptual modeling and coding taking place together. Some groups only documented the conceptual model at the end in order to include a diagram in the report.

In the expert study and both novice studies, the subjects were asked to note any changes in the conceptual model each week and the reason. For the expert project, there was one significant conceptual model alteration toward the end of the project. This involved a scope reduction due to the fact that the collected data were not sufficient to support the model built. In one of the novice groups, one student left in the middle of the project causing a change of project application to make the problem easier to model. Another group attempted to increase the scope from modeling the busy hours of a grocery store to modeling the whole period of the business hours, but gave up after experiencing some difficulties in finding out the right distribution of data and transferring it into the computer model. The other novice groups did not adjust the conceptual models after the coding stage.

3.5 Discussion and Conclusions

This study was designed to provide information to improve the understanding of conceptual modeling. The time spent on different topics had a quite different pattern between the expert project and the nine novice projects. The expert project had much more overlapping of the topics and had a higher proportion of time on conceptual modeling and verification and validation. The novice projects had much more time allocated to data collection because they had to collect the data themselves. One of the difficulties with research on real projects is that all the projects are different. Generally, the student projects are fairly simple problems and so the differences between the expert and novice data could be evidence of different working styles or it could be due the greater complexity of the expert project.

Carrying out this type of research is difficult and a number of problems were encountered. Following (shadowing) experts in real projects has practical difficulties, such as project confidentiality, and finding the projects with right size. Obtaining data in this way also requires a significant time commitment. The number of projects that could be followed was therefore limited. Following real projects also inevitably limits the analysis compared to an artificial laboratory experiment in that all the projects were different. For example, Willemain (1995) was able to compare different modelers tackling the same problem and some of the modelers tackling different problems, to try and identify any modeler and problem effects. With hindsight, using more topic categories for the expert project would have also provided more detailed data. A possible ideal approach in the future would be to follow a real small-scale consultancy project, and then bring it to the classroom to be tackled by novices. However finding an appropriate project is likely to be very difficult. A different approach would be to use a questionnaire to obtain data from a larger sample size of experts on their perceptions of their modeling projects. We have carried out such a questionnaire and some of the results are reported in Wang and Brooks (2007).

Obtaining this sort of information about conceptual modeling and the modeling process is an important step toward a better understanding of the key aspects of successful modeling practice. In the long term, it is hoped that further research in this area will improve the success of simulation and OR projects, and that the information can help in the training of novice modelers.

Acknowledgments

This chapter is reproduced, with minor editing, from: Wang, W., and R. J. Brooks. 2007. Improving the understanding of conceptual modelling. *Journal of Simulation* 1(3): 153–158. © 2007 Operational Research Society Ltd. Reproduced with permission of Palgrave Macmillan.

Some parts of this chapter are based on: Wang, W., and R.J. Brooks. 2007. Empirical investigations of conceptual modeling and the modeling process. In *Proceedings of the 2007 Winter Simulation Conference*, ed. S. G. Henderson, B. Biller, M.-H. Hsieh, J. Shortle, J.D. Tew, and R.R. Barton, 762–770. Piscataway, NJ: IEEE Computer Society Press.

References

Brooks, R. J. 2007. Conceptual modelling: Framework, principles, and future research. Working paper no. 2007/011, Lancaster University Management School, Lancaster, UK.

Brooks, R. J., and S. Robinson. 2001. *Simulation*, with Inventory Control (author C. Lewis), Operational Research Series. Basingstoke: Palgrave.

Hillier, F. S., and G. J. Lieberman. 1967. *Introduction to Operations Research.* San Francisco, Holden-Day.

Law, A. M. 1991. Simulation model's level of detail determines effectiveness. *Industrial Engineering* 23(10): 16–18.

Pidd, M. 2003. *Tools for Thinking, Modelling in Management Science,* 2nd edition. Chichester: John Wiley and Sons.

Powell, S. G., and T. R. Willemain. 2007. How novices formulate models. Part I: Qualitative insights and implications for teaching. *Journal of the Operational Research Society* 58(8): 983–995.

Robinson, S. 1994. Simulation projects: Building the right conceptual-model. *Industrial Engineering* 26(9): 34–36.

Salt, J. D. 1993. Keynote Address: Simulation should be Easy and Fun! In *Proceedings of the 1993 Winter Simulation Conference,* ed. G. W. Evans et al., 1–5. New York: IEEE.

Wang, W., and R. J. Brooks. 2007. Empirical investigations of conceptual modeling and the modeling process. In *Proceedings of the 2007 Winter Simulation Conference,* ed. S. G. Henderson, B. Biller, M.-H. Hsieh, J. Shortle, J.D. Tew, and R.R. Barton, 762–770. Piscataway, NJ: IEEE Computer Society Press.

Ward, S. C. 1989. Arguments for constructively simple models. *Journal of the Operational Research Society* 40(2): 141–153.

Willemain, T. R. 1994. Insights on modeling from a dozen experts. *Operations Research* 42(2): 213–222.

Willemain, T. R. 1995. Model formulation: What experts think about and when. *Operations Research* 43(6): 916–932.

Willemain, T. R., and S. G. Powell. 2007. How novices formulate models. Part II: A quantitative description of behaviour. *Journal of the Operational Research Society* 58(10): 1271–1283.

Part II

Conceptual Modeling Frameworks

4

A Framework for Simulation Conceptual Modeling

Stewart Robinson

CONTENTS

4.1 Introduction

Chapter 1 set out the foundations of conceptual modeling for simulation. It provided an understanding of current thinking on the topic and gave a definition of a conceptual model. It also discussed the requirements for a conceptual model: validity, credibility, utility, and feasibility (Chapter 1, Section 5). Such discussions are useful for informing a simulation modeling project, but they do not answer the question of how to develop a conceptual model. That is the question addressed in this chapter whose key contribution is to provide a framework for developing conceptual models for simulations of operations systems (Wild 2002). This is something that is largely missing from the current literature on simulation.

The framework that is presented provides a sequence of activities required for the development of a conceptual model. For each of these activities there is a set of guidelines and methods for performing them. Some might argue that the approach outlined does not go far enough to be described as a true framework, which would require a much more detailed and structured approach. The structure of the proposed framework, however, is found in the ordered sequence of activities and the guidelines and methods for performing each activity. These go beyond simple guidelines or lists of do's and don'ts for modeling and hence, in the author's view, present a framework for conceptual modeling.

This chapter describes the framework and the guidelines and methods for performing each activity within the framework. It concludes with a discussion on how data requirements can be identified and how the model can be assessed against the four requirements of a conceptual model (Chapter 1, Section 5). The framework is illustrated with the example of the Ford Motor Company (Ford) engine assembly plant model described in Chapter 1 (Section 2).

The framework presented here has been developed based on the author's experience, of nearly 20 years, with developing and using simulation models of operations systems, mainly manufacturing and service systems. By reflecting on the cognitive processes involved in reaching decisions about the scope and level of detail of models developed, a set of guidelines and methods have been devised. The framework aims to be useful for both novice and more expert modelers alike. For novice modelers it provides a guide on how to make decisions about the nature of a simulation model that is to be developed for a specific project. For more experienced modelers, it provides a greater sense of discipline to the conceptual modeling activity. It is hoped that by providing more discipline, greater creativity can be encouraged as the more basic tasks are formalized (Ferguson et al. 1997). At present there appears to be very little discipline in conceptual modeling. Pidd (1999), for instance, sees modeling as a process of muddling through.

Two other groups may benefit from this framework. Teachers may find it useful for giving their students a basis on which to learn about conceptual modeling. Researchers may use the framework as a basis for further and much needed research in this important area of simulation modeling.

4.2 A Framework for Developing a Conceptual Model

Figure 4.1 provides an overview of the conceptual modeling framework that is described in more detail below. In this framework conceptual

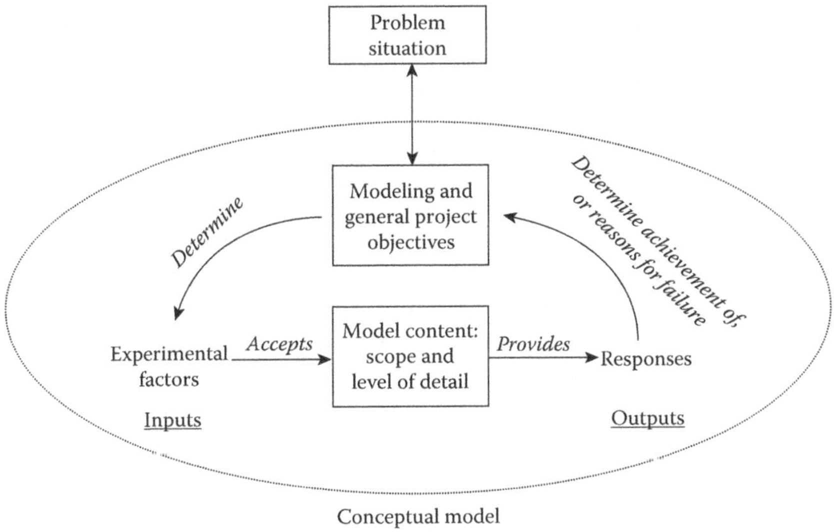

FIGURE 4.1
A framework for designing the conceptual model. (Adapted from Robinson, S., *Simulation: The Practice of Model Development and Use*, Wiley, Chichester, UK, 2004. With permission.)

modeling consists of five key activities that are performed roughly in this order:

- Understanding the problem situation
- Determining the modeling and general project objectives
- Identifying the model outputs (responses)
- Identifying the model inputs (experimental factors)
- Determining the model content (scope and level of detail), identifying any assumptions and simplifications

Starting with an understanding of the problem situation, a set of modeling and general project objectives are determined. These objectives then drive the derivation of the conceptual model, first by defining the outputs (responses) of the model, then the inputs (experimental factors), and finally the model content in terms of its scope and level of detail. Assumptions and simplifications are identified throughout this process.

The order of these activities is not strict as it is expected that there will be much repetition and iteration between them. For instance, the problem situation is rarely static and so continual revision to the conceptual model is required. Further to this, conceptual modeling is not performed in isolation, but is part of simulation study that itself is repetitive and iterative in nature, for instance, work carried out during model coding and experimentation may both lead to alterations in the conceptual model. (Indeed, the simulation study is normally just a part of a wider project that will also involve repetition and iteration.) For the purposes of explaining each of the conceptual modeling activities, however, it is useful to separate them and describe them in the order in which they would generally progress. This is the approach used below. Meanwhile, the reader is reminded to constantly bear in mind the repetitive and iterative nature of the modeling process.

Within this framework the purpose of the model outputs is seen as twofold: first, to determine whether the modeling objectives are being met and second, if the objectives are not being met, to help determine why. As such, the objectives are central to determining the outputs. The experimental factors are also determined by the modeling objectives. Attempts are made to achieve the modeling objectives by changing the values of the experimental factors. Once the model inputs and outputs are determined, the content of the conceptual model must be designed in such a way as to ensure that it can accept the inputs and provide the required outputs, with sufficient accuracy (accuracy is a measure of the correspondence of the model outputs with the real world). Model content consists of two elements. The scope is the boundary of the model in terms of its breadth. The level of detail is the boundary of the model in terms of the depth of detail modeled for each component within the scope. Throughout the process of developing the conceptual model various assumptions and simplifications are made. These should be explicitly recorded alongside the detail of the conceptual model.

It should be apparent from the description above that the modeling objectives are central to the conceptual modeling framework described here. It is for this reason that determining the modeling objectives is described as part of the conceptual modeling process. Since the understanding of the problem situation is central to the formation of the modeling objectives, it also is considered to be part of the conceptual modeling process, although not formally part of the conceptual model (Figure 4.1).

There now follows a more detailed description of the five activities outlined above. Following this, there is a discussion on the identification of data requirements and checking whether the model meets the four requirements of a conceptual model.

4.3 Understanding the Problem Situation

The requirement for a simulation model should always be driven by the need to improve a problem situation. (Here the term *problem situation* is borrowed from Checkland (1981, p. 316): "A nexus of real-world events and ideas which at least one person perceives as problematic.") Indeed, a simulation study would normally be commissioned because the clients perceive a problem and simulation as an aid to addressing that problem. As such, the starting point in any simulation study and, therefore, conceptual modeling for simulation, is to develop an understanding of that problem situation.

It is obviously necessary for the modeler to develop a good understanding of the problem situation if he/she is to develop a model that adequately describes the real world. The approach to this activity depends in large measure on the extent to which the clients and subject matter experts (domain experts) understand, and are able to explain, the problem situation. In this respect, there are three possible scenarios:

- The problem situation is clearly understood and expressed.
- The problem situation is apparently well understood and expressed, although actually it is not.
- The problem situation is neither well understood nor expressed.

In the first case, developing an understanding of the problem situation only requires discussion and careful note-taking. It is also useful for the modeler to confirm his/her understanding by providing descriptions of the problem situation for the clients. This acts as a means of validating the conceptual model as it is developed.

Unfortunately, the first scenario rarely exists. Very often, the clients and domain experts may believe they understand a problem situation and they may express that understanding, but further investigation reveals gaps and discontinuities in their knowledge. This can occur because they do not have a good grasp of cause and effect within the problem domain; hence the need for simulation! In a recent study of a telephone helpline, understaffing (cause) was being blamed for the poor level of customer service (effect). The simulation revealed, however, that extra staff had a negligible effect and that the business process was to blame.

Apart from having a poor grasp of the problem situation, there is the difficulty of each client and domain expert having a different view of the problem (Weltanschauungen [Checkland 1981]). In a recent study of maintenance operators there were as many explanations of working practice as there were staff. This was further confounded when observations of the operators at work did not tie in with any of their explanations. This problem should not be a surprise, especially when dealing with systems involving human activity where the vagaries of human behavior impact upon the performance of the system.

It is apparent that although on the face of it the modeler's role is to learn from the clients and domain experts in order to develop an understanding of the problem situation, the modeler has to play a much more active role. Speaking with the right people and asking searching questions is vital to developing this understanding. The modeler should also be willing to suggest alternative interpretations with a view to unearthing new ways of perceiving the problem situation. Such discussions might be carried out face-to-face in meetings and workshops, or remotely by telephone, email or web conference.

In the third scenario, where the problem situation is neither well understood nor expressed, the job of the modeler becomes all the more difficult. In such situations, there is opportunity to adopt formal problem structuring methods, such as, soft systems methodology (Checkland 1981), cognitive mapping (Eden and Ackermann 2001), and causal loop diagrams (Sterman 2000). Lehaney and Paul (1996) and Kotiadis (2007) are both examples of the use of soft systems methodology for problem structuring prior to the development of a simulation. Meanwhile, Balci and Nance (1985) describe a methodology for problem formulation in simulation.

As an alternative to the formal problem structuring methods listed above, some have recommended the use of simulation itself as a problem structuring approach (Hodges 1991, Robinson 2001, Baldwin et al. 2004). The idea is not so much to develop an accurate model of the system under investigation, but to use the model as a means for debating and developing a shared understanding of the problem situation. Validity is measured in terms of the usefulness of the model in promoting this debate, rather than its accuracy. This idea has been made more feasible with the advent of modern visual interactive modeling systems.

During the process of understanding the problem situation, areas of limited knowledge and understanding will arise. As a result, assumptions about these areas have to be made. These assumptions should be recorded and documented. Indeed, throughout the simulation study areas of limited understanding will be discovered and further assumptions will be made.

The problem situation and the understanding of it are not static. Both will change as the simulation study progresses. The simulation model itself acts as a catalyst for this change because the information required to develop it almost always provides a focus for clarifying and developing a deeper understanding of the real world system that is being modeled. Change is also the result of influences external to the simulation, for instance, staff changes and budgetary pressures within an organization. Such continuous change acts to increase the level of iteration between modeling processes across a simulation study, with adjustments to the conceptual model being required as new facets of the problem situation emerge.

The Ford Motor Company Example: Understanding the Problem Situation

In Chapter 1 (Section 2), the problem situation at the Ford Engine Assembly plant is described. Two models were developed: one for determining the throughput of the plant, the other for investigating the scheduling of key components. In order to illustrate the conceptual modeling framework, the development of a conceptual model for the throughput problem is described. Details of the framework as applied to the scheduling problem are available on request from the author.

The reader is referred to the description of the problem situation at Ford in Chapter 1. In this case there was a clear understanding of the problem among the clients and domain experts; they were uncertain as to whether the required throughput from the production facility as designed could be achieved.

4.4 Determining the Modeling Objectives

Key to the development of an appropriate model are the modeling objectives. They drive all aspects of the modeling process providing the means by which the nature of the model is determined, the reference point for model validation, the guide for experimentation, and a metric for judging the success of the study. The following sections show how the modeling objectives are used to develop the conceptual model.

Before concentrating on specific modeling objectives, it is useful to identify the overall *aims of the organization*. The aims are not so much expressed in terms of what the model should achieve, but what the organization hopes to achieve. Once the organizational aims have been determined, it is possible to start to identify how simulation modeling might contribute to these. In most cases, of course, the simulation model will probably only be able to contribute to a subset of the organization's aims. This subset is expressed through the modeling objectives.

The purpose of a simulation study should never be the development of a model. If it were, then once the model has been developed the simulation study would be complete. Albeit that something would have been learnt from the development of the model, there would be no need for experimentation with alternative scenarios to identify potential improvements. This may seem obvious, but it is surprising how often clients are motivated by the desire for a model and not for the learning that can be gained from the model. The objectives should always be expressed in terms of what can be achieved from the *development* and *use* of the model. As such, a useful question to ask when forming the objectives is "by the end of this study what do you hope to achieve?"

Objectives can be expressed in terms of three components:

- *Achievement*: what the clients hope to achieve, e.g., increase throughput, reduce cost, improve customer service, improve understanding of the system

- *Performance*: measures of performance where applicable, e.g., increase throughput by 10%, reduce cost by $10,000
- *Constraints*: the constraints within which the clients (modeler) must work, e.g., budget, design options, available space

The clients may not be able to provide a full set of objectives. This can be the result of either their limited understanding of the problem situation, or their limited understanding of simulation and what it can provide for them. The latter might lead to the opposite problem, expecting too much from the simulation work. Whichever, the modeler should spend time educating the client about the potential for simulation, what it can and cannot do. The modeler should also be willing to suggest additional objectives as well as to redefine and eliminate the objectives suggested by the clients. In this way the modeler is able to manage the expectations of the clients, aiming to set them at a realistic level. Unfulfilled expectations are a major source of dissatisfaction among clients in simulation modeling work (Robinson 1998, Robinson and Pidd 1998).

As discussed above, the problem situation and the understanding of it are not static. So too, the modeling objectives are subject to change. Added to this, as the clients' understanding of the potential of simulation improves, as it inevitably does during the course of the study, so their requirements and expectations will also change. This only adds to the need for iteration between the activities in a simulation study, with changes to the objectives affecting the design of the model, the experimentation and the outcomes of the project. The two-way arrow in Figure 4.1 aims to signify the iteration between the problem situation and the modeling objectives.

Determining the General Project Objectives

The modeling objectives are not the only concern when designing a conceptual model. The modeler should also be aware of the general project objectives. *Timescale* is particularly important. If time is limited, the modeler may be forced into a more conservative model design. This would help reduce model development time as well as lessen the requirements for data collection and analysis. It would also quicken the run-speed of the model, reducing the time required for experimentation. If the problem situation is such that it requires a large scale model, the modeler may consider the use of a distributed simulation running in parallel on a number of computers. This should improve the run-speed of the simulation, but it may increase the development time.

The modeler should also clarify the nature of the model and its use since this will impact on the conceptual model design. Consideration should be given to some or all of these:

- *Flexibility*: the more it is envisaged that a model will be changed during (and after) a study, the greater the flexibility required.
- *Run-Speed*: particularly important if many experiments need to be performed with the model.
- *Visual Display*: whether a simple schematic (or indeed, no animated display) up to a 3-D animation is required.
- *Ease-of-Use*: ease of interaction with the model should be appropriate for the intended users.
- *Model/Component Reuse*: proper conceptual model design can aid model and component reuse (Balci et al. 2008).

The Ford Motor Company Example: Objectives

Figure 4.2 gives the modeling and general project objectives for the Ford throughput model.

Organizational aim

The overall aim is to achieve a throughput of X units per day from the assembly line. (Note: the value of X cannot be given for reasons of confidentiality.)

Modeling objectives

- To determine the number of platens required to achieve a throughput of X units per day, or
- To identify the need for additional storage (and platens) required to achieve a throughput of X units per day.

The second objective only needs be considered if throughput cannot be achieved by increasing platens only.

General project objectives

- *Time-scale*: 30 working days.
- *Flexibility*: limited level required since extensive model changes beyond changes to the data are not expected.
- *Run-speed*: many experiments may be required and so a reasonable run-speed is important, but not at the forfeit of accuracy.
- *Visual display*: simple 2D animation. (The model is largely required for performing experiments and obtaining results, communication through detailed graphics is not a major need especially as the client is familiar with simulation. Therefore, the level of visual display needs only to enable effective model testing and aid the diagnosis of problems during experimentation.)
- *Ease-of-use*: simple interactive features will suffice since the model is for use by the modeller.

FIGURE 4.2
The Ford throughput model example: Modeling and general project objectives.

4.5 Identifying the Model Outputs (Responses)

Once the objectives are known, the next stages are to identify the outputs and inputs to the model, depicted as the responses and experimental factors in Figure 4.1. It is much easier to start by giving consideration to these, than to the content of the model (Little 1994). It is also important to know the responses and experimental factors when designing the content of the conceptual model since these are the primary outputs and inputs that the model must provide and receive, respectively. In general it does not matter in which order the responses and experimental factors are identified. The responses are placed first because it is probably a little easier to think initially in terms of what the clients want from a model rather than what changes they might make while experimenting with the model.

Identification of the appropriate responses does not generally provide a major challenge. The responses have two purposes:

- To identify whether the modeling objectives have been achieved
- To point to the reasons why the objectives are not being achieved, if they are not

In the first case, the responses can normally be identified directly from the statement of the modeling objectives. For example, if the objective is to increase throughput, then it is obvious that one of the responses needs to be the throughput. For the second case, identification is a little more difficult, but appropriate responses can be identified by a mix of the modeler's past experience, the clients' understanding and the knowledge of the domain experts. Taking the throughput example, reports on machine and resource utilization and buffer/work-in-progress levels at various points in the model would be useful for helping to identify potential bottlenecks. Quade (1988) provides a useful discussion on identifying appropriate measures for the attainment of objectives.

Once the required responses have been identified, consideration should also be given to how the information is reported; this might impact on the required content of the model. Options are numerical data (e.g., mean, maximum, minimum, standard deviation) or graphical reports (e.g., time-series, bar charts, Gantt charts, pie charts). These can be determined through consultation between the simulation modeler, clients and domain experts. Consideration should also be given to the requirements for model use as outlined in the general project objectives.

The Ford Motor Company Example: Determining the Responses

Figure 4.3 shows the responses identified for the Ford throughput model. Daily throughput is selected as the response to determine the achievement of the

Outputs (to determine achievement of objectives)
- Time-series of daily throughput
- Bar chart of daily throughput
- Mean, standard deviation, minimum and maximum daily throughput

Outputs (to determine reasons for failure to meet objectives)
- Percentage machine utilization: idle, working, blocked, and broken

FIGURE 4.3
The Ford throughput model example: Responses.

objectives because it is the measure of performance identified in the modeling objectives. The three reports identified will enable an analysis of the distribution of daily throughput and its behavior over time. Utilization reports are selected as the means for determining the reasons for failing to meet the modeling objectives. This is because the level of disturbance caused by breakdowns (expected to be a key reason for failure to meet throughput) can be identified by the percentage of time each machine is broken, as well as, in part, the time machines spend idle and blocked. Further to this, any system bottlenecks and a shortage or surplus of platens can be identified by idle and blocked machines.

4.6 Identifying the Model Inputs (Experimental Factors)

The experimental factors are the model data that can be changed in order to achieve the modeling objectives. They may either be quantitative data (e.g., number of staff or speed of service) or qualitative (e.g., changes to rules or the model structure). Using this definition, the experimental factors are a limited subset of the general input data that are required for model realization.

As with the responses, identification of the experimental factors is driven by the modeling objectives. The experimental factors are the means by which it is proposed that the modeling objectives will be achieved. They may be explicitly expressed in the modeling objectives, for instance, "to obtain a 10% improvement in customer service by developing effective *staff rosters*," or "to increase throughput … by changing the *production schedule*." Alternatively, they can be obtained by asking the clients and domain experts how they intend to bring about the desired improvement to the real system. The modeler can also provide input to this discussion based on his/her experience with simulation. Altogether, this might lead to a substantial list of factors.

Although the general expectation is that the clients will have control over the experimental factors, this is not always the case. Sometimes, it is useful to experiment with factors over which there is little or no control, for example, the customer arrival rate. Such experimentation can aid understanding of the system or help plan for future events.

Where the objective of the model is, at least in part, to improve understanding, then the list of experimental factors may be a more subtle. The modeler needs to determine, with the clients and domain experts, what factors might be most useful to help improve understanding.

Apart from determining the experimental factors, it is useful to identify the range over which the experimental factors might be varied (e.g., the minimum and maximum number of staff on a shift). The simulation model can then be designed to accept this range of values, potentially avoiding a more complex model that allows for a much wider range of data input. Methods of data entry should also be considered, including: direct through the model code, model-based menus, data files or third party software (e.g., a spreadsheet). The requirement depends upon the skills of the intended users of the model and the general project objectives.

In the same way that the problem situation and modeling objectives are not static, so the experimental factors and responses are subject to change as a simulation study progresses. The realization that changing staff rosters do not achieve the required level of performance may lead to the identification of alternative proposals and, hence, new experimental factors. During experimentation, the need for additional reports may become apparent. All this serves to emphasize the iterative nature of the modeling process.

The Ford Motor Company Example: Determining the Experimental Factors

Figure 4.4 shows the experimental factors identified for the Ford throughput model. Both of these factors are derived directly from the modeling objectives.

Experimental factors

- The number of platens (maximum increase 100%)
- The size of the buffers (conveyors) between the operations (maximum increase of 100%)

FIGURE 4.4
The Ford throughput model example: Experimental factors.

4.7 Determining the Model Content: Scope and Level of Detail

The framework separates the identification of the scope of the model from the model's level of detail. These are logically different, the former identifying the boundaries of the model, the latter the depth of the model. Procedures for selecting the scope and level of detail are described below as well as the identification of assumptions and simplifications made during the modeling process.

Before making decisions about the scope and level of detail of the proposed simulation model, the use of simulation should be questioned. Is simulation the right approach for the problem situation? Robinson (2004) discusses the prime reasons for the selection of simulation as variability, interconnectedness and complexity in the systems being modeled. He also identifies the relevance of discrete-event simulation for modeling queuing systems as a prime reason for its choice. Most operations systems can be conceived as queuing systems. Along side an understanding of these reasons, the definition of the problem situation, the objectives, experimental factors and responses will help to inform the decision about whether simulation is the right approach.

Up to this point, most of the discussion is not specific to conceptual models for simulation. It is possible that another modeling approach might be adopted. It is from this point forward that the conceptual model becomes specific to simulation.

4.7.1 Determining the Model Scope

In general terms, simulation models can be conceived in terms of four types of component: entities, active states, dead states and resources (Pidd 2004). Here these are referred to as entities, activities, queues and resources, respectively. Examples of each component type are as follows:

- *Entities:* parts in a factory, customers in a service operation, telephone calls in a call center, information in a business process, forklift truck in a warehouse.
- *Activities:* machines, service desks, computers
- *Queues:* conveyor systems, buffers, waiting areas, in-/out-trays, computer storage
- *Resources:* staff, equipment

Unlike the first three components, resources are not modeled individually, but simply as countable items. Some substitution is possible between using resources and a more detailed approach using individual components. For instance, a machine could be treated as an activity and modeled at some level of detail, or it could be modeled as a resource (equipment) that needs to be available to support some other activity.

The author's experience suggests that these four component types are sufficient for most simulations of operations systems; at least those that are discrete in nature. This is largely because they can be conceived as queuing systems. Readers may be able to think of additional component types, for instance, transporters and elements of continuous processing systems (e.g., pipes). The framework can quite easily be extended to include additional component types.

Determining the scope of a model requires the identification of the entities, activities, queues and resources that are to be included in the model. The question is, how can a modeler make this decision? The following three-step approach is suggested.

Step 1: identify the model boundary. The experimental factors and responses provide a good starting point for identifying where the edges of the model might lie. The need to experiment with interarrival times provides an obvious entry point into a model. The requirement to report factory throughput strongly suggests that the last operation before work exits the factory needs to be included in the model. Beyond the experimental factors and responses, careful consideration of the system being modeled is important. At this point, the knowledge of the clients and domain experts is vital.

Step 2: identify all the components (entities, activities, queues, and resources) in the real system that lie within the model boundary. It is of particular importance to identify all components that directly connect the experimental factors to the responses, for instance, in a fast food restaurant the number of service staff (an experimental factor and resource) with waiting time (a response related to a queue). The connection between these is the service tasks. This can be thought of as the critical path that must be modeled in order to get the most basic representation that connects the experimental factors with the responses. Apart from direct connections, all interconnections also need to be considered, for the example above this might be the supply of food and drink. Some restraint is required in identifying the potential components of the model so that clearly irrelevant factors (e.g., in the case of a fast food restaurant this might be the cost of food supply) are not taken forward for further consideration.

Step 3: assess whether to include/exclude all components identified. For each component assess whether it is important to the validity, credibility, utility and feasibility of the model. If they are not needed to fulfill any of these requirements, then exclude them from the model. Judgments need to be made concerning the likely effect of each component on the accuracy of the model, and as such its validity. Will removing a component reduce the accuracy of a model below its requirement for sufficient accuracy? These judgments are, of course, confounded by interaction effects between components, for instance, the effect of removing two components may be much greater than the sum of the effects of removing them individually. Past experience will no doubt help in making such judgments. A cautious approach is advised, keeping components in the model where there is some doubt over their effect on validity.

Similarly, the effect on credibility also needs to be considered. It may be that a component is not particularly important to the accuracy of the model, but that its removal would damage the credibility of a model. In this case, it should probably be included. Indeed, a wider scope (and more detail) may be included in a model than is strictly necessary for validity, simply to increase its credibility.

Consideration should be given to the issue of utility. The inclusion of a component may significantly increase the complexity of a model or reduce its run-speed. Both could reduce the utility of the model. The effect of each component on feasibility should also be considered. It may be that the data for modeling a component are unlikely to be available, or the complexity of modeling a component would mean that the simulation study could not meet its timescale.

A careful balance between validity, credibility, utility and feasibility must be sought. For a component, where any one (or more) of these is seen as being of vital importance, then it should be included in the model. If it appears that a component is of little importance to any of these, then it can be excluded. In performing Step 3 the model boundary may well become narrower as components are excluded from the model. In Zeigler's (1976) terms, Steps 1 and 2 are about identifying the base model (at least to the extent that it is known) and Step 3 about moving to a lumped model.

In order to work through these three steps, a meeting or sequence of meetings could be arranged between the modeler, clients and domain experts. This is probably most effective in bringing the differing expertise together rather than holding meetings with smaller groups or relying on telephone or electronic media. Step 2 could consist of a brainstorming session, in which all parties identify potential model components without debate about the need, or otherwise, to include them. It is expected that there will be a number of iterations between the three steps before the model scope is agreed.

The discussions about the scope of the model need to be recorded to ensure that there is agreement over the decisions that are being made. The records also provide documentation for model development, validation, and reuse. A simple table format for documenting the model scope is suggested (see Table 4.1). The first column provides a list of all the components in the model boundary (Steps 1 and 2). The second column records the decisions from Step 3, and the third column describes the reasoning behind the decision to include or exclude each component. Having such a record provides a representation around which the modeler, clients and domain experts can debate and reach an accommodation of views on what should be in the model scope.

TABLE 4.1

The Ford Throughput Model Example: Model Scope

Component	Include/exclude	Justification
Entities:		
Engines	Include	Response: throughput of engines
Platens	Include	Experimental factor
Subcomponents	Exclude	Assume always available
Activities:		
Line A	Include	Key influence on throughput
Head Line	Include	Key influence on throughput
Line B	Include	Key influence on throughput
Hot Test and Final Dress	Exclude	Limited impact on throughput as large buffer between Line B and Hot Test
Queues:		
Conveyors	Include	Experimental factor
Resources:		
Operators	Exclude	Required for operation of manual processes, but always present and provide a standardized service. They cause no significant variation in throughput.
Maintenance staff	Include	Required for repair of machines. A shortage of staff would affect throughput

It may be helpful in some circumstances, particularly where there are differences in opinion, to generate a number of alternative model scopes and then to compare and debate the relative merits of each. Such a debate could focus on the validity, credibility, utility and feasibility of each model version.

Along side the scope table it is probably useful to have a diagram of the system and identify the model scope. A visual representation provides a more accessible view of the decisions being made about model scope, but it can only provide limited information. Meanwhile, the table is able to provide more detail, especially concerning the justification of the model scope.

The Ford Motor Company Example: Determining the Model Scope

Table 4.1 shows the model scope for the Ford throughput model. This is shown diagrammatically in Figure 4.5. The main opportunity for scope reduction comes from the exclusion of the Hot Test and Final Dress areas.

4.7.2 Determining the Model Level of Detail

Determining the level of detail requires decisions about the amount of detail to include for each component in the model scope. That is, determining the

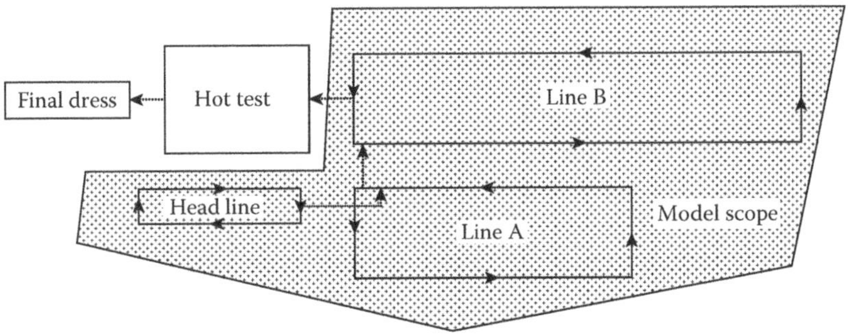

FIGURE 4.5
The Ford throughput model example: Model scope shown as the shaded area.

level of detail for each entity, activity, queue and resource to be included in the model. Table 4.2 provides a list of details that could be considered for each component type. This is not intended to be an exhaustive list, as indicated by the "other" category, but it does provide a useful starting point; although restraint should be used in defining "other" details to avoid unnecessarily long lists of clearly irrelevant details. Again, the reader may be able to think of additional details that could be listed against each component type. These can simply be added to those listed in Table 4.2.

The modeler, clients and domain experts can work through the details in Table 4.2 for each component in the model scope, determining whether the detail should be included or excluded, and also deciding on how each detail should be modeled. In a similar fashion to the model scope, the decision on whether to include a detail or not should be guided by its perceived effect on the validity, credibility, utility and feasibility of the model. These decisions might be made at a meeting between the modeler, clients and domain experts. Decisions about the level of detail can be made with reference to these:

- *Judgment:* of the modeler, clients, and domain experts
- *Past experience:* particularly on behalf of the modeler
- *Data analysis:* analysis of preliminary data about the system
- *Prototyping:* developing part of the model and testing the effect of including and excluding details

Prototyping (Powell 1995, Pidd 1999) is useful for reducing the judgmental aspect of the decisions. In particular, the development of small computer models to test ideas can aid decisions about the level of detail required for a component. Indeed, prototyping can also aid decisions about model scope, particularly through the use of high-level models in which sections of the model can be sequentially included or excluded to determine their effect on the responses.

TABLE 4.2

Template for Level of Detail by Component Type

Component	Detail	Description
Entities	Quantity	Batching of arrivals and limits to number of entities Grouping so an entity represents more than one item Quantity produced
	Arrival pattern	How entities enter the model
	Attributes	Specific information required for each entity, e.g., type or size
	Routing	Route through model dependent on entity type/attributes, e.g., job shop routing
	Other	E.g., display style
Activities	Quantity	Number of the activity
	Nature (X in Y out)	E.g., representing assembly of entities
	Cycle time	
	Breakdown/repair	Nature and timing of breakdowns
	Set-up/change-over	Nature and timing of set-ups
	Resources	Resources required for the activity
	Shifts	Model working and break periods
	Routing	How entities are routed in and out of the activity
	Other	E.g., scheduling
Queues	Quantity	Number of the queue
	Capacity	Space available for entities
	Dwell time	Time entities must spend in the queue
	Queue discipline	Sequence of entities into and out of the queue
	Breakdown/repair	Nature and timing of breakdowns
	Routing	How entities are routed in and out of the queue
	Other	E.g., type of conveyor
Resources	Quantity	Number of the resource
	Where required	At which activities the resource is required
	Shifts	Working and break periods
	Other	E.g., skill levels, interruption to tasks

A simple table format for recording these decisions is suggested, as shown in Table 4.3. This shows the components in the scope and each of the details, as listed in Table 4.2. The third column shows whether the detail is to be included in the model or excluded, while the fourth column provides a justification for the decision. Apart from listing details in the second column, it also provides a brief explanation of how a detail is to be modeled, but only for those details that are included in the model. This table provides a way of showing how the base model (the full list of details) is converted into a lumped model, by outlining what is to be included in the model and how it is to be represented.

TABLE 4.3

The Ford Throughput Model Example: Model Level of Detail

Component	Detail	Include/ Exclude	Justification
Entities:			
Engines	Quantity: produced. Model engines as an attribute of a platen (full/empty) to count engines produced	Include	Response: throughput of engines
	Arrival pattern	Exclude	Assume an engine block is always available to be loaded to the platen
	Attribute: engine derivative	Exclude	No effect on machine cycles and therefore no effect on throughput
	Routing	Exclude	Engines are only modeled as an attribute of a platen
Platens	Quantity: for Line A, Head Line and Line B	Include	Experimental factor
	Arrival pattern	Exclude	All platens are always present on the assembly line
	Attribute: full/empty Needed to count engines produced as platen leaves last operation on the line	Include	Response: throughput of engines
	Routing	Exclude	Routing determined by process not platen
Activities:			
Line A	Quantity: quantity of machines for each operation	Include	Model individual machines as each may have a significant impact on throughput
	Nature	Exclude	Subcomponents are not modeled and so no assembly is represented
	Cycle time: fixed time	Include	Required for modeling throughput. Assume no variation in time for manual processes.
	Breakdown: time between failure distribution	Include	Breakdowns are expected to have a significant impact on throughput
	Repair: repair time distribution	Include	Breakdowns are expected to have a significant impact on throughput
	Set-up/change-over	Exclude	No set-ups in real facility

(Continued)

TABLE 4.3 (Continued)

The Ford Throughput Model Example: Model Level of Detail

Component	Detail	Include/ exclude	Justification
	Resources	Include	Identify number of maintenance staff required to perform repair of machines
	Shifts	Exclude	No work takes place outside of on-shift time
	Routing: next conveyor including routing to rework areas after test stations	Include	Routing of platens defines the key interaction between system components
Head Line	*As for Line A.*		
Line B	*As for Line A.*		
Queues:			
Conveyors	Quantity: 1	Include	All conveyors are individual
	Capacity	Include	Experimental factor
	Dwell time: model as index time for platens	Include	Affects movement time and so throughput
	Queue discipline: FIFO	Include	Affects movement time and so throughput
	Breakdown/repair	Exclude	Failures are rare and so have little effect on throughput
	Routing: to next machine including routing logic to operations with more than one machine	Include	Routing of platens defines the key interaction between system components
	Type: accumulating conveyors	Include	Enables maximum utilization of buffer space and so improves throughput
Resources:			
Maintenance staff	Quantity	Include	Because there are fewer maintenance staff than machines, it is possible for staff shortages to be a bottleneck affecting throughput
	Where required: identify machines that require maintenance staff for repair	Include	Required to allocate work to maintenance staff
	Shifts	Exclude	No work takes place outside of on-shift time
	Skill level	Exclude	Assume all staff can repair all machines

The Ford Motor Company Example: Determining the Level of Detail

Table 4.3 shows the level of detail for the Ford throughput model. Note that an *operation* is the type of activity, while a *machine* is the equipment that performs that operation. There is more than one machine for some operations.

4.8 Identifying Assumptions and Simplifications

In determining the scope and level of detail of the model, various assumptions and simplifications are made. As a reminder, assumptions are made when there are uncertainties or beliefs about the real world being modeled, while simplifications are incorporated into a model to enable more rapid model development and use, to reduce data requirements and to improve transparency (understanding) (Chapter 1, Section 3.1). For the purposes of clarity, it is useful to explicitly list the assumptions and simplifications.

In large measure, the assumptions and simplifications can be identified with reference to those components and details that have been excluded from the model. Indeed, a component or detail will have been excluded on the basis that it is an assumption, simplification or a fact, the latter category referring to truisms about the real system. For instance, in the Ford throughput model (Table 4.3), set-ups/change-overs are excluded because it is known that there are no set-ups or change-overs in the real system. This is a fact. It should be noted that the assumptions and simplifications (indeed, facts) are not listed in the excluded items alone. For instance, under activities for line A in Table 4.3, there is an assumption about the cycle time of manual processes. This suggests that the modeler should not only look under the excluded components and details in the scope and level of detail tables for assumptions and simplifications, but he/she should pay careful attention to those items included in the model as well.

Once all the assumptions and simplifications have been identified it may be useful to assess each of them for their level of impact on the model responses (high, medium, low) and the confidence that can be placed in them (high, medium, low). This should be jointly agreed between the modeler, clients and domain experts. Obviously such assessments can only be based on judgment at this stage. This process, however, can be useful for ensuring that all the assumptions and simplifications seem reasonable and for ensuring all parties agree with the modeling decisions that are being made. Particular attention might be paid to those assumptions and simplifications that are seen to have a high impact and for which the confidence is low. Where necessary, the conceptual model might be changed to mitigate concerns with any of the assumptions and simplifications.

Modeling assumptions

- Capacity of the buffer before hot test and final dress is sufficient to cause minimal blockage to the assembly line from downstream processes.
- Manual operators are always present for manual processes and provide a standardized service.
- An engine block is always available to be loaded to a platen.
- No work is carried out during off-shift periods, therefore shifts do not need to be modeled.
- Conveyor breakdowns are rare and so have little impact on throughput.
- All staff can repair all machines.

FIGURE 4.6
The Ford throughput model example: Modeling assumptions.

Model simplifications

- Subcomponents are always available.
- No variation in time for manual processes.

FIGURE 4.7
The Ford throughput model example: Model simplifications.

One issue that is not discussed here is how to select appropriate simplifications. The identification of opportunities for simplification is largely a matter of the experience of the modeler, although discussion between the modeler, clients and domain experts may also provide ideas for simplification. Beyond this, it is useful to make reference to a standard set of simplifications. A range of simplification methods exist, such as, aggregating model components, replacing components with random variables and excluding infrequent events. These have been the subject of a number of publications (Morris 1967, Zeigler 1976, Innis and Rexstad 1983, Courtois 1985, Ward 1989, Robinson 1994).

The Ford Motor Company Example: Assumptions and Simplifications

Figures 4.6 and 4.7 list the assumptions and simplifications for the Ford throughput model.

4.9 Identifying Data Requirements

Apart from defining the nature of the model, the level of detail table also provides a list of data requirements. Three types of data are required for a simulation study: contextual data, data for model realization and validation data (Pidd 2003). Contextual data are required for understanding the problem situation and as an aid to forming the conceptual model (e.g., a layout diagram of the operations system and preliminary data on service times). Data for model realization can be directly identified from the level of detail table. Data for validation (e.g., past performance statistics for the operations system, if it currently exists) need to be considered in the light of the model that is being developed and the availability of data for the real system. Here, we shall only consider data for model realization.

It is a fairly straightforward task to identify the data for model realization from the level of detail table. This can be done with reference to the components and their details that are to be included in the model. These data split into two types: the experimental factors (inputs) and model parameters. Experimental factors are varied during experimentation but require initial values. Parameters are data that remain unchanged during experimentation. Identifying the data from the level of detail table supports the idea that the model should drive the data and not vice versa (Pidd 1999).

Once the data for model realization are identified, responsibility for obtaining the data should be allocated with clear direction over the time when the data need to be available. Of course, some data may already be available, other data may need to be collected and some may be neither available nor collectable. Lack of data does not necessitate abandonment of the project. Data can be estimated and sensitivity analysis can be performed to understand the effect of inaccuracies in the data. Even where data are available or can be collected, decisions need to be made about the sample size required and care must be taken to ensure the data are sufficiently accurate and in the right format. For a more detailed discussion on data collection see Robinson (2004).

If data cannot be obtained, it may be possible to change the design of the conceptual model so that these data are not required. Alternatively, the modeling objectives could be changed such that an alternative conceptual model is developed that does not require the data in question. During data collection it is almost certain that various assumptions will have to be made about the data; these assumptions should be recorded along with those identified from the conceptual model. This all serves to increase the iteration in the modeling process, with the conceptual model defining the data that are required and the availability of the data defining the conceptual model. In practice, of course, the modeler, clients and domain experts are largely cognizant of the data that are available when making decisions about the nature of the conceptual model.

The Ford Motor Company Example: Data Requirements

Figure 4.8 shows the data that are required for the Ford throughput model. These have been identified from the details of the included components in the level of detail table (Table 4.3).

Data requirements

- Planned quantity of platens on each assembly line
- Machines: quantity for each operation, cycle time, time between failure distribution, repair time distribution, routing rules (e.g., percentage rework after a test station)
- Conveyors: capacity, index time for a platen, routing rules (e.g., split to parallel machines)
- Maintenance staff: quantity, machines required to repair

FIGURE 4.8
The Ford throughput model example: Data requirements for model realization.

4.10 Model Assessment: Meets the Requirements of a Conceptual Model?

Throughout the development of the conceptual model, the extent to which the proposed model meets the requirements for validity, credibility, utility and feasibility needs to be checked and questioned. In doing so this provides an assessment of the conceptual model.

Conceptual model *validity* is "a perception, on behalf of the modeler, that the conceptual model can be developed into a computer model that is sufficiently accurate for the purpose at hand" (Chapter 1, Section 5). It is not possible to measure the accuracy of the conceptual model until at least a full computer representation is available, if it is possible to do so then (Pidd 2003, Robinson 1999). The modeler, however, is able to form an opinion about whether the proposed model is likely to deliver sufficient accuracy for the purpose to which it will be put. This opinion will largely be based on a belief as to whether all the key components and relationships are included in the model. The modeler's opinion must also be based on a clear understanding of the model's purpose (modeling objectives) and the level of accuracy required by the clients. Further to this, input from the clients and especially the domain experts is important in forming this opinion about validity.

Credibility meanwhile is defined as "a perception, on behalf of the clients, that the conceptual model can be developed into a computer model that is sufficiently accurate for the purpose at hand" (Chapter 1, Section 5). Judgment about the credibility of the model relies on the clients' opinions. This is formed by the past experience of the clients and their experience with the current project, much of which is a reflection upon their interaction with the modeler (Robinson 2002). In particular, the clients need to have a good understanding of the conceptual model. A clear description of the conceptual model is therefore required. This can be delivered through a project specification that outlines all the phases of conceptual model development as described above, from the understanding of the problem situation and the modeling objectives through to the scope and level of detail of the model and the assumptions and simplifications. Ultimately the modeler and the clients must have confidence in the conceptual model, reflected in the validity and credibility of the conceptual model, respectively.

The *utility* of the conceptual model is "a perception, on behalf of the modeler and the clients, that the conceptual model can be developed into a computer model that is useful as an aid to decision-making within the specified context" (Chapter 1, Section 5). Issues to consider are the ease-of-use, flexibility, run-speed, visual display, and potential for model/component reuse. These requirements are expressed through the general project objectives. All must be of a sufficient level to satisfy the needs of the project. For instance, if the model is to be used by the modeler for experimentation, then ease-of-use is of less importance than if the model is to be used by the clients or a third party.

The final requirement, *feasibility*, is "a perception, on behalf of the modeler and the clients, that the conceptual model can be developed into a computer model with the time, resource and data available" (Chapter 1, Section 5). Can the model be developed and used within the time available? Are the necessary skills, data, hardware and software available? The modeler, clients and domain experts need to discuss these issues and be satisfied that it is possible to develop and use the conceptual model as proposed.

It may be useful for the modeler to generate several conceptual model descriptions and then to compare them for their validity, credibility, utility and feasibility. The model that is perceived best across all four requirements could then be selected for development.

All of the above is contingent on being able to express the conceptual model in a manner that can be shared and understood by all parties involved in a simulation study. In the terms of Nance (1994), this requires the expression of the modeler's mental conceptual model as a communicative model. The tables derived in the conceptual modeling framework described above provide one means for communicating the conceptual model; see Figures 4.2, through 4.4, 4.6, and 4.7 and Tables 4.1 and 4.3. Beyond this, diagrammatic representations of the model are also useful (Figures 4.5 and 4.9), and possibly more beneficial

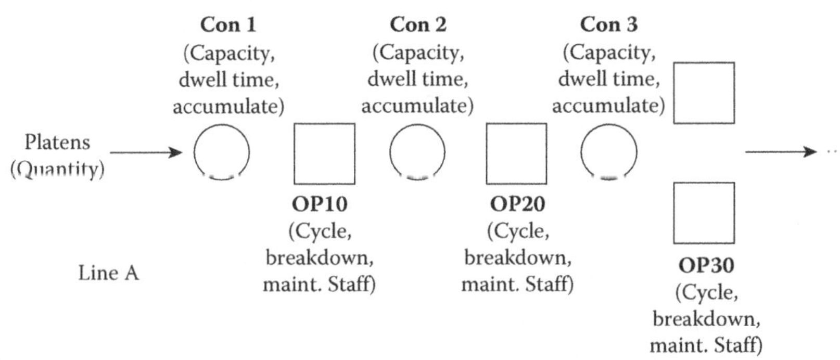

FIGURE 4.9
An illustrative process flow diagram of part of the Ford throughput conceptual model.

as a communicative tool (Crapo et al. 2000). A range of such methods have been used for representing simulation conceptual models, for instance:

- Process flow diagrams (Robinson 2004); see Figure 4.9
- Activity cycle diagrams (Hills 1971)
- Event graphs (Schruben 1983, Som and Sargent 1989)
- Digraphs (Nance and Overstreet 1987)
- UML (the unified modeling language) (Richter and März 2000)
- Object models (van der Zee 2006)
- Simulation activity diagrams (Ryan and Heavey 2006)

Pooley (1991) provides a useful review of diagramming techniques that might support simulation modeling. The conceptual model could, of course, be represented using the visual display facilities of the simulation software, without the need for coding the detail of the model. Figure 4.9 shows a simple process flow diagram for a portion of the Ford throughput model.

4.11 Conclusion

The conceptual modeling framework described above provides a series of iterative activities for helping a modeler to design a conceptual model for a specific problem situation. Each activity is documented with a table summarizing the decisions made. The use of these tables (along with diagrammatic representations of the model), provides a means for communicating and debating the conceptual model with the clients and domain experts. As a result, it provides a route to agreeing upon the nature of the simulation model that is required to intervene in the problem situation.

In conclusion, we consider the question of whether there is a right conceptual model for any specified problem. For two reasons, the answer is "no." First, we have identified conceptual modeling as an art. Albeit that the framework above provides some discipline to that art, different modelers will not come to the same conclusions. Any other expectation would be akin to expecting an art class to paint exactly the same picture of the same subject. There has to be room for creativity in any art, including conceptual modeling. There are, of course, better and worse conceptual models. The four requirements of a conceptual model (validity, credibility, utility, and feasibility) provide a means for distinguishing better from worse.

A second reason why there is no right conceptual model is because the model is an agreement between more than one person (the modeler, clients, and domain experts). Each has his/her own preferences for and perceptions of what is required. These preferences and perceptions are expressed through the four requirements of a conceptual model. The framework provides a means for communicating and debating the conceptual model, with a view to reaching an agreement, or at least an accommodation of views, over the nature of the model. The conceptual model is, therefore, some compromise between alternative preferences and perceptions of the world.

In short, there is no absolutely right conceptual model because the model is dependent on the preferences and perceptions of the people involved in the simulation study. It would seem that the idea of developing conceptual modeling frameworks that will always lead to a single best model is futile. Instead, our aim should be to provide frameworks that provide a means for communicating, debating and agreeing upon a conceptual model, while also releasing the potential for creativity in the modeling process. This is what the conceptual modeling framework described here aims to provide.

Acknowledgments

This chapter is reproduced, with minor editing, from: Robinson, S. 2008. Conceptual modelling for simulation part II: A framework for conceptual modelling. *Journal of the Operational Research Society* 59 (3): 291–304. © 2008 Operational Research Society Ltd. Reproduced with permission of Palgrave Macmillan.

Some sections of this chapter are based on the following:

- Robinson, S. 2004. *Simulation: The Practice of Model Development and Use.* Chichester, UK: Wiley.
- Robinson, S. 2004. Designing the conceptual model in simulation studies. In *Proceedings of the 2004 Operational Research Society Simulation Workshop (SW04)*, ed. S.C. Brailsford, S.C., L. Oakshott, S. Robinson, and S.J.E. Taylor, 259–266. Birmingham, UK: Operational Research Society.
- Robinson, S. 2006. Issues in conceptual modelling for simulation: Setting a research agenda. In *Proceedings of the Operational Research Society Simulation Workshop (SW06)*, ed. J. Garnett, S.C. Brailsford, S. Robinson, and S.J.E. Taylor, 165–174. Birmingham, UK: Operational Research Society.
- Robinson, S. 2006. Conceptual modeling for simulation: Issues and research requirements. In *Proceedings of the 2006 Winter Simulation Conference*, ed. L.F. Perrone, F.P. Wieland, J. Liu, B.G. Lawson, D.M. Nicol, and R.M. Fujimoto, 792–800. Piscataway, NJ: IEEE.

The Ford engine plant example is used with the permission of John Ladbrook, Ford Motor Company.

References

Balci, O., and R.E. Nance. 1985. Formulated problem verification as an explicit requirement of model credibility. *Simulation* 45 (2): 76–86.

Balci, O., J.D. Arthur, and R.E. Nance. 2008. Accomplishing reuse with a simulation conceptual model. In *Proceedings of the 2008 Winter Simulation Conference*, ed. S.J. Mason, R.R. Hill, L. Mönch, O. Rose, T. Jefferson, and J.W. Fowler, 959–965. Piscataway, NJ: IEEE.

Baldwin, L.P., T. Eldabi, and R.J. Paul. 2004. Simulation in healthcare management: A soft approach (MAPIU). *Simulation modelling practice and theory* 12 (7–8): 541–557.

Checkland, P.B. 1981. *Systems Thinking, Systems Practice*. Chichester: Wiley.

Courtois, P.J. 1985. On time and space decomposition of complex structures. *Communications of the ACM* 28 (6): 590–603.

Crapo, A.W., L.B. Waisel, W.A. Wallace, et al. 2000. Visualization and the process of modeling: a cognitive-theoretic view. In *Proceedings of the Sixth ACM SIGKDD International Conference on Knowledge Discovery and Data Mining*, ed. R. Ramakrishnan, S. Stolfo, R. Bayardo, and I. Parsa, 218–226. New York: ACM Press.

Eden, C., and F. Ackermann, F. 2001. SODA: The principles. In *Rational analysis for a problematic world revisited*, 2nd edition, ed. J.V. Rosenhead and J. Mingers, 21–41. Chichester: Wiley.

Ferguson, P., W.S. Humphrey, S. Khajenoori, et al. 1997. Results of applying the personal software process. *Computer* 30 (5): 24–31.

Hills, P.R. 1971. *HOCUS*. Egham, Surrey: P-E Group.

Hodges, J.S. 1991. Six (or so) things you can do with a bad model. *Operations research* 39 (3): 355–365.

Innis, G., and E. Rexstad. 1983. Simulation model simplification techniques. *Simulation* 41 (1): 7–15.

Kotiadis, K. 2007. Using soft systems methodology to determine the simulation study objectives. *Journal of simulation* 1 (3): 215–222.

Lehaney, B., and R.J. Paul. 1996. The use of soft systems methodology in the development of a simulation of out-patient services at Watford General Hospital. *Journal of the operational research society* 47 (7): 864–870.

Little, J.D.C. 1994. Part 2: On model building. In *Ethics in modeling*, ed. W.A. Wallace, 167–182. Amsterdam: Elsevier (Pergamon).

Morris, W.T. 1967. On the art of modeling. *Management science* 13 (12): B707–717.

Nance, R.E. 1994. The conical methodology and the evolution of simulation model development. *Annals of operations research* 53: 1–45.

Nance, R.E., and C.M. Overstreet. 1987. Diagnostic assistance using digraph representation of discrete event simulation model specifications. *Transactions of the society for computer simulation* 4 (1): 33–57.

Pidd, M. 1999. Just modeling through: a rough guide to modeling. *Interfaces* 29 (2): 118–132.

Pidd, M. 2003. *Tools for Thinking: Modelling in management science*, 2nd ed. Chichester: Wiley.

Pidd, M. 2004. *Computer Simulation in Management Science*, 5th ed. Chichester: Wiley.

Pooley, R.J. 1991. Towards a standard for hierarchical process oriented discrete event diagrams. *Transactions of the society for computer simulation* 8 (1): 1–41.

Powell. S.G. 1995. Six key modeling heuristics. *Interfaces* 25 (4): 114–125.

Quade, E.S. 1988. Quantitative methods: uses and limitations. In *Handbook of Systems Analysis: Craft Issues and Procedures*, H.J. Miser and E.S. Quade, 283–324. Chichester: Wiley.

Richter, H., and L. März. 2000. Toward a standard process: The use of UML for designing simulation models. In *Proceedings of the 2000 Winter Simulation Conference*, ed. J.A. Joines, R.R. Barton, K. Kang, and P.A. Fishwick, 394–398. Piscataway, NJ: IEEE.

Robinson, S. 1994. Simulation projects: Building the right conceptual model. *Industrial engineering* 26 (9): 34–36.

Robinson, S. 1998. Measuring service quality in the process of delivering a simulation study: the customer's perspective. *International transactions in operational research* 5 (5): 357–374.

Robinson, S. 1999. Simulation verification, validation and confidence: a tutorial. *Transactions of the society for computer simulation international* 16 (2): 63–69.

Robinson, S. 2001. Soft with a hard centre: Discrete-event simulation in facilitation. *Journal of the operational research society* 52 (8): 905–915.

Robinson, S. 2002. General concepts of quality for discrete-event simulation. *European journal of operational research* 138 (1): 103–117.

Robinson, S. 2004. *Simulation: The Practice of Model Development and Use.* Chichester, UK: Wiley.

Robinson, S., and M. Pidd. 1998. Provider and customer expectations of successful simulation projects. *Journal of the operational research society* 49 (3): 200–209.

Ryan, J., and Heavey, C. 2006. Requirements gathering for simulation. In *Proceedings of the Operational Research Society Simulation Workshop (SW06)*, ed. J. Garnett, S.C. Brailsford, S. Robinson, and S.J.E. Taylor, 175–184. Birmingham: Operational Research Society.

Schruben, L.W. 1983. Simulation modeling with event graphs. *Communications of the ACM* 26 (11): 957–963.

Som, T.K., and R.G. Sargent. 1989. A formal development of event graphs as an aid to structured and efficient simulation programs. *ORSA journal on computing* 1 (2): 107–125.

Sterman, J.D. 2000. *Business Dynamics: Systems Thinking and Modeling for a Complex World.* New York: Irwin/McGraw-Hill.

van der Zee, D.J. 2006. Building communicative models: A job oriented approach to manufacturing simulation. In *Proceedings of the Operational Research Society Simulation Workshop (SW06)*, ed. J. Garnett, S.C. Brailsford, S. Robinson, and S.J.E. Taylor, 185–194. Birmingham: Operational Research Society.

Ward, S.C. 1989. Arguments for constructively simple models. *Journal of the operational research society* 40 (2): 141–153.

Wild, R. 2002. *Operations Management*, 6th ed. London: Continuum.

Zeigler, B.P. 1976. *Theory of Modeling and Simulation*. New York: Wiley.

5

Developing Participative Simulation Models: Framing Decomposition Principles for Joint Understanding

Durk-Jouke van der Zee

CONTENTS

5.1 Introduction

A few decades ago, Hurrion introduced the notion of visual interactive simulation (Hurrion 1976, 1989). Its basic contribution lies in the fact that it brings analysts and stakeholders together by means of an animated display. As such, it facilitates the joint discussion on model validation/verification, and—maybe even more important—alternative and possibly better solutions to the problem that the modeling and simulation project is meant to solve (Bell and O'Keefe 1987, Bell et al. 1999). Refinement of the approach is possible building on principles of object-oriented design (Booch 1994). Object orientation was reembraced as a metaphor for simulation modeling in the 1990s, being developed for the early simulation language Simula™ (Dahl and Nygaard 1966). It foresees in a natural one-to-one mapping of real-world concepts to modeling constructs (Glassey and Adiga 1990, Kreutzer 1993, Roberts and Dessouky 1998).

Visual interaction and object orientation set rough guidelines for building a "conceptual" model for simulation (Balci 1986). Typically, a conceptual model is meant to facilitate the joint search for better-quality solutions, building on a common understanding of the problem and system at hand. This implies the conceptual model being both transparent and complete to all parties involved in the study.

Clearly, model visualization greatly popularized the use of modeling and simulation for systems design. This popularity makes clear that facilitating stakeholders' involvement in the simulation study is crucial to the acceptance of modeling and simulation as a decision support tool. It may be expected that such "facilitation" is even more important nowadays as systems design often involves multiple problem owners as in, for example, supply chains, health-care chains and transportation networks. Moreover, the complexity of suchlike systems makes problem owners *participation* in the search for better solutions indispensable, given their role as domain experts. The challenge for the analyst is to contribute to, and guide this process, aiming to build mutual trust, and fostering the joint creation and acceptance of good quality solutions. Clearly, a visual interactive model that is validated, and understood by all parties involved, may act as an enabler in answering to this challenge.

Unfortunately, many simulation models tend to be limited with respect to transparency and completeness in terms of decision variables. Typically, a subset of relevant system elements is not visualized or distributed over the model. Common causes may be found in the analyst's use of his implicit reference models and (limited) facilities offered by the simulation language, which is not domain specific or is incomplete. In addition to the analyst's skills and tool qualities, the characteristics of the project at hand should be mentioned, such as, the available resources, budget restrictions, time horizon, client's interest in modeling, and modeling efforts as they follow from problem complexity. Taylor and Robinson (2006) mention domains such as health care, services, and business processes as either not being represented or represented in such a way that needs to significantly improve over the next decade. Also, more traditional fields of simulation application may be plagued by similar shortcomings. Several authors report that control structures in manufacturing systems, that is, the managers or systems responsible for control, their activities and their mutual attuning of these activities are often left implicit (Mize et al. 1992, Pratt et al. 1994, Bodner and McGinnis 2002, Galland et al. 2003). Control elements are, for example, dispersed over the model, are not visualized, or form part of the time-indexed scheduling of events.

In our previous work we addressed the above issue for the manufacturing field. We did so by presenting a conceptual modeling framework for manufacturing simulation (van der Zee and Van der Vorst 2005, van der Zee 2006a). The framework offers a high-level description of essential manufacturing elements and relationships as well as their dynamics. It is meant to serve as an *explicit* frame of reference for more disciplined modeling and visualization, and to offer a common conceptual basis for improving model understanding, see Figure 5.1. We studied the relevance and use of the modeling framework for simulation studies on manufacturing planning.

The development of the modeling framework, as it was presented and applied in our previous work, is the net result of (1) the recognition of the aforementioned problem on simulation model development—the *lack of explicit guidance for the analyst in model creation*, and (2) a *domain analysis* for the manufacturing field, which resulted in the identification of a comprehensive set of general and domain-specific *decomposition principles*. Essentially, the framework distinguishes among of three elementary object classes, i.e., agents, flows, and jobs, which are further tailored to the manufacturing domain, using this set of decomposition principles. In this chapter we consider the process of developing the conceptual modeling framework for a specific domain in greater detail. This is motivated by two reasons. The first is our idea of highlighting the underlying *approach toward model engineering*, which is assumed to have a validity that exceeds the manufacturing domain.

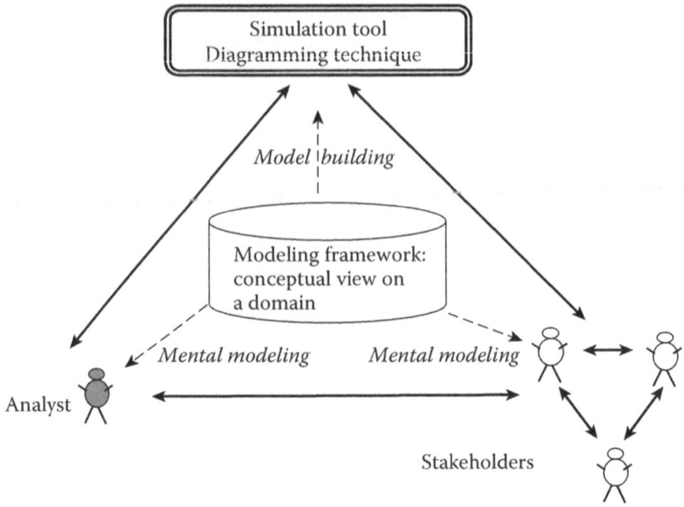

FIGURE 5.1
Modeling framework: Role in supporting simulation model development.

Second, and more specific, we aim to make the proposed modeling framework for manufacturing simulation to an *open architecture* amendable for improvements, extensions, and refinements.

5.2 Literature Review: Seeking Discipline in Model Creation

The thesis underlying this chapter is that transparent simulation models enable an active participation of stakeholders in decision support. In turn this is assumed to facilitate the build up of mutual trust, model validation, creativity in solution finding, and solution acceptance. This brings us to the basic question: how to contribute to model transparency, i.e., enhance model understanding for all stakeholders? In particular, which guidance is available for the analyst fulfilling this task of model development, i.e., the choice, detailing, and visualization of model elements and their workings? Note that our prime focus in this chapter is on the development of the conceptual model. Obviously, it is the embedding of the model in project contents and its organization, which determines its contribution to ultimate project success (Robinson 2002).

The modeling task of the simulation analyst is often considered an art (Shannon 1975). This stresses the importance of the analyst's creativity in model building—trying to capture relevant elements of a system with the right amount of detail. Typically, this creativity is bounded and guided by implicit or explicit guidelines, i.e., good modeling practices and principles

(Pidd 1999, Law and Kelton 2000), domain related insights (Valentin and Verbraeck 2005), and—last but not least—logic and libraries underlying simulation software (Kreutzer 1986). Let us consider these guidelines in somewhat more detail, starting from a recent survey of Robinson (2006, 2008).

Robinson distinguishes between three basic approaches on simulation model development: principles of modeling, methods of simplification and modeling frameworks. Here we characterize each of the approaches. For related references please consult the work of Robinson. *Principles of modeling* refer to the general case of conceptual modeling. Guiding principles include the need for model simplicity, the advocated policy of incremental modeling, and the good use of metaphors, analogies, and similarities in model creation. *Methods of simplification* focus on the possibility of reducing model scope and/or the level of detail for model elements, starting from their relevance for model accuracy. Gains may, for example, be realized by combining model elements, leaving them out or adapting their attributes. Clearly, these methods are helpful in model construction by pointing at possibilities for model pruning. However, they do not address model creation in terms of what is to be modeled. *Modeling frameworks* specify a procedural approach in detailing a model in terms of its elements, their attributes and their relationships. Examples include the general case of systems representation and domain related cases. The general case of systems representation foresees in conceptualization building on elementary system elements, i.e., components, including their variables and parameters, and mutual relationships, see Shannon (1975). Such representations are reflected in basic diagramming techniques for example, Petri Nets, Activity Cycle Diagrams (Pooley 1991), and Event Graphs (Schruben 1983). Domain related cases refer primarily to the military field, see Nance (1994). Outside this domain, examples are scarce. Guru and Savory (2004) propose a framework for modeling physical security systems. Also, our previous work on the modeling framework for manufacturing systems (van der Zee and Van der Vorst 2005, van der Zee 2006a) may be included in this category.

Next to principles, methods and frameworks the analyst may be guided—or restricted—in his conceptual modeling efforts by the simulation software being adopted for the project. Pidd (1998) distinguishes between several types of software ranging from general purpose languages to visual interactive modeling systems (VIMS). While the former category does not provide a conceptual notion or basis for modeling, the latter category is tailored toward simulation use assuming model building to be based on an elaborate library of building blocks, which may be domain related to a certain degree. Further we mention a specific class of simulation modeling tools based on elementary concepts like, for example, DEVS (Zeigler 1990) and Petri Nets (Murata 1989). Where VIMS offer contextual rich libraries, these tools force the user to build models from a small set of elementary components. However, where the logic underlying library set up for VIMS may be found in a pragmatic and evolutionary path, libraries for tools like

DEVS and Petri Nets conform to more rigorous mathematical standards. This enhances model transparency; however, this tends to be only valid for small-scale problems. Embedding real-life systems often leads to representations in terms of large networks of similar building blocks, each adding just little detail. Typically, they are difficult to interpret for nonanalysts (Kamper 1991).

The above discussion leads us to the following observations on available guidance for the analyst:

- Relevant approaches are available for supporting the analyst in somewhat more disciplined modeling. However, they mostly address the general case, or aim at model pruning instead of model creation.
- Much guidance may come from the specialized libraries of so-called visual interactive modeling systems. However, their contextual richness does not a priori guarantee model completeness or transparency. Libraries may be incomplete with respect to modeling requirements of certain domains, for example, consider the modeling of manufacturing control (see section 5. 1). Also, the skills of the analyst in mastering the tool and the availability of insightful documentation are of great significance in this respect.
- Simulation tools building on more fundamental concepts like DEVS or Petri Nets may allow for transparent modeling, given a clear set of well defined building blocks. However, this transparency tends to be restricted to small-scale problems.

The motivation for developing our conceptual modeling framework for manufacturing simulation (van der Zee and Van der Vorst 2005) is the lack of an explicit manufacturing domain related approach for model development. A specific focus concerns the modeling of control logic. See van der Zee and Van der Vorst (2005) and van der Zee (2006a) for a comparison with alternative frameworks. In this article we will elaborate on the setup of the modeling framework. The framework defines a conceptual architecture consisting of a number of component classes, i.e., agents, flow items, and jobs. The architecture is founded on a comprehensive set of general and domain-specific decomposition principles. Identifying and classifying these decomposition principles, as well as their combination, will enable these:

- Possible improvements, extensions, and refinements of the conceptual modeling framework for manufacturing simulation, as its source code is now "open"
- The development of alternative frameworks for other or related domains

5.3 On the Construction of a Modeling Framework

In this section we address the construction of our conceptual modeling framework for manufacturing. Before we do so we will clarify our idea of a model.

5.3.1 Model and Experimental Frame

As a first step in modeling Ören and Zeigler (1979) suggest to distinguish between the "experimental frame" and the "model." Where the model corresponds to a (limited) representation of a real-world system, the experimental frame specifies those circumstances under which the real system is to be observed or experimented with. The separation between model and experimental frame has been implemented in the language SIMAN™ (Pegden et al. 1990, Pidd 1998). Here the model describes the physical elements of the system (machines, workers, goods, etc.) and their logical interrelationships. The experimental frame specifies the experimental conditions under which the model is to run, like initial conditions, type of statistics gathered, etc.

5.3.2 Modeling Framework

The construction of the modeling framework foresees in two phases: (1) a domain analysis resulting in a set of decomposition principles and (2) an engineering process in which decomposition principles are "framed" in terms of high-level class definitions for the field, see Figure 5.2.

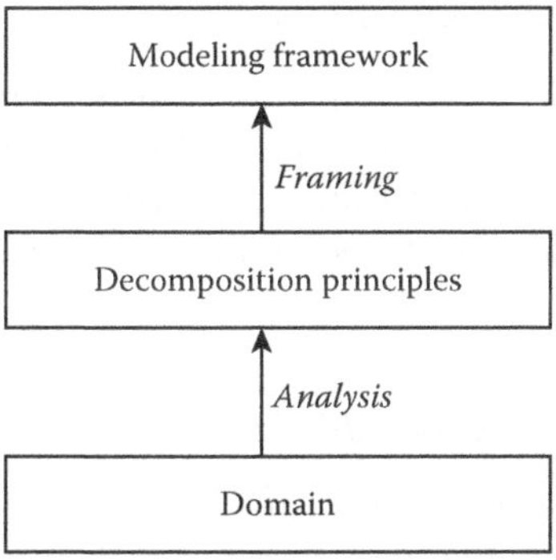

FIGURE 5.2
Construction of a modeling framework.

TABLE 5.1

Decomposition Principles Underlying the Modeling Framework for Manufacturing Simulation

I	External and internal entities (system boundary)
II	Movable and nonmovable entities
III	Queues and servers
IV	Intelligent and nonintelligent entities
V	Infrastructure, flows, and jobs
VI	Modality: physical, information, and control elements
VII	Dynamics: executing jobs

5.3.3 Decomposition Principles

The structure for a model concerns two types of elements (Ören and Zeigler 1979):

- Component models that make up the overall model: They specify the static characteristics for the system. Their attributes characterize states, inputs and outputs.
- Rules of interaction among component models: They specify dynamics for the system.

Let us now consider decomposition principles for identifying and characterizing both types of elements as they resulted from our domain analysis of the manufacturing field, see Table 5.1.

Principles I–IV characterize entities for a wider category than just manufacturing systems, whereas principles V–VII are somewhat more specific for the manufacturing field.

5.3.3.1 I External and Internal Entities

The distinction between external and internal entities follows from the concept of a system boundary. Typically, system design involves the choice, configuration and operation of "internal" entities. "External" entities are only modeled as far as their behavior is relevant for the system. Rather, they act as "sources" or "sinks" for physical flows (goods, resources), or data.

5.3.3.2 II Movable and Nonmovable Entities

The physical and logical infrastructure for a manufacturing system is typically made up of entities like workstations, information systems, and managers. As such the respective entities are considered nonmovable. They communicate by exchanging movable items like goods and messages. Note how the distinction between movable and nonmovable entities may sometimes be subtle. For example, consider an operator being assigned to a single machine vs. an operator being assigned to multiple machines. In the former case the operator may be considered as belonging to the manufacturing infrastructure, whereas the latter case suggests considering the operator as a movable shared resource.

5.3.3.3 *III Queues and Servers*

Nonmovable entities may be classified according to their associated activities. Following the general notion in management science of manufacturing systems being queueing systems, a basic distinction can be made between queues (store items waiting) and servers (service the items being processed).

5.3.3.4 *IV Intelligent and Nonintelligent Entities*

According to Lefrancois and Montreuil (1994) and Lefrancois et al. (1996), a distinction between intelligent and nonintelligent entities permits a more natural and richer presentation and implementation of systems modeled. In such a context, intelligent entities are modeled as agents. Agents are used to implement decision rules inherent to manufacturing system planning and control. Examples include logic for scheduling, dispatching and releasing jobs for a machine or department. According to Lefrancois and Montreuil (1994) and Lefrancois et al. (1996), workstations and work orders are assumed to be nonintelligent. We agree with respect to work orders. However, in this chapter we consider work stations to be intelligent, because of their local control logic for initiating and steering jobs. We come back to this point below; see section 5.3.4.

5.3.3.5 *V Infrastructure, Flows, and Jobs*

Manufacturing systems are built up of infrastructural elements like workstations, information systems, and managers. Flows refer to the objects being exchanged and transformed within this infrastructure as a net effect of jobs being executed.

5.3.3.6 *VI Modality: Physical, Information, and Control Elements*

The separation of physical, information, and control elements is assumed to facilitate a higher degree of model reusability and a more "natural" model building environment (Mize et al. 1992). Mize et al. point at the fact that traditional languages do not provide natural constructs for separately and distinctly modeling the three types of basic functions, i.e., physical transformation, information exchange, and control/decision. In addition, the constructs provided for information exchange and control specification are often dispersed in the model (Pratt et al. 1994). Clearly, this hinders modification and programming. A policy that adheres rather strictly to the "one component–one function" doctrine would suffer less from these drawbacks. In turn this provides a more natural modeling environment as the modeler is forced to think about model elements independently.

5.3.3.7 VII Dynamics: Executing Jobs

An important principle introduced for our modeling framework is: "all activities in the manufacturing system have a common denominator: the job" (van der Zee and Van der Vorst 2005). Identifying each activity as a job is meant to bring two important advantages. First, the use of this common denominator for all activities will provide a clear and natural mechanism for event scheduling, where events are related to the start and the completion of jobs. Second, an explicit notion and allocation of company activities (compare V), increases visibility and traceability of decision variables.

5.3.4 Engineering the Framework: Framing Decomposition Principles

In our focus a *modeling framework* concerns a well-defined conceptual view on a domain. A modeling framework is the outcome of an engineering process in which generic classes of entities and their workings are identified and characterized, building on a comprehensive set of domain related and more general decomposition principles.

Here we will discuss the construction of our modeling framework for manufacturing simulation. More in particular we will show how we "framed" decomposition principles in terms of domain related classes, class hierarchies, and class relationships. As a starting point for discussion we use Tables 5.1 and 5.2. As far as notation is concerned we will conform ourselves to an object-oriented approach, as described in "the object model," see the appendix and Booch (1994) for more details. Application of decomposition principles will be referred to by their Roman numbering, see Table 5.1. The resulting elements of the framework will be addressed by their alphabetical numbering.

5.3.4.1 Main Classes and Their Hierarchies

To represent entities in the manufacturing domain we define three main classes in our modeling framework: *agents, flow items,* and *jobs* (V; A–C). *Agents* represent the infrastructural, *nonmovable* elements of a manufacturing system such as workstations, information systems and managers (II; A). They are assumed to be intelligent to a certain extent (IV; D). Their decision-making capabilities relate to transformations of goods or data. A boundary is recognized between the system under study and its related environment. This is reflected by distinguishing between *internal* and *external* agents in the class hierarchy (I; A). For internal agents such as machines, warehouses, Automatic Guided Vehicles, planners, etc., we consider two subclasses,

TABLE 5.2

Framing Decomposition Principles in Class Definitions, Class Hierarchies, and Class Relationships

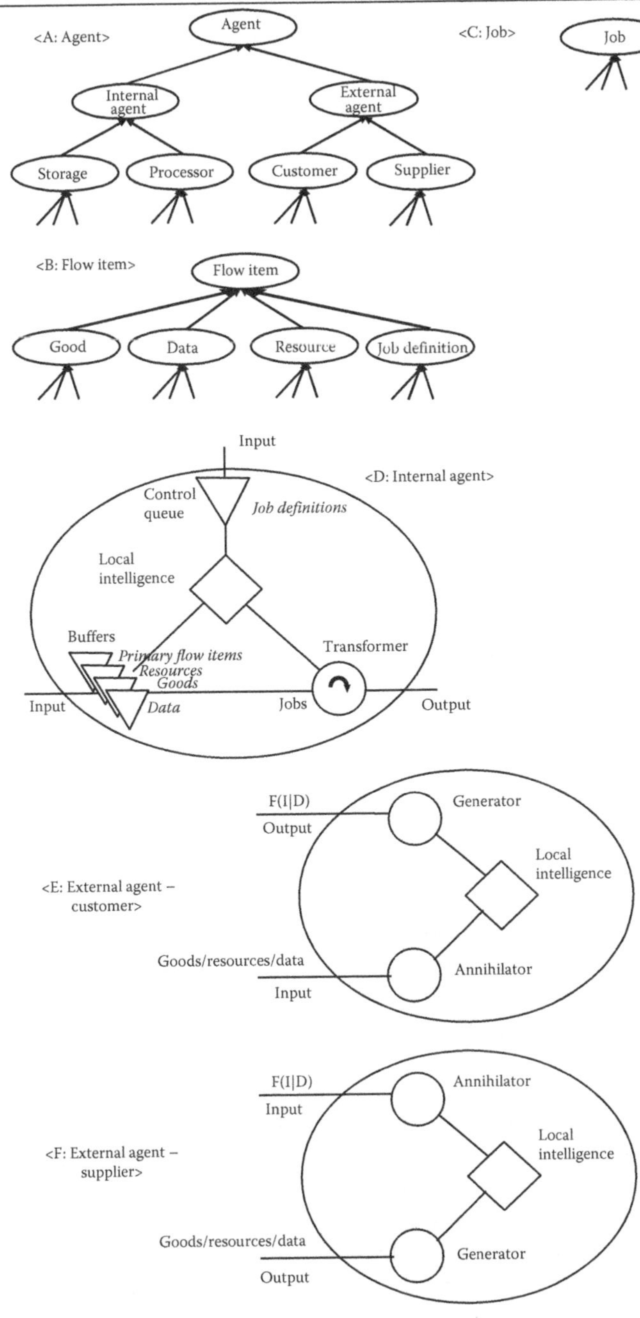

(Continued)

TABLE 5.2 (Continued)

Framing Decomposition Principles in Class Definitions, Class Hierarchies, and Class Relationships

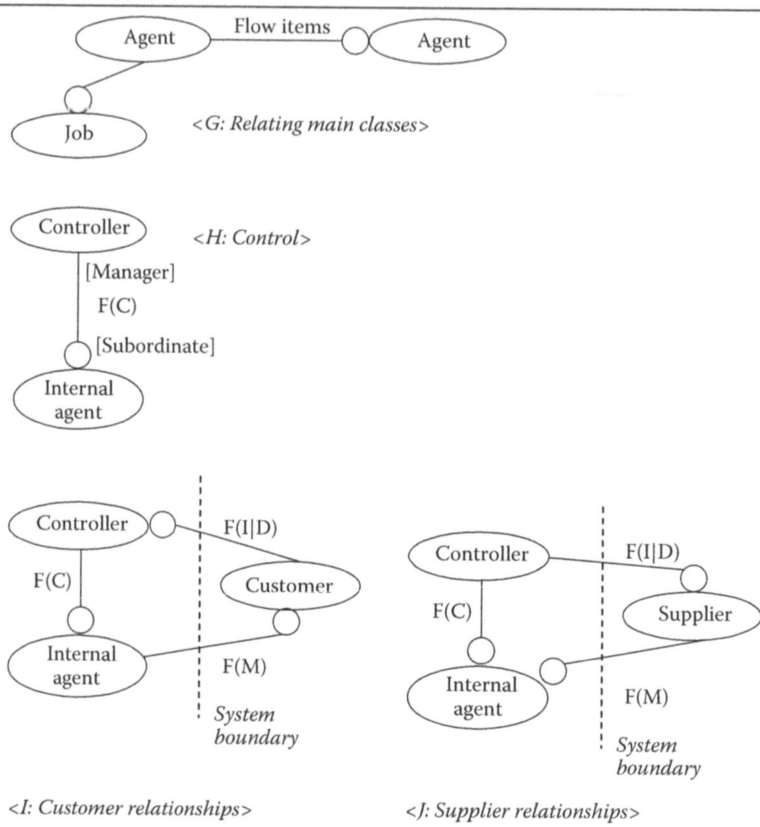

i.e., processors, and storages. They are distinguished by considering the nature of their dominant type of resources (III; A). For external agents, it is more common to consider sub classes suppliers and customers, i.e., processor types that act as "sources" and "sinks," respectively (III; A).

Flow items constitute the *movable* objects within manufacturing systems (II; B). We include four types of flow items in the modeling framework: *goods* (like, for example, materials, parts, semifinished products), *resources* (like, for example, manpower, tools, vehicles), *data* (like, for example, feed back on control decisions, forecasts) and *job definitions* (VI; B). Goods, resources, or data seldom flow spontaneously from one location to another, as mostly some form of control is exercised over agent activities. Typically, agents' jobs are directed by messages. We address this type of messages as *job definitions*. Job definitions act both as a trigger for initiating agent activities and as a carrier of relevant information related to these activities, such as, for example, their input, processing conditions and the agents to whom the resulting output should be sent.

In a manufacturing system agents and flows are linked by *jobs,* which describe manufacturing activities (V; C). In our *job-oriented worldview,* we assume that *each manufacturing activity is referred to as a job,* being the responsibility of a specific agent. In turn, a job concerns a comprehensive set of activities, i.e., transformations linking a set of flow items and agent resources. Note that a job definition is an intrinsic element of this set of flow items, which influences both timing and characteristics of jobs, see above.

It is common practice to think of *agents* in terms of the type of flow items that are the subject of their jobs. In line with practice it is possible to define more specific classes of internal agents, where the type of flow item serves as a parameter. For example, a workstation may be considered an internal agent of a processor type handling goods. In a similar way control systems and *decision-makers may be defined as internal agents producing job definitions.*

5.3.4.2 Class Definitions: Agents

In this subsection we consider the structure for agents. This follows from its role as an intelligent entity (IV; D). The structure for an internal agent was inspired by the atomic model as defined by Zeigler (1976, 1990). Starting from a general view on simulation modeling an atomic model *encapsulates* basic elements and functions of an entity in a formal way. In our discussion of a class definition for agents we will distinguish between internal and external agents.

The state of an agent relates to its attributes and their values. Attributes include buffers and transformers (III; D). Buffers model the temporary storage of those flow items that are the prime subject of a future job or that enable (facilitate) job execution (resources, information). The first category of flow items is addressed in Table 5.2 as primary flow items. For example, a machine job foresees in sheet metal (goods) being its prime subject, while tools, and personnel (resources) facilitate the transformation of the sheet metal. Note how the latter conception of personnel only considers their working skills and "neglects" their reasoning capabilities. Except for the buffer that stores the job definitions for an agent, i.e., the control queue, buffers for facilitative flow items are optional. The transformer reflects the physical or logical location for those flow items that are associated with the agent's set of jobs being executed.

The previous paragraph presented the basic elements of an internal agent. To discuss agent functions we distinguish between input and output operations, and the local intelligence (IV; D). The handling of incoming flow items is dealt with by one or more input operations. An input operation puts flow items in the right buffers. The diagram shows two such input operations: one that puts job definitions in the control queue and one that updates buffers. In a similar way, the *output* operations take care of sending the flow items resulting from a job to the respective output addresses (agents) by calling the respective input operations.

The initiation of a job is enabled by rules comprised in the *local intelligence* (IV; D). As a first rule in initiating a job, the job with the highest priority in the control queue is selected for processing. Before a job may be started, two requirements (preconditions) have to be fulfilled: (1) the availability of a job definition, and (2) the availability of the required input (which can be null) for a job. In accordance with our job-oriented approach (V, VII; D) each job has to be prespecified. This is reflected in the requirement that a job definition should be present in the control queue. The job definition specifies the required input to be withdrawn from the buffers, capacity needed, processing conditions and the identifiers of agents to which the job's output has to be sent.

The notion of local intelligence applies to all agents, including planners and the work stations assigned to them. Where intelligence for work stations may be restricted to elementary rules for timing and release of jobs, decision logic for planners may be comprehensive. Decision logic for planners may refer to a wide range of rules that support, for example, capacity planning, material planning, scheduling and dispatching. Here the output of the one decision job (for example, capacity planning) may determine input for the other decision job (for example, scheduling). In this respect a hierarchy of planners and their jobs may be distinguished (see Relationships between agents).

Let us now consider external agents, i.e., the customers and suppliers that make up the environment for a manufacturing system. Besides the element local intelligence (IV; E, F), which is also found in internal agents, *generators* and *annihilators* are defined for external agents. Generators represent "sources" of flow items, while annihilators model "sinks" in which flow items disappear (I; E, F). Local intelligence may be used to link activities of generator and annihilator. For example, local intelligence may comprise a rule that states that a new order may only be issued if the goods corresponding to the last order have been received. Here an order corresponds to a subclass of data (F(I|D)), which acts as a trigger for a control action of a planner (represented as Controller), so that the flow items (F(M)) are produced to satisfy the customer. Note how notation reflects the entity class (F for Flow Item), the subclass (G for Good, R for Resource, I for Data, C for Job Definition, and M a parameter—in case of no prespecified subclass), and their specific characteristics (D for Trigger). In a similar way, we define the supplier by interchanging the labels for input and output (F(I|D, F(M)), and the generator and the annihilator, so that the supplier is triggered by the agent Controller to supply goods to the system.

5.3.4.3 Relationships between Agents

Agents communicate with other agents by exchanging flow items, being the net result of job execution (V; G). In this subsection we consider two specializations of the basic type of relationship between agents in somewhat more detail:

- The relationship between an internal agent and his controller
- Relationships between external and internal agents

Control is assumed to be effectuated by the sending of job definitions from a controller, such as a planner, dispatcher, etc., to a subordinate internal agent (VI; H). Each agent refers to exactly one controller from which it receives its job definitions, denoted as F(C). Conversely, a subordinate can send information (F(I|D)) about its status to its controller. Such data act as a request for control; it is one of the jobs of the controller to interpret this type of message. Mechanisms like hierarchical control and coordinated control are embedded in this class relationship. Both mechanisms may be considered as important building blocks in planning and control systems and supply chain coordination. Essential choices by the controller include the timing and contents of decision jobs, i.e., planning and control activities.

For external agents we distinguish between customers and suppliers (I; E, F, I, J). Table 5.2 shows how a customer sends an order (F(I|D)) to an internal agent of the type controller. The controller in its turn specifies a job definition (F(C)) for an internal agent who is responsible for the deliverance of the requested items (F(M)), with M a parameter for setting the subclass of flow items. In the case of a supplier, roles have changed: the controller sends an order to a supplier, who has to take care of delivery of the requested items.

5.3.4.4 Dynamics Structure: Agents Executing Jobs

In line with our job-oriented view we assume the execution of jobs by agents as the driving force of manufacturing dynamics (VII; D). Whereas the initiation of jobs is considered a conditional event, relying on the presence of a job definition and its associated inputs, job completion is bound to a specific moment in time.

5.4 Applying the Modeling Framework: Enhancing Participation

The basic contribution of the domain-specific modeling framework is to help guide the analyst in creating transparent conceptual models, which appeal to stakeholders from a specific field of interest. Typically, such models should foster user participation as an answer to the challenges of conceptual modeling that we stressed in the introduction. In this section, we present a case study of a repair shop to illustrate this point of view. First we consider the use of the framework for conceptual modeling, and illustrate the way it may be used for model coding. Next we reflect on the added value of the manufacturing-specific conceptual modeling framework for model creation and its use.

5.4.1 Case: Repair Shop

5.4.1.1 Introduction

The case study refers to a simulation model for a small fictitious company that repairs engines for pleasure yachts. Facing a competitive market, the company considers opportunities for reducing customer delivery times. Here the prime focus is on reducing manufacturing lead times, which make up the largest part of the delivery times. The example has been used for educational purposes and is partly based on van der Zee (2006a, 2006b). The choice of the case study is motivated by reasons of clarity and simplicity of understanding. More elaborate examples on supply chain design, and planning systems design—related to industrial cases—can be found in van der Zee and Van der Vorst (2005), Van der Vorst et al. (2005), van der Zee et al. (2008), and van der Zee (2009).

5.4.1.2 Objectives of the Study

The objectives of the study are to model the dynamic behavior of the repair shop and to evaluate the consequences of alternative shop configurations for manufacturing lead time performance. Shop configurations are variations on the scheduling system, i.e., the choice of priority rule (First Come First Serve [FCFS] vs. Shortest Processing Time rule [SPT]), scheduling frequency (daily, weekly), and the number of work cells available for repair activities.

5.4.1.3 System Description

The activities of the company are driven by yacht owners that need for preventive or corrective maintenance of engines. This is reflected in an irregular pattern of engine arrivals at the shop. The initial activity at the shop is the inspection of the engine to determine the need for replacement parts. The replacement parts are ordered from the companies' internal warehouse (Figure 5.3). Inspection also serves to make a first estimate of

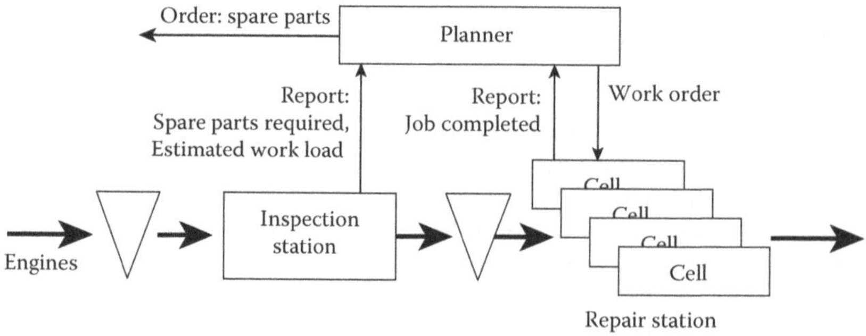

FIGURE 5.3
Repair shop: System description.

the workload associated with an engine, i.e., the time needed to complete repair activities. Estimates on workloads are reported to the shop planner. The repair station consists of a number of identical and autonomous work cells. Each cell is capable of repairing one engine at a time. A planner is made responsible for assigning repair jobs to work cells. He makes a new schedule on a weekly basis. For scheduling he currently applies a FCFS policy. Work orders for new repair jobs are being released by the planner to the shop at the start of each planning period and in response to feed back of the repair shop on jobs being completed. An alternative scheduling policy would be an SPT rule (SPT), whereas planning frequency may be changed from weekly to daily.

5.4.1.4 Conceptual Modeling

As a first step in modeling we define a conceptual model for the repair shop. For building the conceptual model we used the system description as a starting point. In practice, also other sources of information may be relevant, for example, visual observations or drafts of the system under study, and domain knowledge. The conceptual model is set up according to the high-level definitions developed in our manufacturing domain modeling framework. Figure 5.4 displays agent definitions for two key sub systems, the planner, and the repair station. Essentially, the planner is responsible for two types of jobs: (1) the building of a schedule, and (2) the release of job definitions. The first type of job is executed according to a prespecified time interval, and uses messages reporting on arriving engines, the actual schedule—as kept by the planner—and information on shop status as an input. This is reflected in the definition of buffers. The second type of job is triggered by the messages of the repair station reporting jobs' completion. The repair station is responsible for a single type of job—the repair of engines. The repair of an engine is allowed to start in a free work cell, if both a job definition is received from the planner, and the engine identified in the job definition is available.

5.4.1.5 Model Coding

For coding the repair shop a class library is built concerning the class definitions for flow items, agents and jobs (Figure 5.5). These classes are the essential building blocks for the aggregate class RepairShop. In more complex shops it may be worthwhile to introduce more aggregate classes representing hierarchical levels in modeling. Such classes help to improve model overview. All classes are built starting from the basic class library of EM-Plant®* that covers the class definitions contained in the folders MaterialFlow, InformationFlow, UserInterface, and MUs.

* Registered trademark of Siemens Product Lifecycle Management Software II (DE) GmbH.

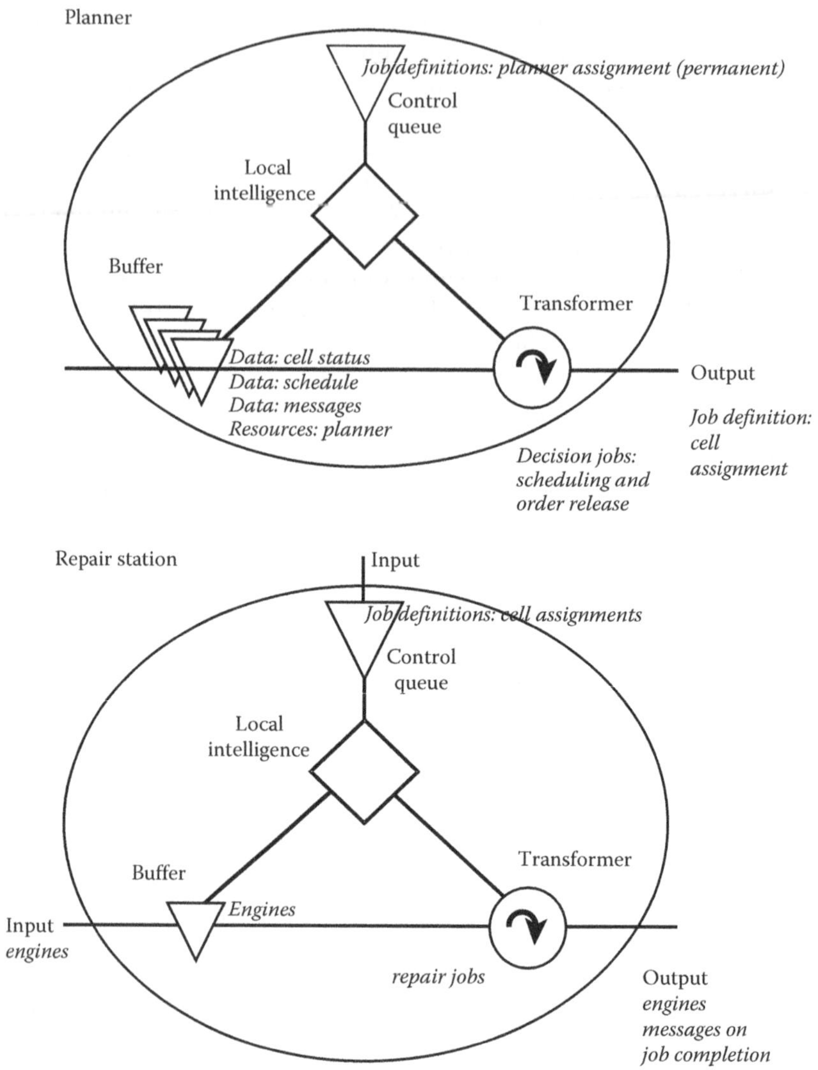

FIGURE 5.4
Conceptual model: Agents Planning and RepairStation.

Let us now discuss the implementation of the three main classes FlowItems, Agents, and Jobs by giving a number of examples. Subsequently, we will relate the classes and model dynamics by considering the internal structure and workings of an agent:

- *Flow items:* Flow items are represented in EM-Plant® by "movable units." We distinguish between four classes of flow items in this model: Engine, StatusUpdate, JobDefinition and Scheduler.

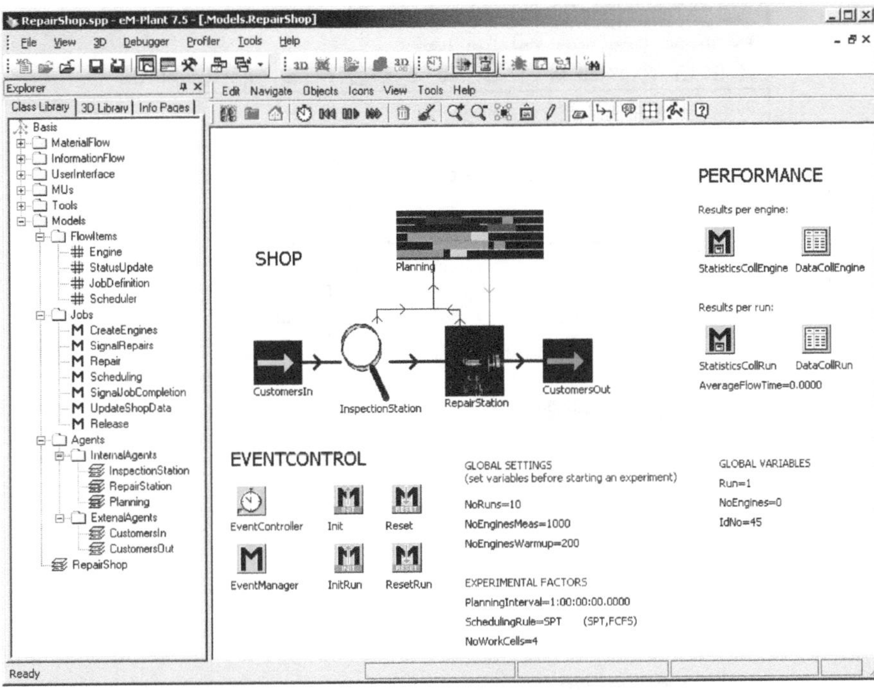

FIGURE 5.5
Class library and class repair shop.

Engines model the physical flows between agents. StatusUpdate and JobDefinition are used to model feedback and control among agents. The Scheduler represents the availability of a person capable of scheduling jobs for the RepairStation. Each flow item has multiple attributes. Next to "header data" needed for identification or routing, they represent the logical or physical contents of the associated object.

- *Agents:* The model distinguishes between *external agents* and *internal agents*. External agents considered are the customers asking for repair of their engines (CustomersIn, CustomersOut). Internal agents model the parties involved in the shop. They are associated with the physical handling of goods (InspectionStation, RepairStation), i.e., engines; data processing (InspectionStation) and control in terms of the scheduling and release of jobs (Planning). Internal structure and workings for an agent are illustrated by Figure 5.6 concerning the agents Planning and RepairStation. Basically, the agents' class definitions cover the functionalities introduced in Table 5.2. For the agent Planning buffers are foreseen for the storage of incoming messages, originating from the InspectionStation estimates of workload) and the RepairStation (job completion).

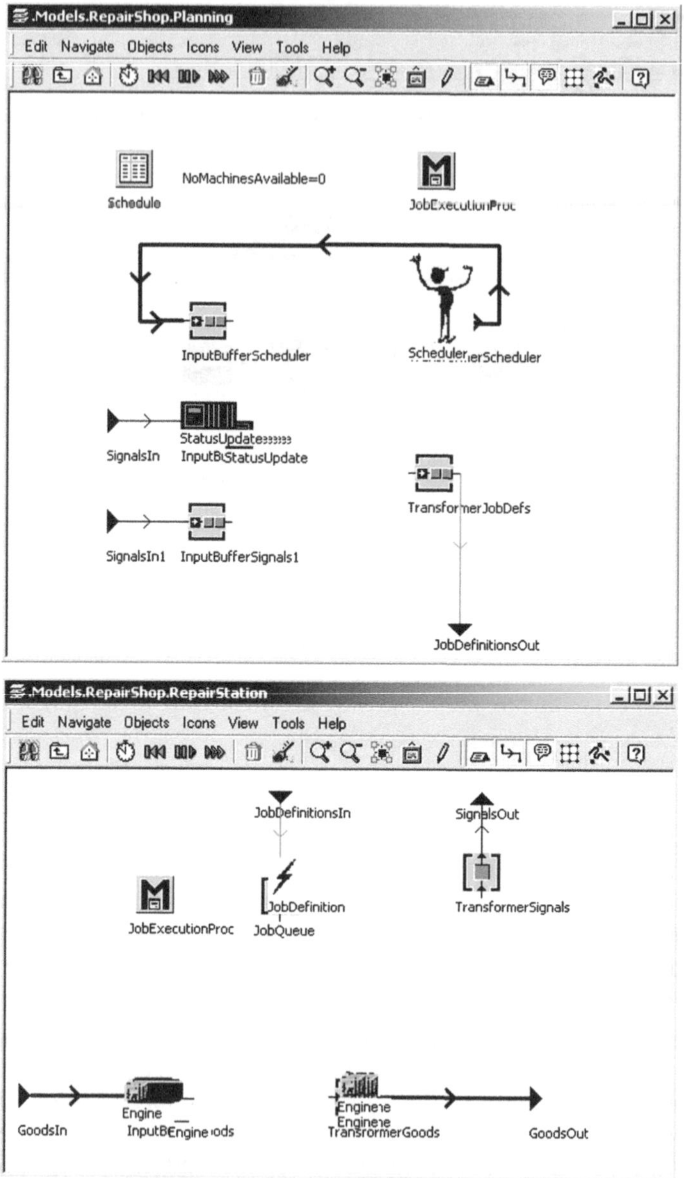

FIGURE 5.6
Coded model: Agents Planning and RepairStation.

Next to these buffers, additional buffers are defined for storing the schedule for the RepairStation (Schedule), and to model availability of the scheduler (InputBufferScheduler). Job execution is taken care of by the local intelligence (JobExecutionProc). This concerns two type of jobs: (1) scheduling for the RepairStation—this is done periodically, and requires availability of the Scheduler—and (2) job release in response to feedback from the shop or the completion of a new schedule. For the RepairStation buffers considered are: InputBufferGoods and JobQueue. Transformers are TransformerGoods and TransformerSignals. Both are linked by local intelligence (JobExecutionProc). The local intelligence takes care of calling on the right jobs for realizing the required transformations. It is activated by the arrival of job definitions and/or goods, i.e., engines.

- *Jobs:* In order to make the shop work jobs have been allocated to agents. To reflect the different nature of jobs we distinguished between several classes of jobs. For example the agent Planning is associated with two classes of jobs: (1) Release, i.e., release repair jobs, and (2) Scheduling, i.e., set up a new schedule of repair jobs on a daily basis. All job classes are implemented in EM-Plant® Methods, i.e., programming code.

5.4.2 Added Value of the Domain-Specific Modeling Framework

5.4.2.1 Guidance in Modeling

We found the modeling framework helpful in model creation for the case study mentioned above, and also for some related cases, see van der Zee et al. (2008) and van der Zee (2009). Helpful, because, instead of having to create an implicit view on a manufacturing system of his own, the analyst could start from a clear point of reference. Advantages of this approach include time and cost savings in model development. Also, it may be easier to express stakeholders' requirements on modeling, as the definition of model elements is closer to their mental models of the system under study. Furthermore, we found that adapting/reusing models to deal with alternative scenarios is relatively easy. For example, consider the following scenarios:

- The number of repair stations: the reduction or increase of the number of stations is realized by adapting attributes for the repair station as well as for the planner.
- Alternative control rules: they can be modeled by adapting the definition of job classes associated with the planning department.
- The choice of another planning period: this is realized by considering the time-related behavior of the planning job.

Model changes like the number of repair stations are facilitated by many simulation languages by means of parameter settings of default building blocks. Implementation of the second and third scenario may imply a somewhat greater appeal to the logic of modeling framework, as they refer to nonstandard language features. Moreover, where the above scenarios do not directly involve the (control) structure for the system, others may do so. This may involve the distribution of job classes over the agents or the number of agents involved. For example, where the default shop model assumes one agent to be responsible for both release and scheduling of the repair station, in an alternative setting there may be two specialized agents each responsible for one task. This separation of tasks resembles different levels of shop control. In a similar way, tasks of, for example, work stations may be redistributed. For a major part model flexibility with respect to the representation of control is the net result of a natural and explicit mapping of concepts—knowing where to look and to make the change.

5.4.2.2 Model Completeness and Transparency

Corner stones for building understanding among stakeholders are model transparency and model completeness. Model transparency should result from the notion of a limited set of elementary manufacturing concepts offered by the domain-specific modeling framework. This set should assist in building model structures that appeal to the imagination of all parties involved in the study. On the other hand, model completeness is related to the explicit notion of relevant manufacturing objects and their workings.

Let us consider the issue of model transparency in somewhat more detail, by comparing modeling as it was done for the case study with ad hoc model development. Typically, ad hoc models may violate elementary decomposition principles, as we found them for manufacturing field. Hence, model transparency may be harmed. Here we concentrate on the somewhat more "advanced" principles IV–VII, see Table 5.1, given our focus on manufacturing systems, including their control. We supply some examples, building on previous modeling experiences in industry, research, and education:

- Intelligent and nonintelligent entities (IV): intelligence, i.e., reasoning mechanisms to be associated with model entities, are often not structured or visualized in a uniform and insightful way. Basic reasons may be found in the free formats offered by, for example, diagramming techniques, and simulation tools' internal languages. They set no a priori restrictions on their use, and the way they should be related to the choice of model components and their workings. Alternatively, in the manufacturing domain modeling framework, we choose to relate "intelligence" to the local intelligence of agents.

- Infrastructure, flows, and jobs (V): in principle, modeling tools for simulation foresee in the possibility for a clear separation of manufacturing infrastructure and flow items. We add the explicit notion of the job, being the common denominator for identifying and describing manufacturing systems' value-adding activities. This is in line with lean manufacturing principles (Womack et al. 1990, Goldman et al. 1995). Typically, simulation models restrict the notion and display of jobs to those activities, involving the processing and/or movement of goods and resources.
- Modality—physical, information, and control elements (VI): starting point for many conceptual models is the representation of the goods flow and its associated resources. In principle, there is nothing wrong with that. However, it should not prevent the analyst from considering information and control elements in sufficient detail. For example, decision logistics, i.e. activities associated with planners and schedulers, and the humans or systems supporting them, are often "hidden" by dispersing them over the model, instead of displaying them in a structured way. Also, modalities may not be identifiable as such, as physical activities, data processing, and/or control activities are integrated in single building blocks and/or programming code. Our modeling framework foresees in a separation of modalities in terms of a class hierarchy of flow items, class definitions for agents, and specialized agents, which fit in the class hierarchy of agents.
- Dynamics—executing jobs (VII): in our view, jobs, i.e., companies' value-adding activities, are key to manufacturing performance. By demanding that jobs are the sole driving mechanism for model dynamics, we strive to appeal to practice, and represent manufacturing dynamics according to simple and understandable rules. Alternatively, ad hoc models may foresee in interventions in event control, which do not follow this basic logic. For example, they may allow for an activity to be created, adapted, or removed without the need for explicit identification as one of the agents' jobs.

In sum, the modeling framework is meant to be of assistance in making models more insightful for users other than the analyst, and even for the analyst, building on its conformation to basic decomposition principles. In turn, model transparency may facilitate model validation, i.e., face validation and creativity in solution finding. It is found that these qualities become even more relevant in case of complex business network configurations, such as supply chains, and health-care chains (van der Zee and Van der Vorst 2005). Obviously, more applications of the framework are required to strengthen these findings.

5.4.2.3 Scope: Field of Application

The modeling framework as described above, and applied to the case study, is designed to deal with modeling dynamic systems from the manufacturing field. However, it does not fully cover the field. Essentially, the framework is "machine-oriented," describing flows as passive objects that are operated upon (Kreutzer 1986). An alternative view, not embedded in the framework, is the "material-oriented" view, which starts from the flow items, which display autonomous behavior in acquiring passive resources. In other words, systems that foresee in intelligent behavior of movable entities may be captured less easily by the modeling framework. Think of, for example, (internal) transportation systems or systems where decision logic and attributes of personnel are of relevance, like team operated manufacturing cells.

5.4.2.4 Choice of Simulation Software

Above we showed how the modeling framework guided model implementation in EM-Plant®. In recent research efforts we also considered the use of the object-oriented simulation language Taylor ED™ (Van der Vorst et al. 2005) and ExSpect™, a tool based on the Petri Nets formalism (van der Zee 1997, 2009). All tools allowed for a straightforward implementation of the elements of the Modeling Framework. In our opinion other choices of a tool would be very well possible. However, the choice for a tool that is not object-oriented may restrict modeling flexibility, as it lacks the availability of concepts such as, for example, inheritance. Furthermore, simulation software that is largely parameter driven and lacks an (internal) language for specifying entity behavior may be unsuited for implementing conceptual models developed using the framework. Typically, this would mean that elementary decomposition principles underlying the modeling framework are violated, also see the examples given above (Model completeness and transparency).

5.5 Conclusions and Directions for Future Research

"The process by which a systems engineer or management scientist derives a model of a system he is studying can best be described as an intuitive art. Any set of rules for developing models has limited usefulness at best and can only serve as a suggested framework or approach" (Shannon 1975). In this chapter we recognize the limitations of guidelines for conceptual modeling—their relevance primarily is in model structuring—less in detailing model elements. In our view, however, it is especially the notion of the model structure and behavior, in terms of its basic elements and their workings, which may make a difference in stakeholders' model understanding and their participation in decision support. This makes the identification and studying of modeling frameworks and the underlying rules for model construction, i.e., decomposition principles, worthwhile.

In this chapter we review a modeling framework for modeling manufacturing systems as we proposed it in our earlier work. So far this framework has been presented and applied without highlighting the way it has been constructed. In this chapter we do so for two reasons:

- To show the underlying *approach toward model engineering*, which is assumed to have a validity that exceeds the manufacturing domain
- To make the proposed modeling framework for manufacturing simulation to an *open architecture* amendable for improvements, extensions, and refinements

Model engineering is related to a two phase approach in defining the modeling framework. The initial phase foresees in a domain analysis for isolating model decomposition principles that apply to a field of application. Next, decomposition principles are applied (framed) in defining the modeling framework, as a comprehensive set of domain related concepts, expressed in terms of object classes, their relationships and their dynamics.

The two phase approach is applied to the manufacturing field. First a set of decomposition principles is defined. Among others, for manufacturing systems it is found important to distinguish between control, information and physical elements, and to isolate jobs. Jobs are found relevant for representing value-adding activities, not only for physical activities, but also in data processing and planning and control. Further, job execution serves as transparent mechanism for model dynamics, which appeals to practice. The application of decomposition principles underpins the set up of the modeling framework in terms of a definition of class hierarchies for agents, flow items, and jobs, their relationships and their dynamics. The notion of the framework's underpinnings makes it amendable for change.

Relevance of a modeling framework should be in model structuring aiming at transparent and complete models, see above. We illustrate this point by a case study. First, we show how the modeling framework is helpful in setting up a conceptual model for the case study, and subsequently, the coded model. Next we consider added value of the framework in somewhat more detail. We find that, next to the initial structuring of the model, the framework is especially helpful in case recoding or restructuring of the model is required. Here one may think of, for example, a redefinition and/or redistribution of shop activities. More in particular we study the advantages of the use of the modeling framework for creating transparent models relative to ad hoc approaches. Basically, ad hoc approaches may violate elementary decomposition principles, like, for example, the separation of the control, information and physical elements, and a well-defined notion of jobs and their dynamic.

Some interesting directions for future research include the detailing and deepening of the engineering approach underlying modeling frameworks and their application. For example, the approach may be related to the concept of reference models (see, for example, Biemans 1990 for the manufacturing field), and principles of system engineering and software engineering. Given the obtained insights in the engineering approach the development of modeling frameworks for domains other than manufacturing may be considered. Another direction may concern the use of the modeling frameworks in facilitating simulation models, which assume higher levels of user participation, such as gaming (van der Zee and Slomp 2005, 2009). Further, the notion of decomposition principles may be helpful in setting up libraries of building blocks, or the validation of existing libraries to support their conceptual renewal and improvement. Last but not least, the modeling framework should be further validated by testing it on real-world models.

Acknowledgment

This chapter is reproduced, with editing, from: van der Zee, D.J. 2007. Developing participative simulation models: Framing decomposition principles for joint understanding. *Journal of Simulation*, 1: 187–202. Reproduced with permission from Palgrave.

Appendix Notation

For defining the modeling framework, we adopt the class diagrams as proposed by Booch (1994). Class diagrams describe the class structure of a system. They consist of two basic elements: object classes (for example, machines and employees) and their mutual associations. Classes are described by their name, attributes, and operations (Figure 5.A1).

Attributes are used to describe the state of the object belonging to a class. A change of an attribute value corresponds to a state change for the object. *Operations* refer to the services provided by a class. Operations may change the state of an object (for example, withdraw one item from

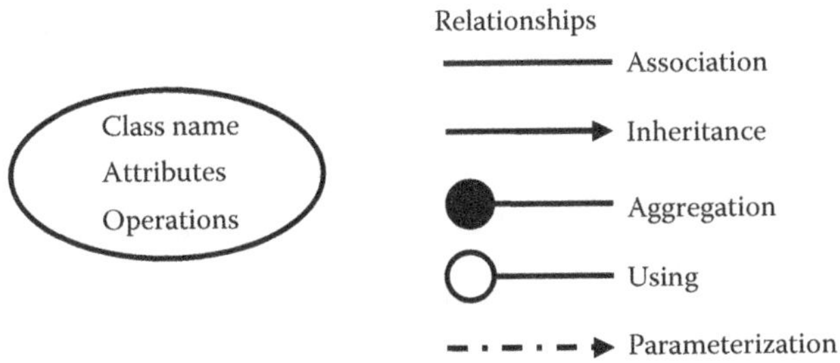

FIGURE 5.A1
Class notation.

a storage) or just access it (for example, determine the serial number of a machine).

The basic type of relationship between two classes is the association, i.e., all other types of relationships are considered as refinements of this relationship. Typically, associations are labelled by noun phrases, which describe the nature of the relationship. Let us now consider the refinements of the association:

Inheritance denotes a relationship among classes, where a subclass shares a part of the structure or behavior defined in one or more superclasses.

Whole/part relationships *(aggregation)* relate classes to an aggregate class.

Using refers to a client/supplier relationship. Whereas an association represents a bidirectional semantic connection, a using relationship makes a clear distinction between the client and the provider of certain services.

Parameterization is represented by a dashed line that connects the parameterized class and its concrete class. Parameterization supposes a using relationship with the parameter class.

References

Balci, O. 1986. Credibility assessment of simulation results. In *Proceedings of the 1986 Winter Simulation Conference,* 209–222. Piscataway, NJ: IEEE.

Bell, P.C., and R.M. O'Keefe. 1987. Visual Interactive Simulation: History, recent developments, and major issues. *Simulation* 49(3): 109–116.

Bell, P.C., C.K. Anderson, D.S. Staples, and M. Elder. 1999. Decision-makers' perceptions of the value and impact of visual interactive modeling. *Omega: The International Journal of Management Science* 27: 155–165.

Biemans, F.P.M. 1990. Manufacturing planning and control: A reference model. *Manufacturing Research and Technology* 10. Amsterdam: Elsevier.

Bodner, D.A., and L.F. McGinnis. 2002. A structured approach to Simulation modelling of manufacturing systems. In *Proceedings of the 2002 Industrial Engineering Research Conference*. Orlando, FL: IIE.

Booch, G. 1994. *Object-Oriented Analysis and Design with Applications*. Redwood City, CA: Benjamin Cummings.

Dahl, O., and K. Nygaard. 1966. SIMULA: An Algol-based simulation language *Communications of the ACM* 9(9): 671–678.

Galland, S., Grimaud, F., Beaune, P., and J.P. Campagne. 2003. M(A)MA-L: An introduction to a methodological approach for the simulation of distributed industrial systems. *International journal of production economics* 85(1): 11–31.

Glassey, C.R., and S. Adiga. 1990. Berkeley Library of Objects for Control and Simulation of Manufacturing (BLOCS/M). In *Applications of Object-Oriented Programming*, ed. L.J. Pinson and R.S. Wiener, 1–27. Reading, MA: Addison-Wesley.

Goldman, S., Nagel, R., and K. Preiss. 1995. *Agile Competitors and Virtual Organizations: Strategies for Enriching the Customer*. New York: Van Nostrand Reinhold.

Guru, A., and P. Savory. 2004. A template-based conceptual modeling infrastructure for simulation of physical security systems. In *Proceedings of the 2004 Winter Simulation Conference*, ed. R.G. Ingalls, M.D. Rossetti, J.S. Smith, and B.A. Peters, 866–873. Piscataway, NJ: IEEE.

Hurrion, R.D. 1976. The design, use and required facilities of an interactive computer simulation language to explore production planning problems, PhD thesis, University of London, UK.

Hurrion, R.D. 1989. Graphics and interaction. In *Computer Modelling for Discrete Simulation*, ed. M. Pidd. Chichester, UK: Wiley.

Kamper, S. 1991. On the appropriateness of Petri Nets in model building and simulation. *Systems analysis modelling simulation* 8(9): 689–714.

Kreutzer, W. 1993. The role of complexity reduction in the development of simulation programming tools: An advanced tutorial. In *Proceedings of European Simulation Conference*. Delft: Society for Computer Simulation.

Kreutzer, W. 1986. *System Simulation: Programming Styles and Languages*. Sydney: Addison-Wesley.

Law, A.M., and W.D. Kelton. 2000. *Simulation Modeling and Analysis*. Singapore: McGraw-Hill.

Lefrancois, P., and B. Montreuil. 1994. An object-oriented knowledge representation for intelligent control of manufacturing workstations. *IIE Transactions* 26(1): 11–26.

Lefrancois, P., Harvey, S., Montreuil, B., and B. Moussa. 1996. Modelling and simulation of fabrication and assembly plants: An object-driven approach. *Journal of intelligent manufacturing* 7: 467–478.

Mize, J.H., Bhuskute, H.C., Pratt, D.B. and M. Kamath. 1992. Modelling of integrated manufacturing systems using an object-oriented approach. *IIE transactions* 24(3): 14–26.

Murata, T. 1989. Petri Nets Properties, analysis and applications. In *Proceedings of the IEEE* 77(4): 541–580.

Nance, R.E. 1994. The conical methodology and the evolution of simulation model development. *Annals of operations research* 53: 1–45.

Ören, T.I., and B.P. Zeigler. 1979. Concepts for advanced simulation methodologies. *Simulation* 32(3): 69–82.

Pegden, C.S., Shannon, R.E., and R.P. Sadowski. 1990. *Introduction to Simulation Using SIMAN*. New York: McGraw-Hill.

Pratt, D.B., Farrington, P.A., Basnet, C.B., Bhuskute, H.C., Kanath, M., and J.H. Mize. 1994. The separation of physical, information, and control elements for facilitating reusability in simulation modelling. *International journal of computer simulation* 4(3): 327–342.

Pidd, M. 1998. *Computer Simulation in Management Science*. Chichester: Wiley.

Pidd, M. 1999. *Tools for Thinking: Modelling in Management Science,*. 2nd edition. Chichester: Wiley.

Pooley, R.J. 1991. Towards a standard for hierarchical process oriented discrete event diagrams. *Transactions of the society for computer simulation* 8(1): 1–41.

Roberts, C.A., and Y.M. Dessouky. 1998. An overview of object-oriented simulation. *Simulation* 70(6): 359–368.

Robinson, S. 2002. General concepts of quality for discrete-event simulation. *European journal of operational research* 138: 103–117.

Robinson, S. 2006. Issues in conceptual modelling for simulation: Setting a research agenda. In *Proceedings of the Operational Research Society Simulation Workshop (SW06)*, ed. J. Garnett, S. Brailsford, S. Robinson, and S. Taylor, 165–174. Birmingham: Operational Research Society.

Robinson, S. 2008. Conceptual modelling for simulation Part I: Definition and requirements. *Journal of the operational research society* 59: 278–290.

Schruben, L.W. 1983. Simulation modeling with event graphs. *Communications of the ACM* 26(11): 957–963.

Shannon, R.E. 1975. *Systems Simulation: The Art and Science*. Englewood Cliffs, NJ: Prentice Hall.

Taylor, S.J.E., and S. Robinson. 2006. So where to next? A survey of the future for discrete-event simulation. *Journal of simulation* 1(1): 1–6.

Valentin, E.C., and A. Verbraeck. 2005. Requirements for domain specific discrete event simulation environments. In *Proceedings of the 2005 Winter Simulation Conference*, ed. M.E. Kuhl, N.M. Steiger, F.B. Armstrong, and J.A. Joines, 654–663. Piscataway, NJ: IEEE.

Van der Vorst, J.G.A.J., Tromp, S., and D.J. van der Zee. 2005. A simulation environment for the redesign of food supply chain networks: Integrating quality and logistics modeling. In *Proceedings of the 2005 Winter Simulation Conference*, ed. M.E. Kuhl, N.M. Steiger, F.B. Armstrong, and J.A. Joines, 1658–1667. Piscataway, NJ: IEEE.

van der Zee, D.J. 1997. Simulation as a tool for logistics management, PhD thesis, University of Twente, The Netherlands.

van der Zee, D.J. 2006a. Modeling decision making and control in manufacturing simulation. *International journal of production economics* 100(1): 155–167.

van der Zee, D.J. 2006b. Building communicative models: A job oriented approach to manufacturing simulation. In *Proceedings of the Operational Research Society Simulation Workshop (SW06)*, ed. J. Garnett, S. Brailsford, S. Robinson, and S. Taylor, 185–194. Birmingham: Operational Research Society.

van der Zee, D.J. 2009. Building insightful simulation models using formal approaches: A case study on Petri Nets. In *Proceedings of the 2009 Winter Simulation Conference,* ed. M.D. Rossetti, R.R. Hill, B. Johansson, A. Dunkin, and R.G. Ingalls, 886–898. Piscataway, NJ: IEEE.

van der Zee, D.J., and J. Slomp. 2005. Simulation and gaming as a support tool for lean manufacturing systems: A case example from industry. In *Proceedings of the 2005 Winter Simulation Conference,* ed. M.E. Kuhl, N.M. Steiger, F.B. Armstrong, and J.A. Joines, 2304–2313. Piscataway, NJ: IEEE.

van der Zee, D.J., and J. Slomp. 2009. Simulation as a tool for gaming and training in operations management: A case study. *Journal of simulation* 3(1): 17–28.

van der Zee, D.J., Pool, A., and J. Wijngaard. 2008. Lean engineering for planning systems redesign: Staff participation by simulation. In *Proceedings of the 2008 Winter Simulation Conference,* ed. S.J. Mason, R.R. Hill, L. Moench, and O. Rose, 722–730. Piscataway, NJ: IEEE.

van der Zee, D.J., and J.G.A.J. van der Vorst. 2005. A modeling framework for supply chain simulation: Opportunities for improved decision-making. *Decision sciences* 36(1): 65–95.

Womack, K., Jones, D., and D. Roos. 1990. *The Machine that Changed the World.* Oxford: Maxwell Macmillan International.

Zeigler, B.P. 1976. *Theory of Modelling and Simulation.* New York: Wiley.

Zeigler, B.P. 1990. *Object-Oriented Simulation with Hierarchical, Modular Models, Intelligent Agents and Endomorphic Systems.* London: Academic Press.

6

The ABCmod Conceptual Modeling Framework

Gilbert Arbez and Louis G. Birta

CONTENTS

6.1 Introduction

The development of a meaningful conceptual model is an essential phase for the successful completion of any modeling and simulation project. Such a model serves as a crucial bridge between the generalities of the project description and the precision required for the development of the simulation program that ultimately generates the data that is required for resolving the project goals. A conceptual model is a careful blending of abstraction and pertinent detail.

In the realm of continuous time dynamic systems, conceptual model development typically relies on the language of differential equations, which is usually colored by the terminology that is specific to the domain in which the underlying dynamic system is embedded (e.g., engineering, thermodynamics, aerodynamics, etc.). However when the system under investigation (SUI) falls in the realm of discrete-event dynamic systems (DEDS) there is, regrettably, no equivalent language that can adequately characterize behavior because of the diversity and complexity that pervades this domain. The most straightforward means for conceptual modeling is therefore absent. The typical consequence, regrettably, is a leap directly into the intricacies of some computer programming environment with the unfortunate result that the program displaces the model as the object of discourse. Essential features of the model quickly become obscured by the intricacies of the programming environment. Furthermore, the resulting artefact (i.e., the simulation program) has minimal value if a change in the programming environment becomes necessary.

In this chapter we outline the ABCmod conceptual modeling framework (Activity-Based Conceptual modeling), which is an environment for developing conceptual models for modeling and simulation projects in the DEDS domain. Its model building artefacts fall into two categories; namely, entity structures and behavior constructs. These relate, respectively, to the structural and the behavioral facets of the SUI. Care has been taken to ensure that all aspects of the modeling requirements are included in a consistent and transparent manner. In addition to structure and behavior, the framework includes a consistent means for characterizing the inputs and the outputs of the SUI that have relevance to the project goals. The conceptual modeling process within this framework is guided by an underlying discipline but the overall thrust is one of informality and intuitive appeal. The constituents of the framework can be easily extended on an ad hoc basis when specialized needs arise.

The underlying concepts of ABCmod framework have been continuously evolving and earlier versions of this environment have been presented in the literature (Arbez and Birta 2007, Birta and Arbez 2007). The presentation in this chapter incorporates several important refinements. Included here is a clearer and more coherent separation between structural and behavioral aspects of the model, as well as an approach for presenting the model at both a high level of abstraction in addition to a detailed level. The latter specification can be easily translated into simulation programs based on either the event scheduling or process-oriented world views.

Characterization of the SUI's structure and behavior without concerns about programming issues and details is the fundamental intent. We note however, that this intent is more restrictive than the perspective adopted by some authors. For example, Robinson (2004) includes project objective, assumptions, and simplifications as part of the conceptual model.

6.2 Overview and Related Work

There is a variety of ways of packaging the discrete events that are the essence of a discrete-event dynamic system. These alternatives lead to distinct modeling approaches (usually called "world views" [Shannon 1975, Overstreet and Nance 2004, Banks et al. 2005, Birta and Arbez 2007]). Furthermore, these approaches can be conditioned to some degree by the intent of the modeling process. Two particular options can be identified in this regard. In one, there is a significant alignment with the "simulation engine" (Pidd 2004a) that will carry out the execution of the simulation program while in the other the predominant concern is with clarity of communication among the stakeholders in the modeling and simulation project. Conceptual modeling falls squarely within the latter option.

The notion of an activity is fundamental to the ABCmod framework. The word *activity* appears in the modeling and simulation literature with a variety of informal meanings (Kreutzer 1986, Pritsker 1986, Pidd 2004a) Within the ABCmod context, however, its meaning is very specific; namely, an *activity* is the following:

- It is an indivisible unit that characterizes an interaction among entities
- It is associated with some purposeful task within the SUI
- It evolves over a nonzero (but normally finite) interval of time.

Furthermore, we regard an activity as having four components:

a. a starting condition expressed in terms of the state of the model that must be satisfied before the activity can start
b. a list of state changes that take place at the instant that the activity starts
c. a duration that indicates how long the activity will take to complete
d. a list of state changes that take place when the activity terminates

We refer to this view of an activity as the inclusive view. The activity construct in the ABCmod framework encapsulates the inclusive view. This perspective of an activity is by no means a standard within the modeling and simulation literature but it has previously appeared (e.g., Hills 1973).

While an (inclusive) activity-oriented modeling approach may have its limitations within the context of simulation program implementation, it needs to be stressed that these difficulties do not extend into a conceptual modeling context. On the contrary, as will be shown in the discussion of the ABCmod framework that follows, this view is very well suited to the conceptual modeling task. It provides a meaningful and intuitively appealing approach for packaging the events that constitute behavior in the DEDS realm. An appreciation for these important properties by Pidd (2004b) can be reasonably inferred from his enthusiasm expressed for the closely related three phase approach that is aligned with simulation program design (see below).

We summarize below some of the features of this framework that will be explored further in the following sections of this chapter.

i. The ABCmod framework provides a comprehensive environment that can accommodate arbitrary complexity within the SUI. There are, for example, integrated mechanisms to handle input, output, and the preemption or interruption of activities. Its basic model building constructs can be easily extended when specialized needs arise.
ii. A clear distinction is maintained between structural and behavioral aspects of the model. Both graphical and textural formats are used to present each of these principal facets of model construction.
iii. Behavior is formulated both at a high level of abstraction (in a graphical format) and at a detailed level (in a structured and intuitively straightforward text-based format).
iv. While the notion of an (inclusive) activity is central to the ABCmod framework, it is acknowledged that there are important aspects of behavior that fall outside the scope of this notion. An associated notion of an "action" extends the ABCmod behavioral modeling landscape.
v. The notion of instances of both entity structures and activity constructs is a prevailing perspective in ABCmod conceptual model development.
vi. Time management is outside the scope of the ABCmod conceptual modeling process.
vii. The notion of "executing" an ABCmod conceptual model is not meaningful (see in particular the previous point), thus there can be no generation of output. It is nevertheless essential for any conceptual model to include a comprehensive specification of the output that is required for purposes of achieving the goals of the project. Mechanisms for dealing with this critical facet of a conceptual model are integrated into the ABCmod environment.

The inclusive activity notion has not been especially useful from the point of view of simulation engine design where the focus is on logic flow, specifically the management of lists of events that must take place in a correct temporal sequence. Strategies have, nevertheless, evolved from this underlying notion. The two most common are the activity-scanning approach (often called the two-phase approach [Buxton and Laski 1962]) and an extension called the three-phase approach (Tocher 1963). In both these approaches the four facets of the inclusive view of an activity are implicitly recognized but are separated and reconstructed into alternate constructs that are more useful from a software perspective. Nevertheless the word "activity" is usually retained but its meaning is often unclear and/or curious (e.g., it is not uncommon to read that "an activity is an event that ______").

We note finally that a correspondence might be assumed between our behavior diagram (see (iii) above) and the Activity Cycle Diagram (ACD) that can be found in the modeling and simulation literature (e.g., Kreutzer 1986, Pidd 2004a). Both of these diagrams are formed from a collection of life-cycle diagrams that are specific to an entity class. A simple example of an ABCmod life-cycle diagram is given in Figure 6.1, which is intended to show that an entity associated with this life-cycle diagram could flow either to activity Act2 or activity Act3 upon completion of its engagement in activity Act1.

The rectangles in our life-cycle diagram represent activities (more correctly activity instances) as per the inclusive view outlined earlier. In the ACD context, rectangles are often called "active states" (Kreutzer 1986, Pidd, 2004a) and at best (depending on the author), they encompass only parts (b) and (c) of the inclusive activity's four constituents. The circle in Figure 6.1 is intended simply to represent a delay (of uncertain length) encountered by an entity instance that arises because the conditions for initialization of a subsequent activity instance in which it will be engaged may not yet be present. In the ACD context, the circle (usually called a "dead state" [Pidd 2004a]) often corresponds to a queue. Furthermore. there is frequently an implicit suggestion that the ACD reflects structural properties of the SUI. There is no structural implication associated with the ABCmod behavior diagram.

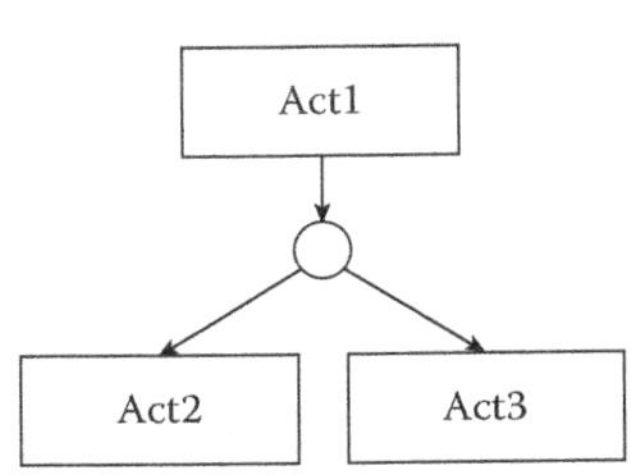

FIGURE 6.1
ABCmod life-cycle diagram.

6.3 Constituents of the ABCmod Framework

6.3.1 Overview

A representation (i.e., abstraction) of the SUI's behavior is the underlying goal of conceptual modeling. This task can be regarded as the task of characterizing the interactions overtime that take place among the collection of elements that populate the space of the SUI. The important implication here is that the characterization of behavior cannot be undertaken until appropriate surrogates for these elements have been identified. In other words there are two main collections of modeling artefacts that are required to carry out the conceptual modeling process. The first deals with the abstraction of the elements that are interacting within the SUI and the second focuses on the nature of these interactions. These two requirements correspond, respectively, to the structural and the behavioral aspects of the model.

Within an ABCmod conceptual model, the surrogates for the elements of interest within the SUI are *entities* that are derived from a collection of *entity structures* that are formulated to accommodate the specific nature of the SUI and the goals of the project. The behavior of the SUI, on the other hand, is formulated in terms of a collection of *behavior constructs*. These fall into two categories called *activity constructs* and *action constructs*. An activity construct in an ABCmod conceptual model represents a specific unit of behavior that is judged to have relevance to the model building task. Such a construct can be viewed as an encapsulation of some relevant dynamic relationships among the entities. These relationships typically take the form of interactions among the entities as they react to, and give rise to, the occurrence of events. The identification of these units of behavior is the task of the conceptual model builder and is largely driven by the project goals.

6.3.2 Exploring Structural and Behavioral Requirements

By way of setting the stage for the presentation of the ABCmod framework, we outline here a facet of a particular discrete-event dynamic system. With some elaboration the outline could evolve into a modeling and simulation project but that is not of concern in this discussion. The intent is simply to illustrate a variety of features that are typical of the DEDS domain with a view toward identifying some possible contents of a toolbox for conceptual model construction.

We consider the operation of a department store. Customers arrive and generally intend to make one or more purchases at various merchandise areas (departments) of the store. At each such area a customer browses/shops, possibly makes one or more selections and if so then pays for them at the service desk located within the area before moving on to the next merchandize area. Upon completion of the shopping task the customer leaves the store.

In this fragment of a DEDS description, each customer corresponds to an entity that we might regard as a type of consumer entity (note that such a generalization is, in fact, an abstraction step). In a similar way we could regard each of the service desks within the various merchandize areas as a resource entity (another abstraction) inasmuch as the consumer entities (the customers) need to access these resource entities in order to complete their purchase transactions (these transactions are, after all, the purpose of their visit to the department store). Because the service function at the resource entity (i.e., service desk) has a finite duration there is a possibility that some consumer entities may not receive immediate attention upon arrival at the resource because it is busy serving other customers. Hence it is reasonable to associate a queue entity with each resource entity where consumer entities can wait for their turn to access the resource.

The merchandize areas where customers evaluate merchandize are distinctive. On one hand, each can be regarded simply as a "holding" area for a collection of consumer entities (i.e., the customers). With this perspective, a merchandize area can be represented as a particular type of aggregate that we call a *group* (an unordered collection of entities). On the other hand, a merchandize area is a prerequisite for an activity called shopping. Hence, it has features of a resource. In effect, then, the merchandize areas within the department store have a dual role. This notion of duality is explored further in the discussions that follow.

As suggested above, the ABCmod framework recognizes two types of aggregate; namely queues and groups. Entities within a group are not organized in a disciplined way as in the case of a queue but rather simply form an identifiable collection. Note furthermore that the discipline that is inherent in a queue introduces two important features. Both of these arise from the fact that there is a natural exit mechanism for the entities in a queue; namely, availability of access to the resource that is associated with the queue. As a consequence the time spent by an entity's membership in a queue is implicitly established and the destination of an entity that departs from a queue is likewise implicitly established. In contrast, neither the time spent by an entity within a group nor its subsequent destination is implicit in the membership property.

In the discussion above we have transformed various elements of the SUI (e.g., customers, merchandise areas, service desks, and customer lines waiting for service) into generic elements that have broader, more general, applicability (consumer entities, group entities, resource entities, queue entities). This is an important abstraction step and lies at the core of the conceptual modeling process. These various generic entities that we have introduced are among the model building artefacts that are explored in greater detail in the discussion that follows.

It needs to be appreciated however that the mapping process from elements in the SUI to generic elements is not always as straightforward as the preceding discussion might suggest. Consider, for example, a set of machines within a manufacturing plant that are subject to failure. A team of maintenance personnel is available to carry out repairs. While the machines are operating, they can certainly be viewed as resource entities in the manufacturing operation but when they fail they become consumer entities because they need the service function of the maintenance team. In other words, the machines can shift from one role to another. Such circumstances are not uncommon and some appropriate generic artefact in our modeling toolbox is essential.

It is usually possible to formulate a graphical representation of the important structural components of the SUI for which a model is being constructed; i.e. a schematic diagram. Some aspects of behavior may also be incorporated. The result can provide a useful integrated view of the various components that need to be incorporated in the model and some insights into how they interact. Figure 6.2 shows such a representation for the department store as outlined above.

The arrows indicate movement of the customers. The dark arrows indicate departure from the department store. The partially shaded arrows indicate movement from one merchandize area to another and the nonshaded arrows show movement within a particular department.

The discussion above has illustrated some structural elements that provide a foundation for the modeling process. We explore this example further, but now from the perspective of behavior. A useful way to begin is to examine how various shoppers in the department store might interact. Figure 6.3 shows a possible interaction scenario for three shoppers called A, B, and C. They arrive at times A_0, B_0, and C_0, respectively, and leave the store at times A_5, B_7, and C_3, respectively.

There are a number of important observations that can be made about Figure 6.3. Notice, in particular, that some type of transition occurs at each of the time points A_0 through A_5, B_0 through B_7 and C_0 through C_3. These transitions, in fact, represent changes that must be captured in the model building process. Notice also that some of these time points are coincident; for example $A_2 = B_2$, $A_3 = C_2$ and $A_5 = B_4$, suggesting that several different changes can occur at the same moment in time. It is also clear from Figure 6.3 that there are intervals of time during which at least some of these three shoppers are engaged in the same activity; for example between B_0 and B_1 all three customers are browsing in Area 1 and between C_1 and A_2 customers A and C are waiting in Queue 1.

We have previously noted that each of the service desks can be regarded as a resource and shoppers need to acquire ("seize") this resource in order to pay for items being purchased before moving on to another merchandise area. The payment activity at a service desk has several noteworthy features:

i. There is a precondition that must be TRUE before this service activity can begin (the server must be available and there must be a shopper seeking to carry out a payment transaction).

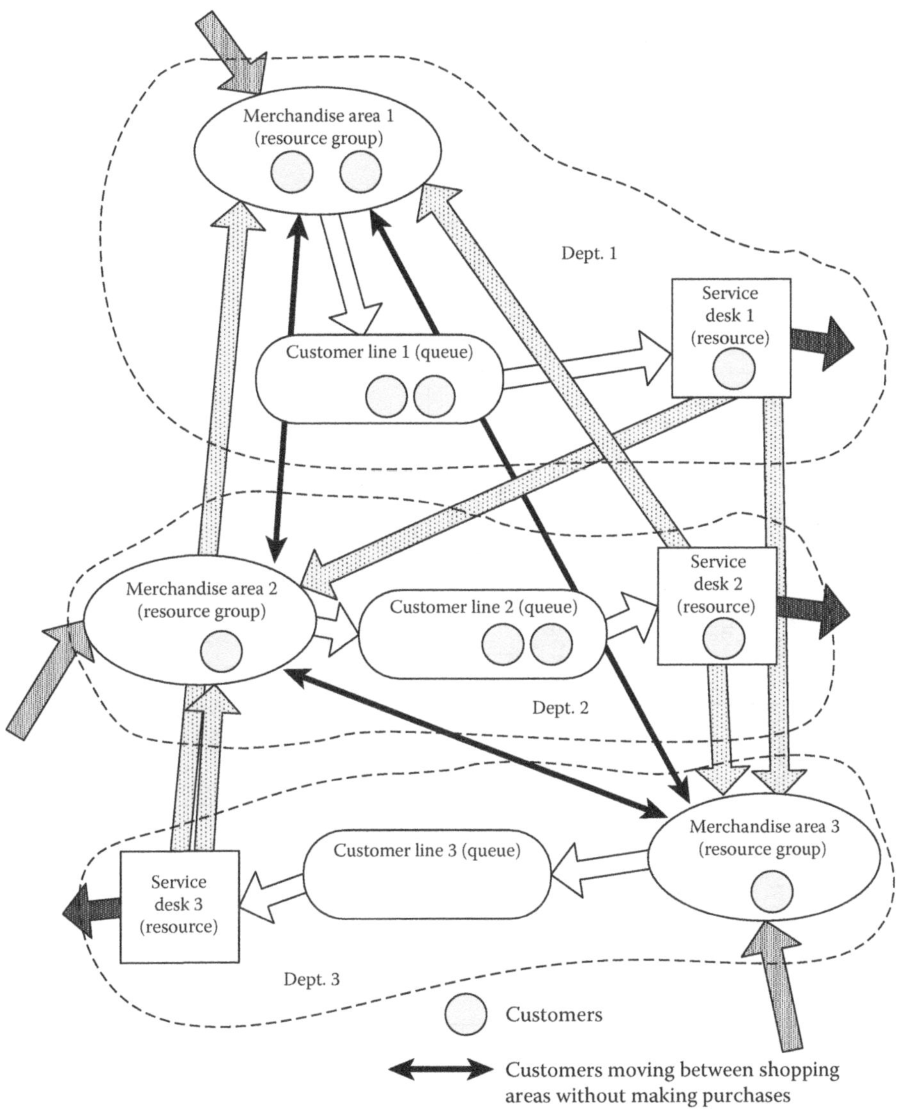

FIGURE 6.2
A schematic view of the conceptual model for department store shoppers. (Based on Birta, L.G. and Arbez, G., *Modeling and Simulation: Exploring Dynamic System Behavior,* Springer, London, Fig. 4.1, p. 99, 2007. With kind permission of Springer Science and Business Media.)

ii. The service activity carries out a purposeful task and has duration; i.e., it extends over an interval of time.

iii. Changes take place when the service function is completed (e.g., at time $A_3 = C_2$ the number of shoppers in merchandize Area 3

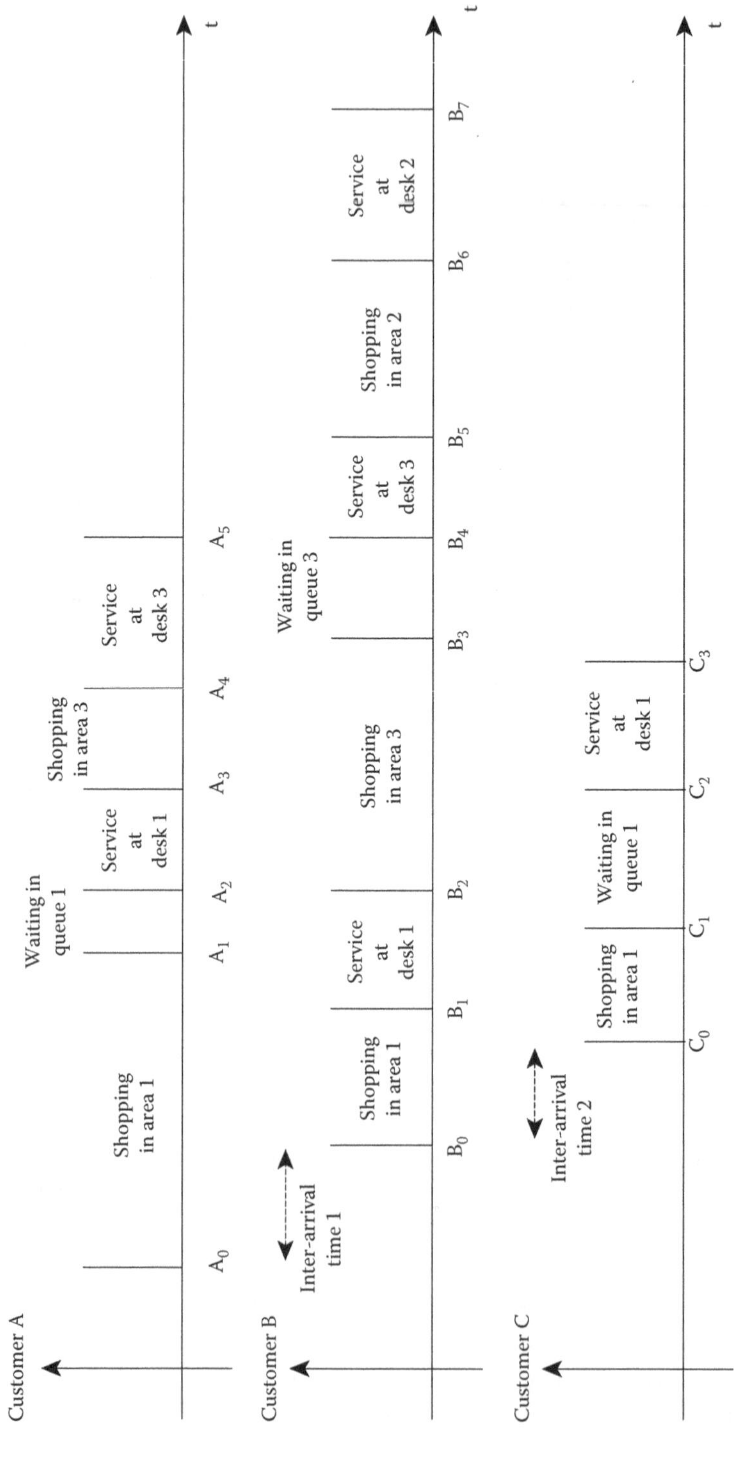

FIGURE 6.3
Behavior of three department store shoppers. (Based on Birta, L.G., and Arbez, G., *Modeling and Simulation: Exploring Dynamic System Behavior*, Springer, London, Fig. 4.2, p. 107, 2007. With kind permission of Springer Science and Business Media.)

increases by one and the number in the queue in front of service desk 1 decreases by one).

These features are directly reflected in one of the main ABCmod constructs used to characterize behavior. This will become apparent in the discussion of Section 6.3.4 below.

From the perspective of the requirements of a modeling and simulation project, the description given above for the department store shoppers is incomplete in several respects. Many details need to be provided; for example, how is the set of merchandise areas that a particular customer visits selected? What is the order of the visitations? And how many servers are assigned to the service desks? Can a particular customer balk; i.e., not make any purchase at one or more of the assigned merchandise areas and if so, then under what circumstances? The information for dealing with these questions is not provided in the descriptive fragment that is given but would most certainly be necessary before a meaningful conceptual model could be formulated. Indeed one of the important functions of the conceptual modeling process is to reveal the absence of such essential details.

Likewise several data models need to be determined. Included here would be the characterization of customer arrival rates and service times at the service desks, allocation of the shopping areas to be visited by the arriving customers and the characterization of the duration of the browsing phase at each merchandise area, etc. It is especially important to observe that these various data models will provide the basis for generating events that give rise to change. For example, the event associated with the end of a particular customer's browsing phase will generally (but not necessarily) result in that customer's relocation into the queue associated with the service desk of that service area.

The intent of the discussion in this section has been to explore some of the important facets of the modeling process within the DEDS domain, at least insofar as they are reflected in the particular problem context that was considered. This overview will serve as a foundation for the discussion that follows.

6.3.3 Model Structure

6.3.3.1 Entity Structures and Entities

As indicated earlier, a necessary constituent of any conceptual modeling process is a collection of entities that serve as surrogates for those elements in the SUI that have relevance to model development. These entities are manipulated in some appropriate manner over the course of the observation interval. While this is most certainly true there remain fundamental questions such as the properties of these entities and the manner in which they acquire their existence. Within the ABCmod framework the notion of an entity has its origins in the more fundamental notion of an entity structure. In this section we outline the relationship between these two notions and explore their important features. A particularly significant outcome is insight that emerges about the state variables of an ABCmod conceptual model.

Each of the entity structures within an ABCmod conceptual model serves as a specification for one or more entities. Such a specification is an *m*-tuple of attribute names together with a description of each attribute. An entity is a named m-tuple of values where the name is derived from the underlying entity structure and the values are assignments to the attributes of that entity structure. Such an entity is said to be derived from the entity structure.

It follows then that one of the important initial steps in the development of a conceptual model for any particular modeling and simulation project is the identification of an appropriate collection of such entity structures; i.e., one that accommodates the modeling requirements of the project. This collection, in effect, defines the structure of the conceptual model being constructed. As will become apparent in Section 6.3.4, the entities that are derived from these entity structures are fundamental in behavior specification.

Each entity structure has two properties, which are called *role* and *scope.* The notion of *role* is intended simply to provide a suggestive (i.e., intuitive) link between the features of the SUI and the conceptual model building environment provided by the ABCmod framework. The value assigned to *role* reflects the model builder's view of the entity structure in question, or more correctly, the entity (or entities) that are derived from that entity structure. There are four basic alternatives (i.e., values for *role*) that align with a wide variety of circumstances; namely the following:

- *Resource*: when a derived entity provides a service.
- *Consumer*: when a derived entity seeks one or more services.
- *Queue*: when a derived entity serves as the means for maintaining an ordered collection of other entities. (The number of such entities that are accommodated at any point in time normally varies and often there is a maximum capacity; these values are typically maintained in attributes associated with the underlying Queue entity structure.)
- *Group*: when a derived entity serves as the means for maintaining an unordered collection of other entities. (The number of such entities that are accommodated at any point in time normally varies and often there is a maximum capacity; these values are typically maintained in attributes associated with the underlying Group entity structure.)

There is no reason to believe, however, that these four alternatives will necessarily encompass all possible circumstances. Note, for example, that it is often the case that an entity structure's *role* may exhibit duality. Consider, for example, an entity intended to represent a machine that periodically breaks down and requires repair. While it is operating, the machine provides a service and hence can be viewed as a resource, but when it breaks down it requires the service of maintenance personnel and hence becomes a consumer. Such situations where the value of *role* can vary over the course of the observation interval are not uncommon and we view the *role* in such a case as having sequential duality. The value assigned to *role* in this example would be Resource Consumer.

There is likewise the possibility of *role* having simultaneous duality. This would occur, for example, in the case of a service counter at fast food outlet. The counter can be regarded as a Group because it maintains a collection (unordered) of customers being served. But, at the same time, it can be regarded as a Resource because customers must acquire a position at the counter (i.e., within the Group) as a prerequisite for the activity of getting served. In this situation the dual value of Resource Group would be assigned to *role*.

In the discussions that follow we shall frequently use phrases such as: "the resource entity structure called *X*" to mean: "the *role* of the entity structure called *X* is Resource." In a similar way, the phrase "consumer entity" is a reference to an entity derived from an entity structure whose *role* is Consumer.

The *scope* of an entity structure reflects upon the number and permanence of the entities that are derived from it. In the case where exactly one entity is derived from an entity structure, that entity structure is said to have *scope* = Unary. When an entity structure yields a finite number, N > 1, of derived entities then its *scope* is Set[N]; i.e., *scope* = Set[N]. If the number of derived entities of an entity structure is indeterminate, then *scope* = Class.

In the case where the scope of an entity structure is either Unary or Set[N], the derived entities remain within the realm of the model over the entire observation interval. In the case where *scope* = Class, the derived entities typically have a transient existence (i.e., they can not only be created but they can likewise be eliminated). These entities have no explicit identifier but can, nevertheless, be referenced (as discussed in the following section). They are usually called instances of the underlying entity structure.

Each entity structure in an ABCmod conceptual model has an identifier that reflects aspects of its two properties (namely, *role* and *scope*). Likewise each entity can be accessed by either an identifier (in the case where *scope* = Unary or Set[N]) or by reference (in the case where *scope* = Class). These identifiers/references have a format that is derived from the identifier of the underlying entity structure. These naming conventions are outlined in the following section.

6.3.3.2 Identifiers for Entity Structures and Entities

A particular format that incorporates pertinent information has been adopted for creating identifiers for entity structures. The intent here is to facilitate interpretation when references to the entity structure occur. The identifier appends type information to a name for the entity structure that is meaningful to the model builder. The general format for this identifier is:

Type: *Name*

where: **Type** = {*role*} {*scope*} and

role either has a value from the set S_r = {Resource, Consumer, Queue, Group} or has a composite value of the form* R_1R_2 where each of R_1 and R_2 is a member of S_r with $R_1 \neq R_2$

scope has a value from the set S_b = {Unary, Set[N], Class} where $N > 1$.

Name is some meaningful name assigned to the entity structure

For example, the entity structure identifier: "Resource Set[2]: Tugboat" would indicate a resource entity structure called Tugboat from which two entities are derived. Alternately, "Resource Unary: Tugboat" indicates a Resource entity structure (called Tugboat) from which a single entity is derived. Continuing with this example, we shall frequently use phrases such as "a Tugboat entity" to imply an entity derived from an entity structure called Tugboat. Note however that this reference does not reveal the *role* or *scope* of the underlying entity structure.

The identifier for an entity has a format that reflects the properties of the underlying entity structure. For the case where *scope* = Unary the unique entity derived from the entity structure has the identifier *X.Name* where *X* is one of R, C, Q, G (or some combination of these alternatives) depending on the value of *role*; i.e., *X* = R if *role* = Resource, *X* = C if *role* = Consumer, *X* = RG if *role* = Resource Group, etc. and *Name* is the name assigned to the underlying entity structure. When the underlying entity structure has *scope* = Set[N], we use *X.Name[j]* where j ($0 \leq j \leq N-1$) designates the jth entity derived from the entity structure. When *scope* = Class, i*X.Name* is simply a reference to some particular instance of the entity structure that is relevant to the context under consideration. It does not serve as a unique identifier.

We note finally that within the ABCmod framework entity identifiers are regarded as having global scope. This means that they can be referenced from all behavior construct instances.

6.3.3.3 Attributes

The identification of appropriate attributes for an entity is governed to a large extent by the requirements that emerge in the process of characterizing behavior. This characterization, within the ABCmod framework, is carried out using a collection of "behavior constructs" that react to and manipulate

* In section 6.3.3.1 it was pointed out that there are situations where *role* may have either sequential or simultaneous duality. The composite value indicated here accommodates such a possibility. In principle, the form could have three components and the intent would be analogous.

entities. Inasmuch as entities reflect attribute values, it follows that the selection of the attributes themselves (in the formulation of the underlying entity structures) is the fundamental issue. Some important insight about the selection of appropriate attributes can be obtained by examining typical attribute requirements for several entity categories.

We begin with an examination of consumer entity instances (cei's); i.e., entities derived from an entity structure with *role* = Consumer and *scope* = Class. In many circumstances, such cei's can be viewed as flowing among the various aggregate entities (Queue entities and Group entities) and the Resource entities that exist within the model. An essential requirement therefore is to track both the existence and the status of these entities to ensure that they can be processed correctly by the rules that govern the model's behavior. In addition, there may be a particular trail of data produced by the cei's that is relevant to the output requirements that are implicit in the project goals. These various requirements suggest typical attributes for entity structures having *scope* = Class.

For example, the cei's derived from a particular entity structure may have properties or features that have direct relevance to the manner in which they are treated by the rules of behavior. In this regard a possible attribute for the entity structure could be "Size," which may have a one of three values (SMALL, MEDIUM, or LARGE) or alternately, "Priority," which may have one of two values (HIGH or LOW).

Observe also that output requirements arising from the project goals often need data that must be collected about the way that cei's have progressed through the model. Frequently this requirement is for some type of elapsed time measurement. For example, it may be required to determine the average time spent waiting for service at a particular resource entity by cei's that utilize that resource. An attribute introduced for this purpose could function as a time stamp storing the value of time, t, when the waiting period begins. A data value placed in a prescribed data set would then be computed as the difference between the value of time when the waiting period ends and the time stamp.

As previously suggested, a perspective that is frequently appropriate is one where cei's flow from Resource entity to Resource entity accessing the services that are provided by them. At any particular point in time, however, access to a particular Resource entity may not be possible because it is already engaged (busy) or is otherwise not available (e.g., out of service because of a temporary failure). Such circumstances are normally handled by connecting the entity to an aggregate entity that is associated with the Resource entity where they can wait until access to the Resource entity becomes possible.

The most common aggregate entity is a Queue entity (i.e., an entity derived from an entity structure for which *role* = Queue). Connecting a cei to a Queue entity corresponds to placing the cei in that Queue entity. From this observation it is reasonable to suggest two particular attributes for any Queue entity structure within the model; namely, List and N. Here List serves to store the cei's that are enqueued in a Queue entity derived from that Queue entity structure and N is the number of entries in that list.

It needs to be stressed that the above selection of attributes for characterizing a Queue entity structure is intended simply to be suggestive and is not necessarily adequate for all situations. In some cases, for example, it may be appropriate to include an attribute that permits referencing the specific Resource entity with which an entity (or entities) derived from the Queue entity structure are associated.

The characterization of a Group entity structure is similar to that of a Queue entity structure but there is an important difference. Consumer entity instances are often placed into a Group entity as in the case of a Queue entity, however there is no intrinsic ordering discipline. On the basis of the observations above, the attributes for a Group entity structure could reasonably include List and N where List is the list of the cei's connected to the Group entity and N is the number of entries in that list. In some situations it may be useful to include an attribute that allows some variation in the capacity of the Group entity. This is very much context dependent and provides a further illustration of the need to tailor the characterizing attributes of entity structures to the specific requirements of a project.

Consider now a Resource entity. One perspective that could be taken is to regard a cei that is being serviced by a Resource entity as being incorporated into it. To support such a perspective the underlying Resource entity structure would have to have an attribute for this purpose (possibly called: Client). In many circumstances it is relevant to have an attribute that reflects the status of an entity that is derived from an underlying Resource entity structure. Such an attribute might, for example, be called Busy where the implication is that the assigned binary value indicates whether or not the Resource entity is busy, i.e., is carrying out its intended function. When the status of a Resource entity may assume more than two values, it may be convenient to introduce an attribute called Status that can acquire these multiple values. For example, Status could assume the values IDLE, BUSY, or BROKEN.

A tabular format is used for the specification of all entity structures in an ABCmod conceptual model. The template for this specification is given in Table 6.1 where **Type** is: {*role*} {*scope*} as outlined earlier.

As will become apparent in section 6.3.4, the behavior constructs that capture the behavior of the SUI react to and modify the attribute values that are encapsulated in entities. A means for referencing these values is therefore essential. Our convention in this regard is closely related to the convention described above for identifying entity structures. In particular, the convention endeavors to clearly reveal the entity structure from which the entity in question is derived. Consider an entity structure with *scope* = Class. By our previously outlined convention, the identifier for an entity instance derived from this entity structure has the generic form:

iX.*Name*

TABLE 6.1

Template for Specifying an Entity Structure

Type: *Name*	
A description of the entity structure called Name	
Attributes	**Description**
AttributeName1	*Description of the attribute called AttributeName1*
AttributeName2	*Description of the attribute called AttributeName2*
. . *AtributeNamen.*	. . *Description of the attribute called AttributeNamen*

where X is the value of *role* and is one of (R, C, Q, G, YZ), where each of Y and Z can assume one of (R, C, Q, G) and $Y \neq Z$. If Attr is an attribute of this entity, then we use

$$\text{i}X.\mathit{Name}.\text{Attr}$$

as a reference to that particular attribute within the entity instance.

Alternately suppose we consider an entity structure with *scope* = Set[N]. The generic identifier for the jth member of this entity structure is:

$$X.\mathit{Name}[j]$$

Again, if Attr is an attribute of this entity structure, then we use

$$X.\mathit{Name}[j].\text{Attr}$$

as a reference to that particular attribute within the entity member in question.

6.3.3.4 State Variables

References to the state of a model are an important and integral part of the discussions surrounding the model development process. Inasmuch as the model's state at time t is simply the value of its state variables at time t, a prerequisite for such discussions is a clear understanding of what constitutes the set of state variables for the model. If the model's state variables are incorrectly identified then aspects of the model's development can become muddled and vague and hence error prone.

Notwithstanding the above, it needs to be recognized that an ABCmod conceptual model lacks a feature that precludes the specification of a fully inclusive set of state variables. We have previously pointed out (Section 6.1) that the ABCmod environment does not address the issue of time management. It is simply assumed that there is a mechanism that moves the time variable, *t*, across the observation interval, starting at its left boundary. While such a traversal mechanism is essential for creating an executable model it is not essential for characterizing the relevant behavior properties of the SUI, which is, after all, the intent of the conceptual modeling process. Nevertheless, there are necessarily aspects of any time management procedure that will impact upon the correct identification of all state variables and because of this our discussion in this section is restricted to the identification of a set of variables that meets most, but not all, the requirements of a complete state variable collection.

The entity attributes can, in fact, be aligned with the model's state variables. This follows from the observation that the information embedded in them is needed in order to satisfy the classical requirements for state variables (Padulo and Arbib 1974, Birta and Arbez 2007). Accordingly all references to the state variables of an ABCmod conceptual model in the discussions that follow should be interpreted as a reference to the attributes of the model's entities. Correspondingly, a reference to the state of the model at time *t* should be interpreted as the value of the attributes of its entities at time *t*.

6.3.4 Model Behavior

Our concern now is with presenting the constructs used within the ABCmod framework to characterize behavior (i.e., the model's evolution over time). We begin by noting that the behavior of an ABCmod conceptual model is, for the most part, aligned with changing attribute values. These changes take place in concert with the traversal of the time variable, *t*, across the observation interval and are a consequence of the occurrence of specific conditions. The identification of these conditions and the attribute value changes that they precipitate are fundamental facets of behavior characterization. Recall also that the collection of entities that exist within the conceptual model can vary over the course of the observation interval as entities enter and leave the model. When such changes take place they clearly introduce another source of changing attribute values.

An important but implicit assumption relating to the variable *t* is the assumption that within all sections of any ABCmod conceptual model, the units associated with *t* are the same, for example seconds, days, years, and the like.

The characterization of behavior in the ABCmod framework is carried out using a collection of *behavior constructs*. These fall into two categories called *activity constructs* and *action constructs*. These are described in the discussion that follows.

6.3.4.1 Activity Constructs

The ABCmod activity construct provides the main modeling artefact for characterizing change, or equivalently, behavior. In general, each activity construct within an ABCmod conceptual model serves to encapsulate a unit of behavior that has been identified as having relevance from the perspective of the project goals. The notion of unit here is intended to suggest minimality; in other words, an activity construct should be viewed as atomic in the sense that it captures an aspect of the model's behavior that is not amenable to subdivision (at least from the perspective taken by the model builder). An activity construct can also be regarded as an abstraction of some purposeful task that takes place within the SUI. Its achievement invariably requires at least one resource entity and usually involves interaction with other entities. The key consequence of both the initiation and the completion of this task generally take the form of changes in the value of some of the state variables (i.e., attributes) within the model.

It is important to note that each activity construct captures a particular type of task. Many instances of a particular activity construct can be simultaneously in progress. For example, in the department store it would be feasible for more than one customer to be served at the same desk. In this case several instances of the activity construct formulated to encapsulate the payment task could occur simultaneously.

An activity construct generally has three phases; namely, an initial phase, a duration, and a terminal phase. Both the initial phase and the terminal phase unfold instantaneously (i.e., they consume no [simulated] time). The duration phase carries the important implication that once an instance of an activity construct (i.e., an activity instance) has become energized, it cannot end until there has been an elapse of some number of time units. Note, however, that this duration need not map onto a contiguous time interval but may instead correspond to a collection of disjoint intervals.

The notion of an event is fundamental in any model building discussion within the DEDS domain. In spite of this importance, a universally accepted definition for this notion has proven to be elusive. Within the ABCmod framework, we regard an event simply as a change in the status of the model that is characterized by a Status Change Specification (SCS). An SCS generally includes, but is not restricted to, a change in the model's state. An event begins and ends at the same point in (simulated) time and consequently all changes specified in its associated SCS occur simultaneously. In most cases there is an event associated with both the initial phase and the terminal phase of an activity construct.

The circumstances that cause an event's occurrence are clearly of fundamental importance to its proper characterization. Such circumstances fall into two broad categories. An event is said to be *conditional* if its occurrence depends on the value of one or more state and/or input variables. On the other hand, if the event's occurrence takes place at some predefined value of time, t, independent of the model's state or its inputs, then the event is said to be *scheduled*.

The initial phase of an activity construct generally includes a precondition and a starting event. The precondition is a prescribed logical expression. Although there are some important exceptions, the precondition is generally formulated in terms of the various state variables and/or input variables within the model. The starting event occurs when the precondition acquires a TRUE value. Hence the starting event of an activity construct corresponds to a conditional event. Furthermore, the SCS for the starting event always includes a state variable change that inhibits an immediate reactivation of that instance of the activity construct (in other words, a change that gives that precondition a FALSE value). Notice that the implication here is that when a precondition is present, a starting event is a mandatory constituent for an activity construct. Note also that nothing in the above precludes the possibility of simultaneous multiple instances of an activity construct.

The event that is associated with the terminal phase of an activity instance (i.e., its terminating event) occurs immediately upon the completion of its duration (hence it can be regarded as a scheduled event).

The state changes embedded in the SCS of either a starting event or a terminating event may cause preconditions of several activity constructs to become TRUE thereby initiating instances of them. This demonstrates that multiple activity instances within the model can be simultaneously in progress. Note that although a terminating event is typically present, it is not a mandatory component of an activity construct.

When an activity instance is initiated it has a tentative duration whose length is specified in the underlying construct's specification. This length is frequently established via a data module (see Section 6.3.7), which, therefore, implies that a data modeling stage has been completed. In the most common circumstance, the duration, Δ, of an activity instance does not change once it is initiated. Furthermore it typically maps onto a continuous time interval. In these circumstances the termination time, t_{end}, of that instance is predetermined when it begins; that is, $t_{end} = (t_{start} + \Delta)$ where t_{start} is the value of time, t, when the activity construct's precondition acquired a TRUE value and an instance was initiated. The terminating event (if present) occurs at time $t = t_{end}$.

Several types of activity constructs are provided in the ABCmod framework. They share most of the features that have been outlined above but nevertheless have distinctive aspects that accommodate special requirements. Each type of activity construct is formulated using a template and these templates, together with a brief outline of distinctive aspects, are presented in the discussion that follows.

Activity: This is the most fundamental of the activity constructs. Each occurrence of this construct in the model has a name and is organized according to a template whose format is given in Table 6.2. Recall that a SCS usually includes (but is not restricted to) the identification of required changes in value to some collection of state variables.

Our convention of regarding an activity construct as an atomic unit of behavior precludes embedding within it a secondary behavior unit even

TABLE 6.2
Template for an Activity

Activity: *Name*	
A description of the Activity called Name	
Precondition	*Boolean expression that specifies the condition for initiation*
Event	*SCS associated with initiation*
Duration	*The duration (typically acquired from a Data Module)*
Event	*SCS associated with termination*

TABLE 6.3
Template for the Triggered Activity

Triggered Activity: *Name*	
A description of the Triggered Activity called Name	
Event	*SCS associated with initiation*
Duration	*The duration (typically acquired from a Data Module)*
Event	*SCS associated with termination*

when it may be closely related. One such situation occurs when one behavior unit directly follows upon completion of another without the need to "seize" a further resource. Our notion of a Triggered Activity provides the means for handling such situations.

As an example, consider a port where a tugboat is required to move a freighter from the harbor entrance to an available berth where a loading (or unloading) operation can immediately begin. Here the berthing and the loading operations each map onto activity constructs but the latter is distinctive because the required resource (i.e., the berth) is already available when the berthing is completed and hence the loading can immediately begin. It is because of this absence of a precondition that the loading operation maps onto a Triggered Activity in our ABCmod framework.

Triggered Activity: The distinguishing feature of a Triggered Activity is that its initiation is not established by a precondition but rather by an explicit reference to it within the terminating event of some other activity construct. Such a reference has the form: TA.*Name* where the "TA" prefix emphasizes that ***Name*** is a reference to a Triggered Activity. Note that this shows that an SCS can be more than simply a collection of specifications for state variable changes inasmuch as it can also include a reference to a particular Triggered Activity, which, in turn, serves to initiate an instance of that construct. The template for the Triggered Activity is given in Table 6.3.

We have previously indicated that an activity construct encapsulates a unit of behavior within the SUI. The flow of this behavior, in the context of a specific activity instance may, however, be subjected to an intervention that disrupts the manner in which behavior unfolds. Such an intervention can have a variety of possible effects; for example, (a) the initial (tentative) duration of the activity instance may be altered, (b) the duration may no longer map onto a continuous time interval but may instead map onto two or more disjoint intervals, possibly in combination with (a), (c) the behavior intrinsic to the activity instance may be stopped and may never be resumed.

Two possible types of intervention are possible; namely, preemption and interruption. We examine each of these in turn. Preemption typically occurs in a situation where two (or more) activity instances require the same resource that cannot be shared. Consider for example the circumstance where the initiation of one activity instance called ActP disrupts the flow of another activity instance called ActQ because a resource that is required by both activities instances must be taken from ActQ and reassigned to ActP because ActP has higher priority access to the resource. The ABCmod presentation of such a circumstance requires that ActQ be formulated as an Extended Activity (see Table 6.4) with a preemption subsegment within its Duration segment. A directive of the form "PRE.ActQ" in the starting SCS of ActP initiates the preemption. This directive links directly to the preemption subsegment of ActQ where the consequences of the preemption are specified.

In other words, an activity instance can disrupt the duration of some lower priority instance that is currently accessing the resource. There is however an implication here that some entity (e.g., a consumer entity instance) that is connected to the resource will be displaced. When this occurs, the completion of the service function for the displaced entity is suspended and consequently the duration of the activity instance, from the perspective of the displaced entity, becomes distributed over at least two disjoint time intervals, or in the extreme case may never even be completed.

An interruption accommodates the impact that changes in the value of an input variable can have on one or more of the activity instances within the model. For example, in response to a change in value of an input variable

TABLE 6.4

Template for the Extended Activity

Extended Activity: *Name*	
A description of the Extended Activity called Name	
Precondition	*Boolean expression that specifies the conditions for initiation*
Event	*SCS associated with initiation*
Duration	*The duration (typically acquired from an attribute)*
Preemption Event	*SCS associated with preemption*
Interruption Precondition	*Boolean expression that specifies the conditions under which an interruption occurs*
Event	*SCS associated with interruption*
Event	*SCS associated with termination*

(see Section 6.3.5), an activity instance may undergo a change in the manner in which it completes the task that was initially undertaken. An interruption can be treated as an event inasmuch as it is associated with a set of changes as reflected in an SCS within an interruption subsegment. The subsegment also provides the means for formulating the condition that defines the occurrence of the interruption.

To accommodate the requirements involved in handling an intervention, a more general activity construct is necessary. This construct is called an Extended Activity.

Extended Activity: As its name suggests, this construct can accommodate more general behavior and is the most comprehensive of the activity constructs. Its template is given in Table 6.4

The notion of interruption is equally relevant to a Triggered Activity. This gives rise to a generalization of the Triggered Activity construct that we call an Extended Triggered Activity.

Extended Triggered Activity: Like its basic counterpart, the distinguishing feature of an Extended Triggered Activity is that its initiation is not established by a precondition but rather by an explicit reference to it within the terminating event of some activity construct. The template for an Extended Triggered Activity is given in Table 6.5.

Table 6.6 summarizes several important features of the various activity constructs.

6.3.4.2 Action Constructs

Action constructs are the second category of behavior constructs. While activity constructs serve to capture the various relevant tasks that are carried out within the SUI, the action constructs provide the means for characterizing relevant events—events that are not embedded within activity constructs. The implication here is that an action construct does not have

TABLE 6.5

Template for the Extended Triggered Activity

Extended Triggered Activity: *Name*	
A description of the Extended Triggered Activity called Name	
Event	*SCS associated with initiation*
Duration	*The duration (typically acquired from an attribute)*
Preemption Event	*SCS associated with preemption*
Interruption Precondition	*Boolean expression that specifies the conditions under which an interruption occurs*
Event	*SCS associated with interruption*
Event	*SCS associated with termination*

TABLE 6.6
Features of the Activity Constructs

Feature	Activity	Triggered Activity	Extended Activity	Extended Triggered Activity
Precondition	Yes	No	Yes	No
Starting Event	Yes	Optional	Yes	Optional
Duration	Yes	Yes	Yes	Yes
Intervention	No	No	Yes	Yes
Terminating Event	Optional	Optional	Optional	Optional

TABLE 6.7
Template for the Conditional Action

Conditional Action: *Name*	
A description of the Conditional Action called Name	
Precondition	*Boolean expression that specifies the condition for initiation*
Event	*The associated SCS*

duration, i.e., it unfolds at a single point in time. There are two direct consequences of this feature; namely, an action construct has a single SCS and the concept of instances of an action construct is not meaningful.

There are two types of action construct and they are called the *Conditional Action* and the *Scheduled Action*. Since action constructs correspond to events, a fundamental requirement is the characterization of the condition that causes the occurrence of the underlying event. In the case of the Conditional Action we retain a parallel with the activity constructs and refer to this characterization as the precondition for the Conditional Action. The template for the Conditional Action has the form shown in Table 6.7.

The Conditional Action is frequently used to accommodate a circumstance where the current state of the model inhibits a particular state change that needs to take place. In effect, the need for a delay of uncertain length is thus introduced. In this circumstance the Conditional Action serves as a sentinel that awaits the development of the conditions that permit the state change to occur.

The Scheduled Action corresponds to a scheduled event and hence its occurrence is autonomous in the sense that it depends only on time, t, and is independent of the model's state. Often the event in question is reoccurring and the requirement therefore is to characterize the points in time (the "time set") when the underlying event occurs. The template for the Scheduled Action is shown in Table 6.8.

As will become apparent in the discussion of Section 6.3.5, the Scheduled Action provides the means for handling the notion of input within the ABCmod framework.

TABLE 6.8

Template for the Scheduled Action

Scheduled Action: *Name*	
A description of the Scheduled Action called Name	
TimeSet	*Characterization of the points in time where the underlying event occurs*
Event	*The associated SCS*

6.3.5 Input

Our particular interest now is with characterizing input within the context of formulating an ABCmod conceptual model. The perspective that we adopt is that the notion of input in the DEDS domain has three constituents; namely, the following:

a. E-input variables that reflect the influence of relevant aspects of the SUI's environment upon the behavior that is of interest; e.g., the occurrence of storms that disrupt the operation of a port
b. Independent variables that provide the means for characterizing time-varying features of the conceptual model that influence its behavior but are not themselves effected by it; e.g., .the work schedule for part-time servers at the counter of a fast food outlet
c. Input entity streams that represent the flow of entities into the domain of the ABCmod conceptual model; e.g., the cars that arrive to purchase gas at a service station.

Any particular ABCmod conceptual model may have many inputs; however, there is no requirement for representation from all of these categories.

Consider a variable, u, that represents an input from either category (a) or (b). This variable is, in fact, a function of time; i.e., $u = u(t)$ and the essential information about it is normally provided by a sequence of ordered pairs of the form: $< (t_k, u_k): k = 0,1,2, ---- >$ where t_k is a value of time and $u_k = u(t_k)$ (we assume that $t_i < t_j$ for $i < j$). Each of the time values, t_k, in this sequence identifies a point in time where there is a noteworthy occurrence in the input, u (e.g., a change in value). We refer to this sequence as the *characterizing sequence* for u and denote it as CS[u]; i.e.,

$$\mathrm{CS}[u] = < (t_k, u_k): k = 0, 1, 2, \ldots > . \tag{6.1}$$

The specifications that allow the construction of CS[u] are part of the data modeling task associated with model development. In this regard, however, note that there are two separate sequences that can be associated with CS[u]. These are:

$$\mathrm{CS_D}[u] = < t_k: k = 0, 1, 2, \ldots > \quad \mathrm{CS_R}[u] = < u_k: k = 0, 1, 2, \ldots > \tag{6.2}$$

which we call, respectively, the *domain sequence* for u and the *range sequence* for u. It is almost always true that the domain sequence for u has a stochastic characterization; i.e., a stochastic data model. Generally, this implies that if t_j and $t_{j+1} = t_j + \Delta_j$ are successive members of $CS_D[u]$, then the value of Δ_j is provided by a stochastic model. The range sequence for u may or may not have a stochastic characterization.

From the perspective of developing inputs we assume that the data modeling task has been completed. This, in particular, means that valid mechanisms for creating the domain sequence and the range sequence for each input variable are available.

In some circumstances the input variable, $u(t)$ being considered, falls in the class of piecewise constant (PWC) time functions. An example of this case is shown in Figure 6.4. Here $u(t)$ could represent the number of electricians, at time t, included in the maintenance team of a large manufacturing plant that operates on a 24-hour basis but with varying levels of production (and hence varying requirements for electricians). The behavior of the model over the interval $[t_j, t_{j+1})$ likely depends directly on the value $u_j = u(t_j)$ hence the representation of $u(t)$ as a PWC function is not only meaningful but is, in fact, essential. The characterizing sequence for $u(t)$ as shown in Figure 6.4 is:

$$CS[u] = < (t_0,1), (t_1,2), (t_2,4), (t_3,4), (t_4,3), (t_5,1), (t_6,2) > \quad (6.3)$$

Observe also that with the interpretation given above this particular input is somewhat distinctive inasmuch as neither its domain sequence nor its range sequence will likely have a stochastic characterization.

As an alternate possibility consider a case where, $u(t)$, represents the number of units of a particular product P requested on orders received (at times $t_\eta, t_{\eta+1}, \ldots t_j \ldots$) by an Internet-based distributing company ($\eta = 0$ if the first

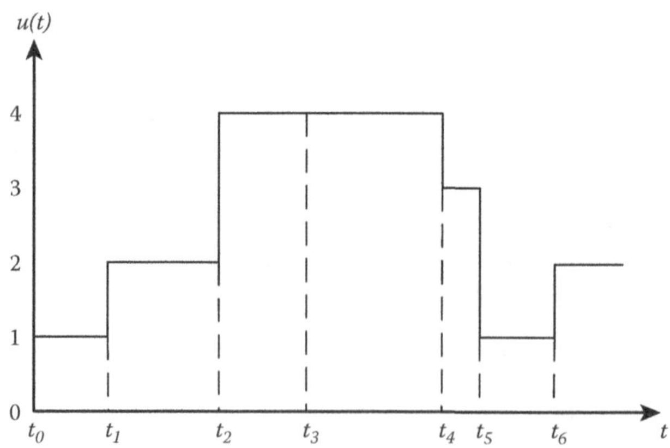

FIGURE 6.4
A piecewise constant time function. (From on Birta, L.G. and Arbez, G., *Modeling and Simulation: Exploring Dynamic System Behavior*, Springer, London, Fig. 4.3, p. 115, 2007. With kind permission of Springer Science and Business Media.)

order arrives at the left boundary of the observation interval, otherwise $\eta = 1$ where $t_1 > t_0$). The characterizing sequence would be written as:

$$CS[u] = < (t_\eta\ , u_\eta), (t_{\eta+1}, u_{\eta+1}), \text{---}\ (t_j, u_j)\ \text{---} > \tag{6.4}$$

Note, however, that only the specific values $u_\eta = u(t_\eta), u_{\eta+1} = u(t_{\eta+2}), \ldots u_j = u(t_j)$ are relevant. In other words, representation of this particular input as a PWC time function is not appropriate because the value of u between order times has no meaning. Note also that the data model for this input would need to provide a specification for both the domain sequence $CS_D[u]$ of order times and the range sequence $CS_R[u]$ of order values as shown in (6.5). Both would likely be in terms of specific probability distribution functions.

$$CS_D[u] = < t_\eta, t_{\eta+1} \ldots t_j, \ldots > \quad CS_R[u] = < u_\eta, u_{\eta+1}, \ldots u_j, \ldots > \tag{6.5}$$

Consider now a variable $s = s(t)$ that represents an input from category (c); i.e., an input entity stream. Recall that the entities in question here would necessarily be instances of some particular entity structure.* The characterizing sequence s can be written as:

$$CS[s] = < (t_\eta, 1), (t_{\eta+1}, 1), (t_{\eta+2}, 1), \text{---}\ (t_j, 1)\ \text{---} > \tag{6.6}$$

Here each value in the domain sequence $< t_\eta, t_{\eta+1}, t_{\eta+2}, \ldots t_j \ldots >$ is the arrival time of an instance of the entity structure in question. Each element of the range sequence has a value of 1; i.e., $s(t_j) = 1$ for all j because we generally assume that arrivals occur one at a time. As above $\eta = 0$ if the first arrival occurs at the left boundary of the observation interval, otherwise $\eta = 1$. The domain sequence is constructed from the arrival process associated with the entity structure in question.

All three categories of input have a characterizing sequence and hence a domain sequence. The impact of inputs from each of the categories is captured in the ABCmod framework by a Scheduled Action whose time set is the domain sequence. It should be emphasized that it is only in limited circumstances that the domain sequence is deterministic; generally the values in the domain sequence evolve in a stochastic manner.

The salient features of the inputs for any ABCmod conceptual model are summarized in a template. The format of this template is shown in Table 6.9. The general format of the associated Scheduled Actions is shown in Table 6.10.

6.3.6 Output

The output of a simulation experiment can be identified with the information that is either explicitly or implicitly required for achieving the goals of the

* The notion of an input entity stream carries the implication of transient existence; hence the entity structures that we associate with this notion always have *scope* =Class.

TABLE 6.9

Template for Inputs

Inputs		
Variable	**Description**	**Scheduled Action**
e-Inputs		
u(t)	*Description of the input variable u(t)*	*Name of the associated Scheduled Action*
Independent Variables		
u(t)	*Description of the input variable u(t)*	*Name of the associated Scheduled Action*
Input Entity Streams		
s(t)	*Description of the input entity stream which that the input variable s(t) represents*	*Name of the associated Scheduled Action*

TABLE 6.10

Templates for the Scheduled Actions for Inputs

Scheduled Action: *uName*	
TimeSet	*t = tk ∈ CSD [u] as defined by DM.uDomain*
Event	*Typically the assignment to the variable u of the value that it acquires at time t = tk ∈ CSD [u] as prescribed by CSR[u], which is provided by a designated data module; e.g., DM.uRange.*

(a) Case where the Scheduled Action corresponds to the e-input variable, u(t)

Scheduled Action: *uName*	
TimeSet	*t = tk ∈ CSD [u] as defined by DM.uDomain*
Event	*Typically the assignment to the variable u of the value that it acquires at time t = tk ∈ CSD [u]] as prescribed by CSR[u], which is provided by a designated data module; e.g., DM.uRange.*

(b) Case where the Scheduled Action corresponds to an independent variable, u(t)

Scheduled Action: *sName*	
TimeSet	*t = tk ∈ CSD [s] as defined by DM.sDomain*
Event	*An entity instance from the entity structure of interest(called ESname) is first established via:* *iX.ESname ← SM.Derive(ESname)* *where the entity structure in question has role = X.* *The Derive operation is typically followed by appropriate attribute value assignments. Where relevant, the newly arriving entity is typically positioned (connected) to some appropriate entity within the model*

(c) Case where the Scheduled Action corresponds to an input entity stream simulation project. The implication here is that a simulation experiment that generates no output information serves no practical purpose. Strictly speaking, such information is outside the scope of both structural and behavioral aspects of a conceptual model. Nevertheless it is fundamental and appropriate steps to capture the relevant data are essential in the development of any simulation model. Inasmuch as a conceptual model can be viewed as a design document for a simulation model, this data requirement needs to be an integral part of conceptual model's formulation.

For the most part, output from a simulation experiment flows from the entities within the conceptual model. Nevertheless there are circumstances where considerable convenience can be realized by manipulating, within the body of a behavior construct, a variable that is unrelated to any entity but whose final value provides information relevant to the project goals; e.g., a counter of the number of customers that wait longer than five minutes in a queue leading to a service desk. In some respects, such an output variable is the most fundamental of the several types that are outlined in this discussion. It is called a Simple Scalar Output Variable (SSOV).

As indicated above, output generally flows from the entities within the conceptual model. In fact, output data is linked to values of specific attributes that have special relevance to project goals. These entity attributes become output variables. Over the course of the observation interval, these output variables generate data sets; i.e., collections of discrete data values and it is these sets of values that are of special interest.

There are two categories of such data sets, one is called a trajectory set and the other a sample set. A trajectory set, denoted by TRJ[y], is a collection of values generated by the output variable y, which is viewed as a time function; i.e., $y = y(t)$. The collected values are a set of ordered pairs of the form (t_k, y_k) where the t_k are points in time that have some special significance (e.g., occurrence of a change in value of the output variable). A sample set on the other hand, is an accumulation of data values deposited by entities flowing within, or through, the conceptual model. The output variable here is again a specific entity attribute but the deposited values generally relate to different entities. A sample set associated with an output variable, y, is denoted by PHI[y] and y is often called a sample variable.

The entire set of data values within a Trajectory Set or a Sample Set is not normally of interest in its entirety. (The notable exception here is the case where a graphical presentation of the data in these sets is desired.) Typically it is some property of the data that has special relevance; e.g., minimum, maximum, average. Normally the scalar value obtained by carrying out such an operation is assigned to an output variable called a Derived Scalar Output Variable (DSOV).

The documentation for outputs is organized in terms of a template that incorporates the various types of output variables that are relevant to the project. The template is shown in Table 6.11.

TABLE 6.11
Template for Summarizing Outputs

Outputs			
Simple Scalar Output Variables (SSOVs)			
Name	**Description**		
Y	*Description of the simple scalar output variable Y*		
Trajectory Sets			
Name	**Description**		
TRJ[y]	*Description of the time variable y(t)*		
Sample Sets			
Name	**Description**		
PHI[y]	*Description of the sample variable y whose values populate the sample set PHI[y]*		
Derived Scalar Output Variables (DSOV's)			
Name	**Description**	**Data Set Name**	**Operator**
Y	*Significance of the value assigned to Y*	*The name of the data set from which the value of Y is derived*	*The operation that is carried out on the underlying data set to yield the value assigned to Y*

6.3.7 Data Modules

It is rarely possible to formulate an ABCmod conceptual model without the need to access data. The simplest such requirement is the case where there is a need for a sample from a prescribed distribution function. Alternately, the requirement might be for a sample from one or several specified distributions according to some prescribed rule. The convention we have adopted in our ABCmod framework is to encapsulate any such data delivery requirement within a named *data module,* which serves as a wrapper for the data specification. The rational here is simply to facilitate modification of the actual source of the data if that need arises. The collection of such data modules that are required within an ABCmod conceptual model is summarized in a table whose template is shown in Table 6.12. Note that by convention, we highlight the name of a data module referenced in the body of behavior constructs with the prefix "DM."

6.3.8 Standard Modules and User-Defined Modules

A variety of standard operations recur in the formulation of the SCS's within the various behavior constructs that emerge during the development of any ABCmod conceptual model. We assume the existence of modules to carry out these operations and each of these is briefly outlined below.

- InsertQue(QueueName, Item): Inserts Item into a queue entity called QueueName according to the declared queuing protocol associated with QueueName.

TABLE 6.12
Template for Summarizing Data Modules

Data Modules		
Name	**Description**	**Data Model**
ModuleName(parameter list)	*Statement of the purpose of the data module called ModuleName*	*Details of the mechanism that is invoked in order to generate the data values provided by the data module called ModuleName. Typically involves sampling values from one or more distributions.*

- InsertQueHead(QueueName, Item): Inserts Item at the head of a queue entity called QueueName.
- Ident ← RemoveQue(QueueName): Removes the item that is at the head of the queue entity called QueueName. Ident is the identifier for the returned item.
- InsertGrp(GroupName, Item): Inserts Item into the group entity called GroupName.
- RemoveGrp(GroupName, Ident): Removes an item from the group called GroupName. Ident is the identifier for the item to be removed from the group.
- Ident ← RemoveGrpAny(GroupName): Removes an arbitrary item from the group called GroupName. Ident is the identifier for the item removed from the group.
- Put(*PHY[y]*, Val): Places the value Val into the sample set called *PHY[y]*
- Ident ← Derive(EntityStructureName): Derives an entity with identifier Ident from the entity structure called EntityStructureName.
- Leave(Ident): It frequently occurs that a specific entity's existence within the model comes to an end. This module explicitly indicates such an occurrence and its argument is the identifier of the entity in question. The module is typically invoked within the SCS of the terminating event of an activity instance.
- Terminate: An instance of a (Triggered) Extended Activity construct that undergoes an intervention must necessarily terminate. This is made explicit by ending the SCS of each intervention subsegment with a reference to the Terminate module.

Typically modules are needed to carry out specialized operations that are distinctive to the specific conceptual model being developed. These can be freely defined wherever necessary to augment the ABCmod framework and ease the conceptual modeling task. They are called User-Defined Modules and they are summarized in a table whose template is given in Table 6.13.

TABLE 6.13

Template for Summarizing User-Defined Modules

User-Defined Modules	
Name	**Description**
ModuleName(parameter list)	*Purpose of the user-defined module called ModuleName*

Note also that by convention, we highlight references to Standard Modules and User-Defined Modules within the body of behavior constructs with the prefix "SM." or "UM.", respectively.

6.4 Methodology for Developing an ABCmod Conceptual Model

The discussion in the preceding sections outlines a framework (namely the ABCmod framework) for capturing and organizing the detail that appropriately characterizes the behavior of some particular SUI within the DEDS domain. However, except for the simplest of cases, the product that emerges (i.e., an ABCmod conceptual model) can become complex. To deal with this complexity, a two stage hierarchical approach has been incorporated to facilitate the model building process. This consists of a high-level formulation followed by a detailed formulation. We briefly summarize each of these below. Further detail is provided in the example that is presented in section 6.5.

The high-level formulation has three constituents:

- *Structural Overview:* The structural overview provides a brief description of each of the entity structures that will be required together with a graphical presentation of the various entities that are derived from them. This structural diagram is constructed from a predefined collection of graphical symbols that are associated with the various *role* values for entity structures (for example, see Figure 6.6).
- *List of Data Models:* This is a list of the data models that will be needed in the development of the data modules that are used in the conceptual model's formulation. Initially the specification of specific stochastic distributions is not necessary since the initial intent is simply to establish what models will be required. Typically establishing the appropriate data models is not a trivial task. An examination of this process is beyond the scope of this presentation. We note, however, that the process can be undertaken in parallel with the development of the detailed conceptual model.

- *Behavioral Overview*: The behavioral overview provides a list of the behavior constructs that will be needed together with a brief description of the task that each carries out. In addition, a behavior diagram is presented that is a collection of life-cycle diagrams (for example, see Figure 6.7). Often an entity participates in more than one behavior construct instance; i.e., it flows from one instance of a behavior construct to another. This flow can take place in any one of a variety of ways and the purpose of the life-cycle diagram is to clarify this behavioral feature.

The detailed-level formulation is organized into five sections:

- *Structural Components:* The specification for each of the required entity structures is presented in the tabular format shown in Table 6.1.
- *Data Components:* The constants and parameters for the model are summarized. Also included here is the summary of the data modules (see Table 6.12) that are needed in the development of the behavior constructs.
- *Input Components:* The inputs to the model (e-inputs, independent variables, and input entity streams) are summarized (see Table 6.9).
- *Output Components:* The various outputs of the model (SSOV's, trajectory sets, sample sets, DSOV's) are summarized (see Table 6.11).
- *Behavioral Components:* This section begins with the identification of the time units that are used and the observation interval. The assumptions relating to initial conditions are summarized in an Initialize Table; i.e., the conditions at the left-hand boundary of the observation interval (usually $t = 0$). By convention all queues and groups are assumed to be empty at $t = 0$ unless otherwise indicated in the Initialize Table. Note also that all behavior constructs are predicated by the assignments (explicit or implicit) of the Initialize Table. User-modules that are introduced to aid in the formulation of behavior constructs are presented in the tabular format shown in Table 6.13. The section ends with the collection of behavior constructs that are pertinent to the project (see Table 6.2 through Table 6.8).

6.5 Example Project: The Bigtown Garage

The ABCmod conceptual modeling process is illustrated in this section. We begin with an outline of a modeling and simulation project and then develop for it an ABCmod conceptual model. The requirements for the model's development demonstrate many (but not all) the features that are available.

6.5.1 SUI Key Features

6.5.1.1 SUI Overview

Bigtown is a large city that operates its own garage to service a range of city owned vehicles. It carries out routine maintenance and breakdown repair work on vehicles assigned to a variety of city departments; e.g., bylaw enforcement, building inspection services, environmental services, and as well, the police department. The police vehicles are by far the largest component of the clientele. Furthermore, police vehicles always receive priority service at the garage because of the key function that they provide.

Because of severe budget cutbacks, the renewal of the fleet of police vehicles has been postponed for several years. The fleet is aging and is placing an increasing burden upon the garage's operation. The impact is becoming an increasing irritant to the ancillary departments because their vehicles are often unavailable due to congestion problems in the garage.

6.5.1.2 General Project Goals

To address this problem the city manager has decided to explore, via a simulation study, the impact of continued restrictions on the renewal of the police car fleet. The specific interest is with gaining some insights into how the congestion problem will develop if funding restrictions continue and, as well, when the problem will likely reach an intolerable level.

6.5.1.3 SUI Details

The garage has four service bays (see Figure 6.5) and is open 24 hours a day, seven days a week. There are three eight-hour shifts: the day, the evening,

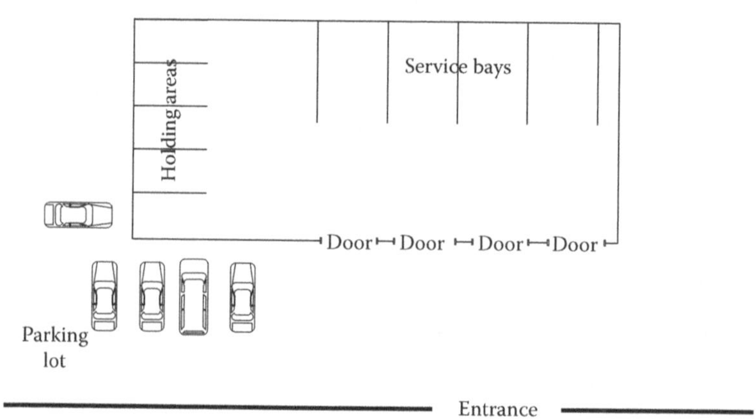

FIGURE 6.5
Bigtown Garage schematic.

and the night, and currently there are two mechanics working during each of these shifts. Although the two extra service bays currently provide some passive utility (see description below) they do offer the possibility for increased throughput if additional mechanics were hired for the one or more of the shifts.

Vehicles arrive for service either because of routine maintenance requirements or because they require repair due to mechanical failure. The vehicles scheduled for routine maintenance on any particular day can be assumed to arrive at the beginning of the day shift; i.e., 8:00 a.m. When a vehicle arrives, a work order is filled out. This summarizes the nature of the service requirement. The time of the vehicle's arrival at the garage is also noted because vehicles are serviced in order of their arrival but with due recognition of the priority of police vehicles. In the case where vehicles have the same arrival time stamp, the vehicles are serviced according to the order in which the work orders were filled out.

The priority given to police vehicles implies that no ancillary vehicle is moved into a service bay from the parking lot until there is no remaining police vehicle waiting to be serviced.

A police vehicle in the parking lot is moved into a service bay if either (a) there is at least one mechanic who is idle, or (b) there is at least one mechanic carrying out a servicing task on an ancillary vehicle; where (b) is applied only when there are no idle mechanics.

In the case of (a), there is at least one unoccupied service bay and the police vehicle is moved into one of them. Both the choice of the bay (if there is more than one that is empty) and the allocated mechanic are random selections. In the case of (b), work on an ancillary vehicle is stopped thereby releasing a mechanic to work on the police vehicle. If there is an empty service bay, the police vehicle is moved there and the freed mechanic moves to that bay to begin the servicing work. If, on the other hand, there is no empty service bay,* then the ancillary vehicle in the service bay of the freed mechanic is moved into one of four holding areas within the garage thereby releasing a service bay for the police vehicle. It may occur that a group of displaced ancillary vehicles is thus created. Work on these vehicles is resumed and completed (provided there are no waiting police vehicles).

6.5.1.4 Detailed Goals and Output

Several measures of performance of the garage's operation have been identified as being of interest. Included here are: (a) the average time spent by vehicles waiting for service to begin (separately for each of the two categories of vehicle); (b) the average "total time" spent from arrival to completion

* This will only occur when there are more than two mechanics working during a shift which is a situation that arises when solution options are explored in the simulation study.

of service (again separately for each of the two categories of vehicle); (c) the average number of busy mechanics

The city manager intends to explore alternate scenarios that correspond to a deteriorating police car fleet. Three scenarios are of particular interest. They correspond to the cases where the distributions of interarrival time for breakdown repair for police vehicles are scaled so that their mean values are decreased first by 20%, then by 40%, and finally by 60% from their current operational values. The effect of increasing the number of mechanics working at the garage to 3 and 4 is also of interest for each of the three scenarios.

6.5.2 ABCmod Conceptual Model

6.5.2.1 High-Level Conceptual Model

6.5.2.1.1 Structural Overview

1. *Consumer Class: PoliceVeh*—Derived entities represent the police vehicles that require servicing.
2. *Consumer Class: AncillaryVeh*—Derived entities represent the ancillary vehicles that require servicing.
3. *Queue Unary: ParkedPV*—The derived entity provides the means for accumulating the police vehicles waiting for servicing. This queue establishes an order of servicing that reflects the arrival time stamps on the work orders.
4. *Queue Unary: ParkedAV*—The derived entity provides the means for accumulating the ancillary vehicles waiting for servicing. This queue establishes an order of servicing that reflects the arrival time stamps on the work orders.

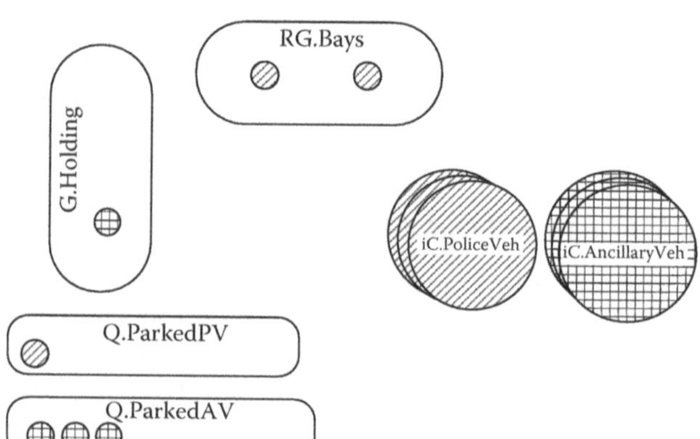

FIGURE 6.6
Bigtown Garage structural diagram.

5. *Resource Group Unary: Bays*—The derived entity represents the service bays used for servicing vehicles.
6. *Group Unary: Holding*—The derived entity represents the holding area where an ancillary vehicle is placed in order to make a service bay available to service a police vehicle.

Notes

- Mechanics are not explicitly modeled. Instead, RG.Bays will have the attribute RG.Bays.freeMechanics to indicate the number of idle mechanics (that is mechanics that are not currently servicing a vehicle). The number of mechanics present during each of the shifts is a model parameter.

6.5.2.1.2 Data Model Overview

- Interarrival times for vehicles requiring routine maintenance (this is constant since vehicles requiring routine maintenance all arrive at 8:00 a.m.). The following values relating to arriving vehicles are random:
  - Number of vehicles that arrive for routine maintenance on any particular day.
  - Fraction of vehicles requiring routine maintenance that are police vehicles.
- Interarrival times of police vehicles requiring breakdown repair. The associated distributions are expected to be dependent on the shift.
- Interarrival times of ancillary vehicles requiring breakdown repair. The associated distributions are expected to be dependent on the shift.
- Service time for breakdown repair (assumed to be the same for both police vehicles and ancillary vehicles).
- Service time for routine maintenance (assumed to be the same for both police vehicles and ancillary vehicles).

6.5.2.1.3 Behavioral Overview

- Scheduled Actions:
  - RMArr: Routine maintenance arrivals of both police and ancillary vehicles.
  - BRArrPV: Breakdown repair arrivals of police vehicles.
  - BRArrAV: Breakdown repair arrivals of ancillary vehicles.
- Activities:
  - ServiceAV: The servicing task for an ancillary vehicle. This activity may be preempted by the ServicePV activity.
  - ServicePV: The servicing task for a police vehicle. This activity may preempt the ServiceAV activity.

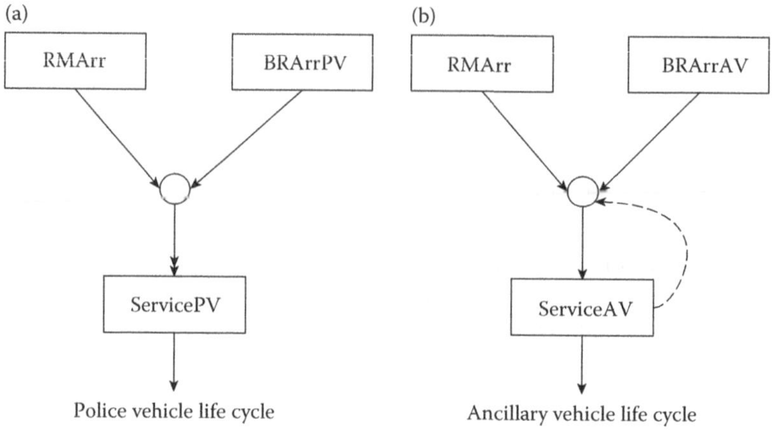

FIGURE 6.7
Bigtown Garage behavioral diagram.

6.5.2.2 Detailed Conceptual Model

Structural Components

Consumer Class: PoliceVeh	
The police vehicles that need servicing.	
Attributes	**Description**
state	Set to SERVICING when the vehicle is being serviced, NOTSTARTED when servicing has not yet started.
serviceType	Indicates type of service required; values are: BR for breakdown repair and RM for routine maintenance.
arrivalTime	The time at which the police vehicle arrived at the garage.

Consumer Class: AncillaryVeh	
The ancillary vehicles that need servicing.	
Attributes	**Description**
state	Set to SERVICING when the vehicle is being serviced, NOTSTARTED when servicing has not yet started, MIDSERVICE when servicing has been preempted.
serviceType	Indicates type of service required; values are: BR for breakdown repair and RM for routine maintenance.
arrivalTime	The time at which the ancillary vehicle arrived at the garage.
start	The time at which the servicing started (or resumed after preemption).
timeToService	The time required to complete the service (this value changes when the servicing is pre-empted).

Queue Unary: ParkedPV	
FIFO queue of police vehicles waiting for servicing.	
Attributes	**Description**
n	Number of police vehicle entities in list.
list	List of police vehicle entities.

Queue Unary: ParkedAV	
FIFO queue of ancillary vehicles waiting for servicing.	
Attributes	**Description**
n	Number of ancillary vehicle entities in List.
list	List of ancillary vehicle entities.

Resource Group Unary: Bays	
This Resource Group represents the service bays.	
Attributes	**Description**
n	Number of vehicles in this Resource Group. Note some vehicles may be in MIDSERVICE. Note that n is less than or equal to four.
list	The list of the vehicle entities in this Resource Group.
freeMechanics	Gives the number of idle mechanics.

Group Unary: Holding	
This group represents the holding areas within the garage.	
Attributes	**Description**
n	Number of ancillary vehicle entities in this Group. Note these vehicles will be in MIDSERVICE. Note that n is less than or equal to four.
list	The list of the ancillary vehicle entities in this Group.

Data Modeling Components

Constants		
Name	**Role**	**Value**
NumBays	Number of service bays.	4
NumRMMin	Minimum number of vehicles that arrive on any particular day for routine maintenance.	TBD
NumRMMax	Maximum number of vehicles that arrive on any particular day for routine maintenance.	TBD
FractionPV	Fraction of vehicles arriving for routine maintenance that are police vehicles.	2/3
MeanAV_Day	Mean interarrival time of ancillary vehicles requiring breakdown repair during the day shift.	TBD
MeanAV_ Evening	Mean interarrival time of ancillary vehicles requiring breakdown repair during the evening shift.	TBD

(*Continued*)

Constants (Continued)		
Name	**Role**	**Value**
MeanAV_Night	Mean interarrival time of ancillary vehicles requiring breakdown repair during the night shift.	TBD
MeanBR_ ServiceTime	Mean service time for breakdown repair.	TBD
MeanRM_ ServiceTime	Mean service time for routine maintenance.	TBD
Parameters		
Name	**Role**	**Value**
MeanPV_Night	Mean interarrival time of police vehicles requiring breakdown repair during the day shift.	MP1, 0.8*MP1, 0.6*MP1, 0.4*MP1
MeanPV_ Evening	Mean interarrival time of police vehicles requiring breakdown repair during the evening shift.	MP2, 0.8*MP2, 0.6*MP2, 0.4*MP2
MeanPV_Night	Mean interarrival time of police vehicles requiring breakdown repair during the night shift.	MP3, 0.8*MP3, 0.6*MP3, 0.4*MP3
NumMechanics	Number of mechanics working at the garage.	2, 3, 4

Data Modules		
Name	**Description**	**Data Model**
NumRMVehicles()	Gives the number of vehicles that arrive for routine maintenance. A fraction (FractionPV) of this number are police vehicles.	Uniform(NumRMMin, NumRMMax)
InterArrivalPV_BR( )	Gives the interarrival times of police vehicles arriving for breakdown repair.	If Shift = DAY: Exponential(MeanPV_Day) If Shift = EVENING: Exponential(MeanPV_Evening) If Shift = NIGHT: Exponential(MeanPV_Night)
InterArrivalAV_BR( )	Gives the interarrival times of ancillary vehicles arriving for breakdown repair.	If Shift = DAY: Exponential(MeanAV_Day) If Shift = EVENING: Exponential(MeanAV_Evening) If Shift = NIGHT: Exponential(MeanAV_Night)
ServiceTime (serviceType)	Gives the time to service a vehicle according to the value of serviceType.	If serviceType = RM: Exponential(MeanRM_ ServiceTime) If serviceType = BR: Exponential(MeanBR_ ServiceTime)
RMArrivals()	Gives the arrival times of vehicles for routine maintenance.	Every 24 hours starting at t = 0; i.e., t = 24k, k = 0,1,2 ...
ShiftChangeTimes()	Gives the points in time when a shift change occurs.	Every 8 hours starting at t = 8; i.e., t = 8k, k = 1,2. 3 ...

Input Components

Inputs		
Variable	**Description**	**Scheduled Action**
Independent Variables		
Shift	Reflects the current shift; values are: DAY (day shift), EVENING (evening shift) or NIGHT (night shift).	ShiftChange
Input Entity Streams		
RMInput	Vehicles requiring routine maintenance.	RMArr
BRInputPV	Police vehicles requiring breakdown repair.	BRArrPV
BRInputAV	Ancillary vehicles requiring breakdown repair.	BRArrAV

Output Components

Outputs	
Trajectory Sets	
Name	**Description**
TRJ[NumBusyMechanics]	NumBusyMechanics = NumMechanics – RG.Bays.freeMechanics
Sample Sets	
Name	**Description**
PHI[WaitServiceAV]	Each value is the time spent by some ancillary vehicle waiting for service to begin.
PHI[WaitServicePV]	Each value is the time spent by some police vehicle waiting for service to begin.
PHI[TotalTimeAV]	Each value is the elapsed time from arrival to completion of service for some ancillary vehicle.
PHI[TotalTimePV]	Each value is the elapsed time from arrival to completion of service for some police vehicle.

Derived Scalar Output Variables (DSOV's)			
Name	**Description**	**Output Set Name**	**Operator**
AvgWaitSrvAV	Average time spent by ancillary vehicles waiting for service to begin.	PHI[WaitServiceAV]	MEAN
AvgWaitSrvPV	Average time spent by police vehicles waiting for service to begin.	PHI[WaitServicePV]	MEAN
AvgTotalTimeAV	Average elapsed time from arrival to completion of service for ancillary vehicles.	PHI[TotalTimeAV]	MEAN
AvgTotalTimePV	Average elapsed time from arrival to completion of service for police vehicles.	PHI[TotalTimePV]	MEAN
AvgBusyMech	Average number of busy mechanics.	TRJ[NumBusyMechanics]	MEAN

Behavioral Components

Time units: hours

Observation interval: t = 0 corresponds to 8:00 a.m.; steady state study, hence right-hand boundary to be determined by experimentation.

User Defined Modules	
Name	**Description**
GetAVBeingServiced(iC. AncillaryVeh)	Returns TRUE if an ancillary vehicle being serviced in RG.Bays (i.e., iC.AncillaryVeh.status = SERVICING). Furthermore, sets iC.AncillaryVeh to reference an AncillaryVeh entity being serviced (random selection).
ServiceBayAvailablePV()	Returns TRUE if an additional police vehicle can be accommodated in RG.Bays. This occurs if either there is an idle mechanic (RG.Bays.freeMechanics ≠ 0) or if there is an ancillary vehicle being serviced in RG.Bays. Note that a free mechanic implies a free bay.
AVMidService(iC. AncillaryVeh)	Returns TRUE if an ancillary vehicle is in mid-service in RG.Bays (i.e., iC.AncillaryVeh.state = MIDSERVICE). Furthermore, sets iC.AncillaryVeh to reference an AncillaryVeh entity in mid-service (random selection).

Initialize
RG.Bays.freeMechanics ← NumMechanics Shift ← DAY

Scheduled Action: RMArr	
Arrival of vehicles requiring routine maintenance.	
TimeSet	t = tk ← CSD [RMInput] as defined by DM.RMArrivals( )

Event	NumVeh ← DM.NumRMVehicles()
	NumPVVeh ← NumVeh * FractionPV
	NumAVVeh ← NumVeh - NumPVVeh
	FOR i = 1 TO NumPVVeh
	    iC.PoliceVeh ← SM.Derive(PoliceVeh)
	    iC.PoliceVeh.state ← NOTSTARTED
	    iC.ServiceType ← RM
	    iC.PoliceVeh.arrivalTime ← t
	    SM.InsertQue(Q.ParkedPV, iC.PoliceVeh)
	ENDFOR
	FOR i = 1 TO NumAVVeh
	    iC.AncillaryVeh ← SM.Derive(AncillaryVeh)
	    iC.AncillaryVeh.state ← NOTSTARTED
	    iC.AncillaryVeh.serviceType ← RM
	    iC.AncillaryVeh.arrivalTime ← t
	    SM.InsertQue(Q.ParkedAV, iC.AncillaryVeh)
	ENDFOR

Scheduled Action: BRArrPV	
Arrivals of police vehicles requiring breakdown repair.	
TimeSet	t = tk ← CSD [BRInputPV] as defined by DM.InterArrivalPV_BR( )
Event	iC.PoliceVeh ← SM.Derive(PoliceVeh) iC.PoliceVeh.state ← NOTSTARTED iC.PoliceVeh.serviceType ← BR iC.PoliceVeh.arrivalTime ← t SM.InsertQue(Q.ParkedPV, iC.PoliceVeh)

Scheduled Action: BRArrAV	
Arrivals of ancillary vehicles requiring breakdown repair.	
TimeSet	t = tk ← CSD [BRInputAV] as defined by DM.InterArrivalAV_BR( )
Event	iC.AncillaryVeh ← SM.Derive(AncillaryVeh) iC.AncillaryVeh.state ← NOTSTARTED iC.AncillaryVeh.serviceType ← BR iC.AncillaryVeh.arrivalTime ← t SM.InsertQue(Q.ParkedAV, iC.AncillaryVeh)

Scheduled Action: ShiftChange	
Assigns appropriate values to the input variable Shift.	
TimeSet	t = tk ← CSD [Shift] as defined by DM.ShiftChangeTimes()

Event	IF Shift = DAY
	    Shift ← EVENING
	ELSE IF Shift = EVENING
	    Shift ← NIGHT
	ELSE // night shift
	    Shift ← DAY
	ENDIF

Activity: ServicePV	
Servicing a police vehicle.	
Precondition	UM.ServiceBayAvailable() AND Q.ParkedPV.n ≠ 0

Event	iC.PoliceVeh ← SM.RemoveQue(Q.ParkedPV)
	SM.Put(PHI[WaitServicePV], t – iC.PoliceVeh.arrivalTime)
	IF RG.Bays.freeMechanics ≠ 0 // Free bay exists in RG.Bays
	    Decrement RG.Bays.freeMechanics
	ELSE // Need to preempt service on ancillary vehicle
	    UM.GetAVBeingServiced(iC.AncillaryVeh)
	    PRE.ServiceAV(iC.AncillaryVeh)

(Continued)

Activity: ServicePV (Continued)	

	IF RG.Bays.n = NumBays // Bay full, need to remove AV
	SM.RemoveGrp(RG.Bays, iC.AncillaryVeh)
	SM.InsertGrp(G.Holding, iC.AncillaryVeh)
	ENDIF
	ENDIF
	SM.InsertGrp(RG.Bays, iC.PoliceVeh)
	iC.PoliceVeh.state ← SERVICING

Duration	DM.ServiceTime(iC.PoliceVeh.serviceType)
Event	SM.RemoveGrp(RG.Bays, iC.PoliceVeh) SM.Put(PHI[TotalTimePV], t – iC.PoliceVeh.arrivalTime) Increment RG.Bays.freeMechanics SM.Leave(iC.PoliceVeh)

Extended Activity: ServiceAV	
Servicing an ancillary vehicle.	
Precondition	((Q.ParkedPV.n = 0) AND ((RG.Bays.freeMechanics ≠ 0 AND (UM.AVMidService(iC.AncillaryVeh) = TRUE OR Q.Holding.n ≠ 0 OR Q.ParkedAV.n ≠ 0))

Event	Decrement RG.Bays.freeMechanics
	IF UM.AVMidService(iC.AncillaryVeh) = TRUE
	// re-initiate servicing
	ELSE IF Q.Holding.n ≠ 0
	iC.AncillaryVeh ←SM.RemoveGrpAny(G.Holding)
	SM.InsertGrp(RG.Bays, iC.AncillaryVeh)
	ELSE // Starting a new service
	iC.AncillaryVeh ← SM.RemoveQue(Q.ParkedAV)
	iC.AncillaryVeh.start ← t
	iC.AncillaryVeh.timeToService ← DM.ServiceTime(iC.AncillaryVeh.serviceType)
	SM.Put(PHI[WaitServiceAV], t – iC.AncillaryVeh.arrivalTime)
	SM.InsertGrp(RG.Bays, iC.AncillaryVeh)
	ENDIF
	iC.AncillaryVeh.state ← SERVICING
	iC.AncillaryVeh.start ← t

Duration	iC.AncillaryVeh.timeToService
Pre-emption Event	iC.AncillaryVeh.timetoService – ← t – iC.AncillaryVeh.start iC.AncillaryVeh.state ← MIDSERVICE SM.Terminate
Event	SM.RemoveGrp(RG.Bays, iC.AncillaryVeh) SM.Put(PHI[TotalTimeAV], t – iC.AncillaryVeh.arrivalTime) Increment RG.Bays.freeMechanics SM.Leave(iC.AncillaryVeh)

6.6 Conclusions

A meaningful conceptual model is essential for a successful simulation study. The development process is driven by the goals that have been identified for the modeling and simulation project and focuses on capturing the structural and behavioral features of the SUI that are relevant to the achievement of those goals. Furthermore the process itself serves as a vehicle that allows all project stake holders to participate in the identification of these structural and behavioral features. The model that evolves serves as the blueprint for the development of the program code for carrying out the simulation study. The ABCmod framework outlined in this chapter provides an environment designed specifically to facilitate the achievement of these fundamental objectives of the conceptual modeling task. It has been extensively used for several years in a senior undergraduate/junior graduate course where students carried out group projects. The wide range of nontrivial student projects that have been completed provide a convincing body of evidence that the framework does achieve its intended purpose very effectively.

Dealing with detail and complexity is an essential requirement of any conceptual modeling environment. This is accommodated in the ABCmod conceptual modeling framework by providing a two stage hierarchal approach. Included at the initial high-level stage is the identification of both the modeling artifacts that map onto objects within the SUI that have relevance to the model development process and, as well, the modeling constructs that capture the behavioral features of these artifacts. These correspond, respectively, to the identification of the entity structures and the behavior constructs pertinent to the model. The second stage is concerned with specifying an appropriate level of detail for the entity structures and the behavior constructs that have been identified.

The high-level model is presented using a graphical format. This includes structural diagrams and a collection of life-cycle diagrams that show how entities move among the behavior constructs. The detailed level model is presented using tables with predefined formats. The text-based tabular format provides the important advantage of accommodating arbitrary complexity. In particular, straightforward mechanisms are provided for dealing with the disruption of entity flow through an activity (e.g., either interruption or preemption).

The complexity inherent in large systems is best handled by formulating an interacting subsystem perspective often organized in a hierarchical manner. Such a perspective naturally is reflected into the conceptual modeling process. Extensions to the ABCmod framework that will conveniently accommodate such a hierarchical perspective are currently under way. A software tool that supports the creation of ABCmod conceptual models is also currently under development.

References

Arbez, G., and L.G. Birta. 2007. ABCmod: A conceptual modelling framework for discrete event dynamic systems. In *Proceedings 2007 Summer Computer Simulation Conference*. San Diego.

Banks, J., J. S. Carson II, B.L. Nelson, and D.M. Nicol. 2005. *Discrete Event System Simulation*, 4th ed. New Jersey: Prentice-Hall.

Birta, L.G., and G. Arbez. 2007. *Modelling and Simulation: Exploring Dynamic System Behaviour*. London: Springer.

Buxton, J.N., and J.G. Laski. 1962. Control and simulation language. *Computer journal* 5: 194–199.

Hills, P.R. 1973. *An introduction to simulation using SIMULA*. Publication No. S55. Oslo: Norwegian Computing Center.

Kreutzer, W. 1986. *System Simulation: Programming Styles and Languages*. Sydney, Wokingham: Addison-Wesley.

Overstreet, C.M., and R. E. Nance. 2004. Characterizations and relationships of world views. In *Proceedings 2004 Winter Simulation Conference*, 279–287. Washington, DC.

Padulo, L., and M.A. Arbib. 1974. *System Theory: A Unified State–Space Approach to Continuous and Discrete Systems*. Philadelphia: W.B. Saunders.

Pidd, M. 2004a. *Computer Simulation in Management Science*, 4th ed. Chichester: John Wiley.

Pidd, M. 2004b. Simulation world views: So what? In *Proceedings 2004 Winter Simulation Conference*. Washington, D.C.

Pritsker, A.B. 1986. *Introduction to Simulation and SLAM II*, 3rd ed. New York: Hallstead Press (John Wiley).

Robinson, S. 2004. *Simulation: The Practice of Model Development and Use*. Chichester: John Wiley.

Shannon, R.E. 1975. *System Simulation: The Art and the Science*. New Jersey: Prentice-Hall.

Tocher, K.D. 1963. *The Art of Simulation*. London: English Universities Press.

7

Conceptual Modeling Notations and Techniques

N. Alpay Karagöz and Onur Demirörs

CONTENTS

7.1 Introduction

Conceptual modeling is a tool that provides a clear understanding of the target domain or problem. In the simulation system development life cycle, conceptual models should be captured early based on project objectives defining what is intended and then should serve as a frame of reference for the subsequent development phases. The conceptual model can be interpreted as part of a problem-specification process and defined as a simplified representation of the real system having the following features: (a) includes structural and behavioral capabilities, assumptions and constraints, (b) provides an implementation independent representation by utilizing a common language for both the client and the modeler, (c) can be used as a basis for early verification and validation of the simulation system and, (d) is produced as a result of an iterative process.

Although there are various approaches offering useful insights on the usage of the conceptual models in the simulation system development lifecycle, they do not provide a systematic guidance on how to develop a conceptual model Robinson (2007a). A systematic guide should include a structured method for developing conceptual models, a notation for representing conceptual models and preferably a software tool that supports this notation. Systematic guidelines provide an orderly way for establishing and maintaining conceptual models and create the link between the conceptual analysis phase with the other activities of the simulation system development lifecycle. Modeling notations are the communication media among the different stakeholders in a simulation system development project. It is essential to use a syntactically and semantically well-defined notation for developing consistent, verifiable, and easy-to-understand conceptual models.

This chapter first outlines the uses of conceptual modeling in the simulation system development lifecycle and describes some of the existing approaches, frameworks and methods related with conceptual modeling. Although these approaches, frameworks, and methods may include extensive facilities, this chapter focuses specifically on their conceptual modeling-related aspects. FEDEP, CMMS (FDMS), DCMF, Robinson's framework, and KAMA framework are described, and a comparison of these methods is provided together with future work opportunities.

7.1.1 Uses of Conceptual Modeling

As Lacy has stated in Lacy et al. (2001), the *conceptual model* is an overloaded term. There exist many close but different definitions, which will not be elaborated in this chapter. Most of these definitions agree that it is essential to develop conceptual models at the early stages of simulation system development life cycle (Balci 1994, DMSO 1997, Pace 1999, IEEE 2003, Mojtahed et al. 2005). Conceptual models are utilized for different purposes in simulation system development lifecycle. Conceptual models are developed to better understand the intended system, depict the requirements of the simulation system or as a basis for verification and validation of simulation systems.

Conceptual modeling is related with the problem domain and can be used as a tool to identify the components of the problem in terms comprehensible to both modelers and domain experts (Sargent 1987, Johnson 1998, Mojtahed et al. 2005, Robinson 2007b). Robinson highlights the importance of following a structured problem definition process in order to define a combined view effectively. He describes possible scenarios in understanding the problem situation and offers simple methods. In cases where the "problem situation is neither well understood nor [clearly] expressed," he suggests that formal problem structuring methods should be utilized (Balci and Nance 1985, Robinson 2007b) define a methodology for problem formulation in simulation system development. Mojtahed et al. (2005) take this approach one step further and treat the conceptual analysis phase as a knowledge engineering activity.

Problem definition in simulation system development is a part of the requirements analysis phase; therefore conceptual models are the products of this phase. Specifications, assumptions, and constraints related with the domain of interest should be included in the conceptual model. These specifications may include the entities, tasks, actions, and interactions among the entities, which will form a basis for the design phase (DMSO 1997). Pace (1999) describes the conceptual model as a bridge between requirements analysis and design phases. Since the boundaries of these phases cannot be sharply defined, there is confusion over whether the conceptual model is a product of the user or the designer (Haddix 1998). In order to reduce this confusion, Haddix defines a conceptual model as "the ultimate definition of the requirements" and uses another term, conceptual design, to mean "initial descriptions of the system's implementation." However, SISO (2006a) disagree with these definitions in its BOM (Base Object Model) standard, stating that the BOMs are defined to "provide an end-state of a simulation conceptual model and can be used as a foundation for the design of executable software code and integration of interoperable simulations."

Being a product of the requirements analysis phase, conceptual models should be independent of the software design and implementation decisions (Sheehan 1998, Pace 1999a, IEEE 2003). This aspect of the conceptual model is based on a software development viewpoint. Johnson (1998) introduces a slightly different aspect of the conceptual model as providing a "simulation-neutral view of the real world." He suggests that the simulation system–specific attributes, even if they are not related with the design phase, should be kept out of a conceptual model. Thus, the conceptual model should include the definitions of a simulation system and it can be realized by different simulation implementations.

It is an established practice in the software engineering field to initiate verification and validation activities as early as possible in the software development life cycle. Software requirements specification is used for ensuring that the developers are producing what the customer really wants. Similarly, early validation of a simulation system is essential for the success of a simulation system development project. Conceptual models can be used as a basis for verification, validation and accreditation activities (Sargent 1987, Haddix 1998). Sargent underlines that the conceptual model should be structured enough to provide means for validation. However, a more thorough validation will be possible using experimentation after the simulation system has been completed. Any defects found during verification and validation activities should be corrected by revisiting the prior phases including the conceptual modeling phase. Hence, conceptual modeling is not a one shot process but rather an iterative one that should be performed in many cycles throughout a simulation system development study (Balci 1994, Willemain 1995).

7.2 Conceptual Modeling Frameworks, Notations, and Techniques

Numerous framework definitions have been proposed covering the various perspectives established in the modeling and simulation community on conceptual modeling. The CMMS (Conceptual Models of the Mission Space) project originated by the US Department of Defense is one of the first initiatives providing detailed guidance on conceptual model development activities. Prior to this work, there were detailed framework definitions (Shannon 1975, Balci and Nance 1985, Zeigler et al. 2000) regarding the whole simulation development lifecycle, but with less guidance on the conceptual modeling phase. The latest framework definition, which belongs to Robinson, includes detailed definitions for developing conceptual models Robinson (2007b). Robinson's work is distinct from others in that it is based on the business-oriented rather than the military domain.

The conceptual modeling phase begins with understanding the problem situation and defining the context (DMSO 1997, Pace 1999a, IEEE 2003, Mojtahed et al. 2005, Robinson 2007b). The context definition phase includes determining the simulation system objectives, identifying the authoritative information sources and defining the assumptions and constraints used to develop the conceptual model to satisfy the simulation system objectives. This context definition is called as mission space (DMSO 1997, Pace 1999a, 1999b, IEEE 2003, Mojtahed et al. 2005) underlines the importance of recording a history of changes made on the assumptions and constraints. The next phase is developing the content, which includes defining the entities, tasks, interactions, inputs, outputs and relationships among all these elements. The output of this phase may be structured information represented in plain text, tables or diagrams that will provide a basis for understanding, sharing, and reviewing.

Although a well-defined representation technique is necessary, some frameworks prefer not to impose a specific conceptual modeling notation (Pace 1999b, IEEE 2003, Robinson 2007b) describes four approaches for documenting conceptual models, which are ad hoc method, design accommodation, CMMS paradigm, and scientific paper approach, all based on free text descriptions. Robinson (2007b) uses a tabular structure for representing conceptual models, while also mentioning the usefulness of the diagrammatic representation such as process flow diagrams (Robinson 2004). However, the free text notation causes ambiguous and recurrent definitions, is unsuitable for machine interpretation, and does not provide adequate guidance to the modeler (Sudnikovich et al. 2004). Recent studies promote the utilization of UML and SysML for conceptual modeling as stated in Borah (2007) and Globe (2007). BPMN and IDEF1X are two other alternative notations that are explained in BPMI (2004) and IEEE (1998). Ryan and Heavey (2006) use simulation activity diagrams. These notations provide different approaches; UML follows a more object-oriented approach, BPMN is more process-oriented and IDEF1X is a more data-oriented approach. Mojtahed et al. (2005) introduced the KnowledgeMetaMetaModel (KM3) as both a language and a tool to construct conceptual models. They state that their intent in developing KM3 was not to construct a unified model description language, but rather provide a way to "capture system structures and behavior in an object-oriented and rule-based way." The abstract syntax of KM3 is defined as a class-diagram and the concrete syntax is textually represented. Mojtahed et al. (2005) state that various graphical representations can be used; however, they do not define a method explaining how to associate these graphical representations with the concrete syntax.

Shannon (1975) claims that modeling is more of an "art" than "science"; therefore, it is generally assumed difficult to define methodical ways to develop conceptual models. However, following disciplined and systematic methods leave more room for creative skills by reducing the amount of routine work. The evolution of newer engineering fields, such as systems and software engineering has shown that using well-defined modeling notations, following defined processes and utilizing software tools definitely improve effectiveness. Hence, conceptual modeling frameworks should include these three elements.

Robinson (2007b) mentions a range of other methods and notations that are used to represent conceptual models. Process flow diagrams, activity cycle diagrams, Petri nets, event graphs, digraphs, the UML (Unified Modeling Language), object models and simulation activity diagrams are examples of these notations. UML is one of the most popular modeling languages used for analysis and design of software. As a result of its success in the software modeling field, UML has been utilized in domains other than software such as systems modeling, business process modeling, data modeling and software process modeling. Although UML and SysML (System Modeling Language), which is an extension of UML, have also been used for developing conceptual models, they have not been extensively used (Richter and Marz 2000, Globe 2007).

7.2.1 KAMA Conceptual Modeling Framework

KAMA is a framework for developing conceptual models of the mission space that includes a process definition (method) for guiding the conceptual modelers and the domain experts, a notation for representing the conceptual models and a tool for supporting the process and the notation (Karagöz and Demirörs 2007, Karagöz 2008). It was developed as part of a research project performed with the collaboration of the academia, industry and military and the framework was validated through case studies (Karagöz 2008) and real life simulation system development projects (Karagöz et al. 2008).

7.2.1.1 KAMA Method

The KAMA method consists of four tasks that are depicted as a task flow diagram in Figure 7.1. The ellipses represent the tasks to be performed by the roles, which are represented as stick figures connected to the tasks with dashed arrows. The inputs to and outputs from the tasks are represented by work products that are shown as rectangles connected to the tasks with dotted arrows. The straight arrows represent the control flow, and the variations in this flow are shown by diamond-shaped decision points. The modeler can identify variations at these decision points; however, the outgoing control flows must have mutually exclusive guard conditions. These tasks are performed with the collaboration of the conceptual modeler, the domain expert, the sponsor and the reviewer.

The first task is knowledge acquisition (KA) about the mission space for which a conceptual model will be developed. As the modeler will reflect her knowledge and experience about the domain onto the conceptual model, it is essential that she use the right and accredited knowledge resources.

The first activity in this task is to identify the high-level simulation system objectives that should be defined in terms of units as measurable as possible.

The second activity is to define the boundaries of the mission space aligned with the simulation system objectives. This may include the high-level needs, assumptions and constraints of the sponsor, the fidelity requirements of the simulation system, and the risks related with this information. Identifying the sources of authoritative information, analyzing these information sources and searching for similar conceptual models are other activities performed in this task.

The second task is defining the context via mission space diagrams, which include missions, roles, objectives, measures and the relationships among them. The missions are high-level tasks that define the boundaries of the conceptual model, which are then detailed by task flow diagrams. Each mission is related with an objective to assure that there is not any irrelevant concept within the mission space. The achievement of an objective is determined by evaluating the quantifiable measures that are linked to an objective.

After the modeler acquires information about the mission space and specifies the context with the assistance of the domain expert, she develops the content. The behavioral features of the conceptual model are shown by using task flow diagrams and entity state diagrams. The structural features are shown by using entity-ontology, entity relationship, command hierarchy

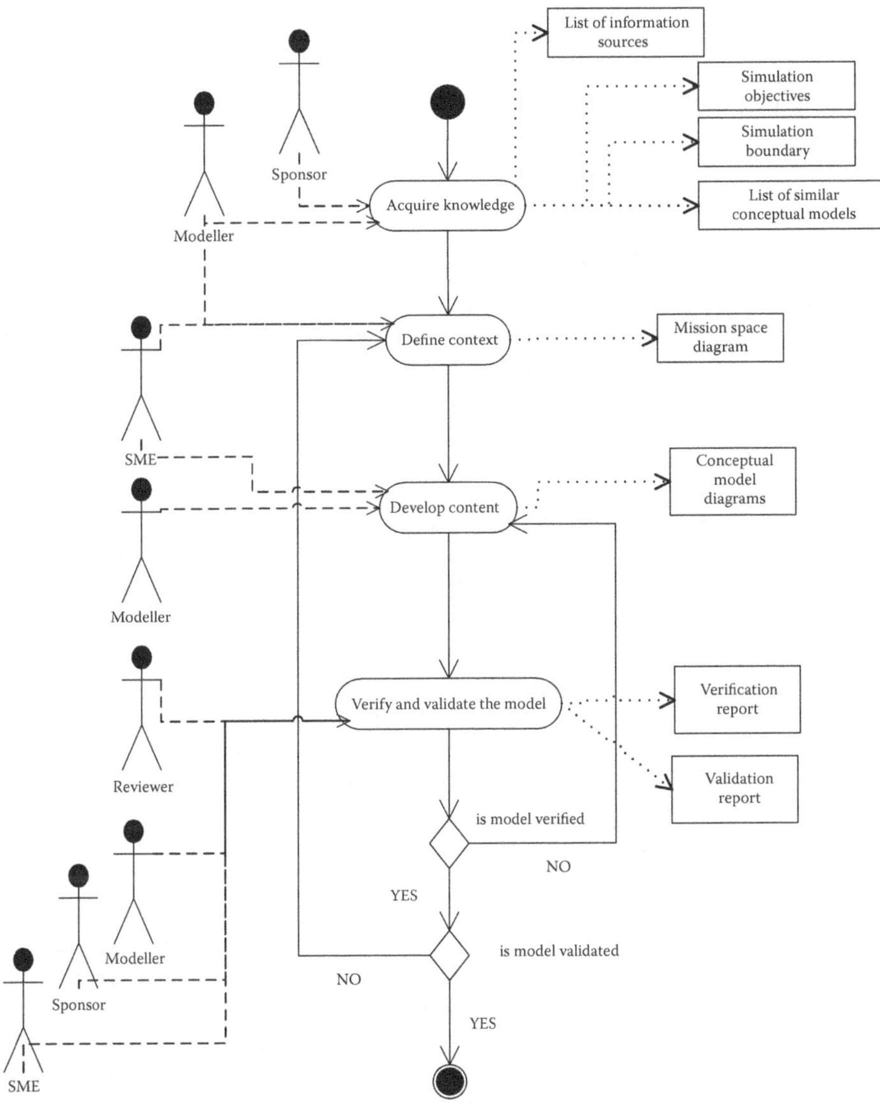

FIGURE 7.1
Flow diagram for the KAMA method. (Based on Karagöz, N.A., A Framework for Developing Conceptual Models of the Mission Space for Simulation Systems, PhD thesis, Middle East Technical University, Department of Information Systems, 2008.)

and organizational structure diagrams. These diagram types are explained in the following section.

After the conceptual model has been finalized, the model is verified and validated with the cooperation of the modelers and domain experts. Both syntactic and semantic rules are checked during the verification process.

Tanriover and Bilgen (2007) has proposed a three-step inspection approach for verifying the conceptual models developed using the KAMA framework (see chapter 15). Usually, the validation of a conceptual model is accomplished via walkthroughs in which the conceptual model diagrams are presented in a step-by-step manner and anomalies are recorded for further analysis and resolution.

7.2.1.2 KAMA Notation

The deficiencies of existing notations, such as the lack of domain-specific modeling approach, the inadequacy of the notations for representing the conceptual models and difficulties the domain experts face in understanding ad hoc notations have all led to the design of the KAMA notation.

The conceptual models should serve the needs of the analysis phase and should be used as an input to the design phase of the simulation system development life cycle. In order to comply with the first need, the modeling notation should be usable and understandable by the domain experts and for the second need the developed models should be transferable to the design phase by some means. KAMA notation constitutes a domain-specific graphical modeling language designed for the conceptual modelers. The jargon of the conceptual modelers is taken into account in defining the language elements. It is based on UML to facilitate the transfer of the knowledge embodied in the conceptual model to the simulation system design phase. As a further development on the framework, an extension for design is provided in Aysolmaz (2007). An easy-to-use domain-specific notation has been provided by modifying the syntax and semantics of some UML elements and by omitting complex structural and behavioral features.

The metamodel diagram of the KAMA notation includes all of metamodel elements that can be used within a KAMA conceptual model as shown in Figure 7.2. The KAMA metamodel uses packages and a hierarchical package structure to reduce complexity, promote understanding, and support reuse. The dependency of KAMA packages on the UML metamodel is encapsulated with the Foundation package, which includes metamodel elements that are directly inherited or derived from the UML metamodel. It is a subset of the UML and provides basic constructs for creating and describing metamodel classes for other KAMA packages.

The Mission Space package extends the Foundation package and includes metamodel elements that are used to represent the missions and tasks in a mission space. The Structure package is based on the Foundation package and includes the metamodel elements that are used to represent the static structure of a conceptual model. The metamodel elements and their relationships in these packages are shown in Figure 7.3 and described in detail in Karagöz (2008).

There are seven types of diagrams used for representing the structural and behavioral views of KAMA conceptual models, which are mission space,

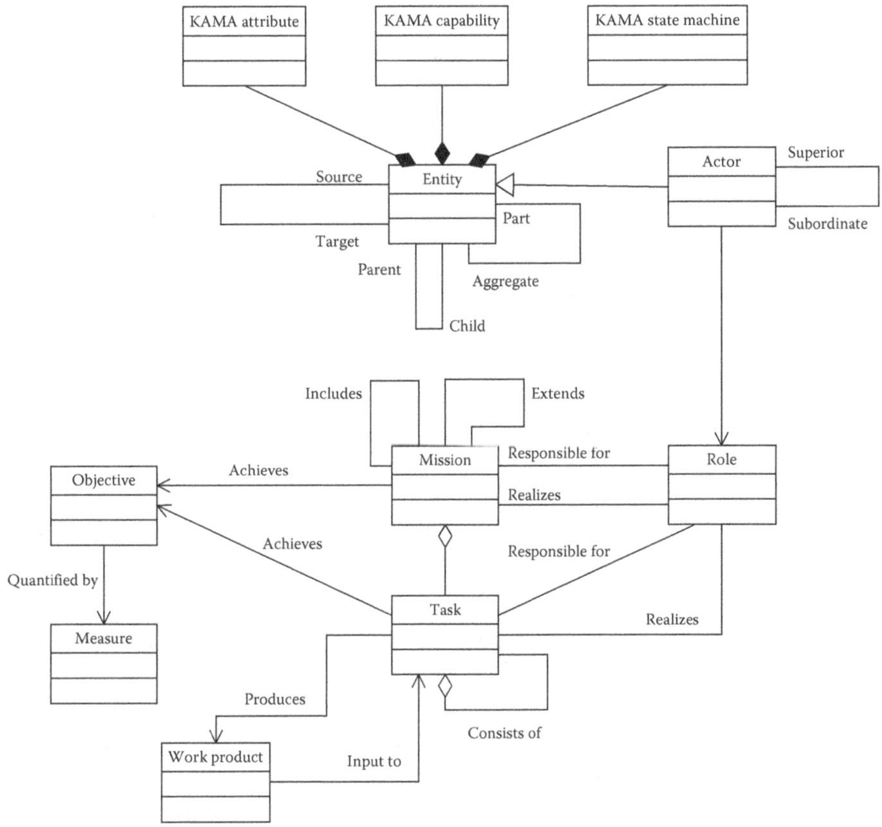

FIGURE 7.2
KAMA metamodel elements. (From Karagöz, N.A., A Framework for Developing Conceptual Models of the Mission Space for Simulation Systems, PhD thesis, Middle East Technical University, Department of Information Systems, 2008.)

task flow, entity ontology, entity relationship, entity state, command hierarchy, and organization structure diagrams.

A sample mission space diagram presented in Figure 7.4, which looks similar to a UML use case diagram, shows the high-level missions of a package in a simulation system. Three roles have been specified that are responsible for or in charge of realizing the missions. Roles may stand for real life people such as commander in our sample or actively participating entities such as a sensor or a platform. Perform Mine Hunting mission includes the Detect Mines mission and is extended by two different missions. The extending missions are (a) Hunt Mine With Unmanned Undersea Vehicle (UUV) and (b) Hunt Mine With Acoustic Mine. These missions share the same objective but use different techniques or tools. UUVs are used to destroy mines by remote operations. Acoustic mines are used to trigger and destroy other mines by making them explode using acoustic waves. For each mission to be

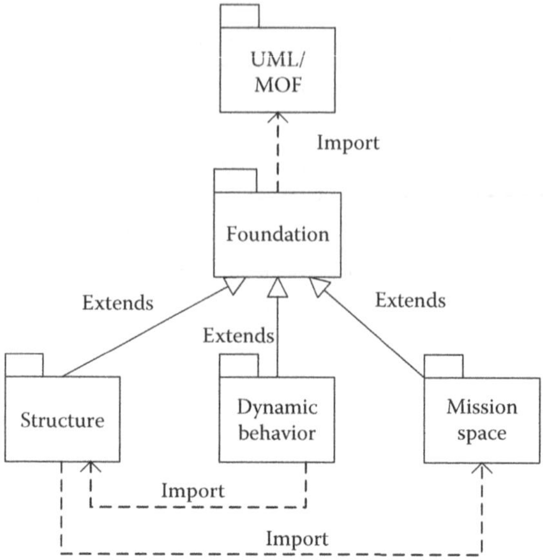

FIGURE 7.3
KAMA package hierarchy. (From Karagöz, N.A., A Framework for Developing Conceptual Models of the Mission Space for Simulation Systems, PhD thesis, Middle East Technical University, Department of Information Systems, 2008.)

accomplished mines should be detected first, which means the inclusion of the Detect Mines mission. The "includes" relationship shows that the including mission requires the execution of the included mission. At some point during the flow of the including mission, the Detect Mines mission will be called for execution. The KAMA framework requires that all of the missions be detailed using task flow diagrams. The two extending missions extend the Perform Mine Hunting mission at the task numbered 5 and specified as "extensionId" on the diagram. The "extensionId" is used when the extended mission has more than one extension points.

The missions are assigned "Objectives" that should be achieved for the successful execution of the mission. An objective can be shared among more than one mission, as is the case with the specified_duration objective. The achievement of an objective is determined by evaluating the quantifiable measures related with an objective. The measures elapsed time, detected area, and the number of mines detected are used to determine the success of the achievement of the specified_area objective. The unit of measure information is stored as an attribute of each measure and is used as a parameter in the "performanceCriteria" attribute of the related objective. An example to this attribute may be "(number of mines detected > = 6) AND ((detected area > = 10 acres) OR (elapsed time = 2 hours))."

The sample task flow diagram presented in Figure 7.5 shows the details of the mission Perform Mine Hunting. The flow of the tasks begins with the

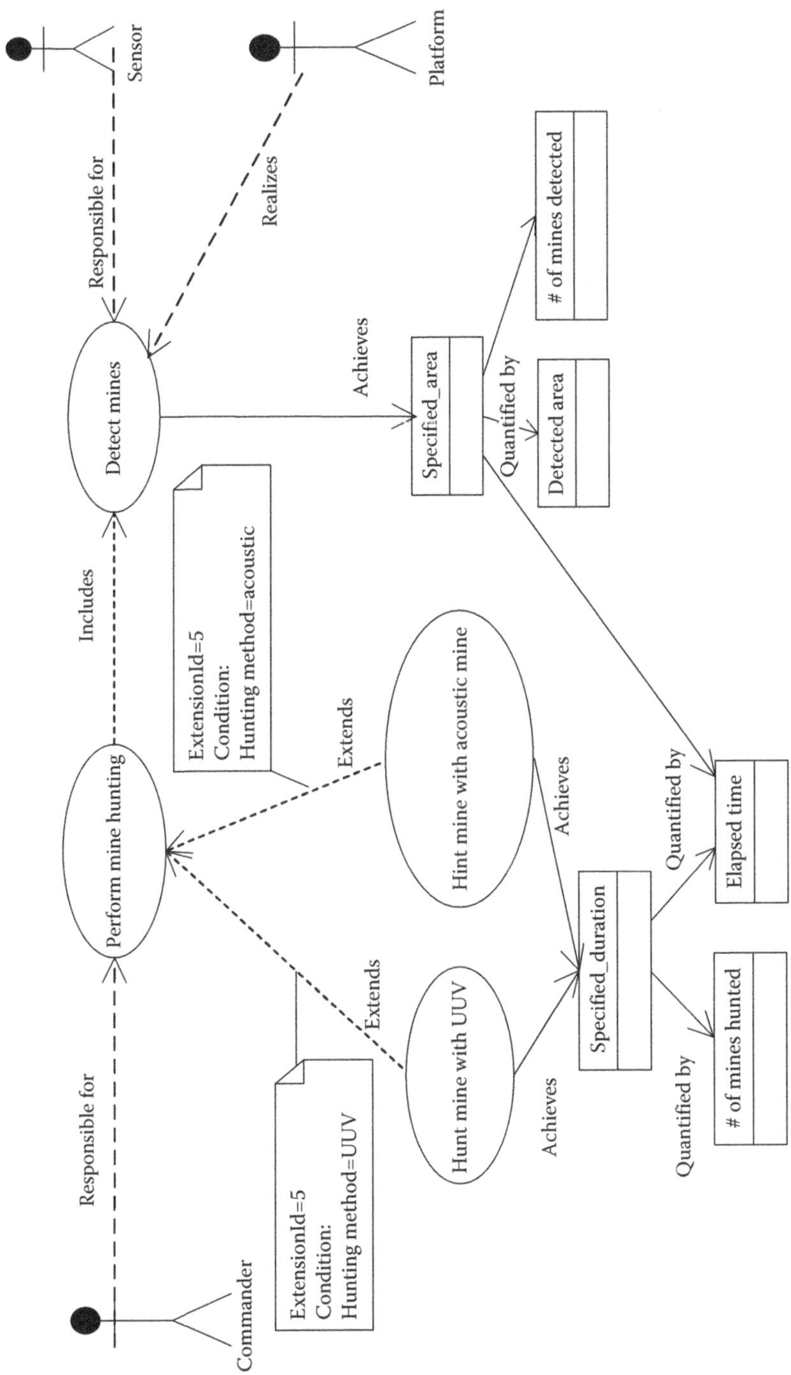

FIGURE 7.4
Example mission space diagram. (From Karagöz, N.A., A Framework for Developing Conceptual Models of the Mission Space for Simulation Systems, PhD thesis, Middle East Technical University, Department of Information Systems, 2008.)

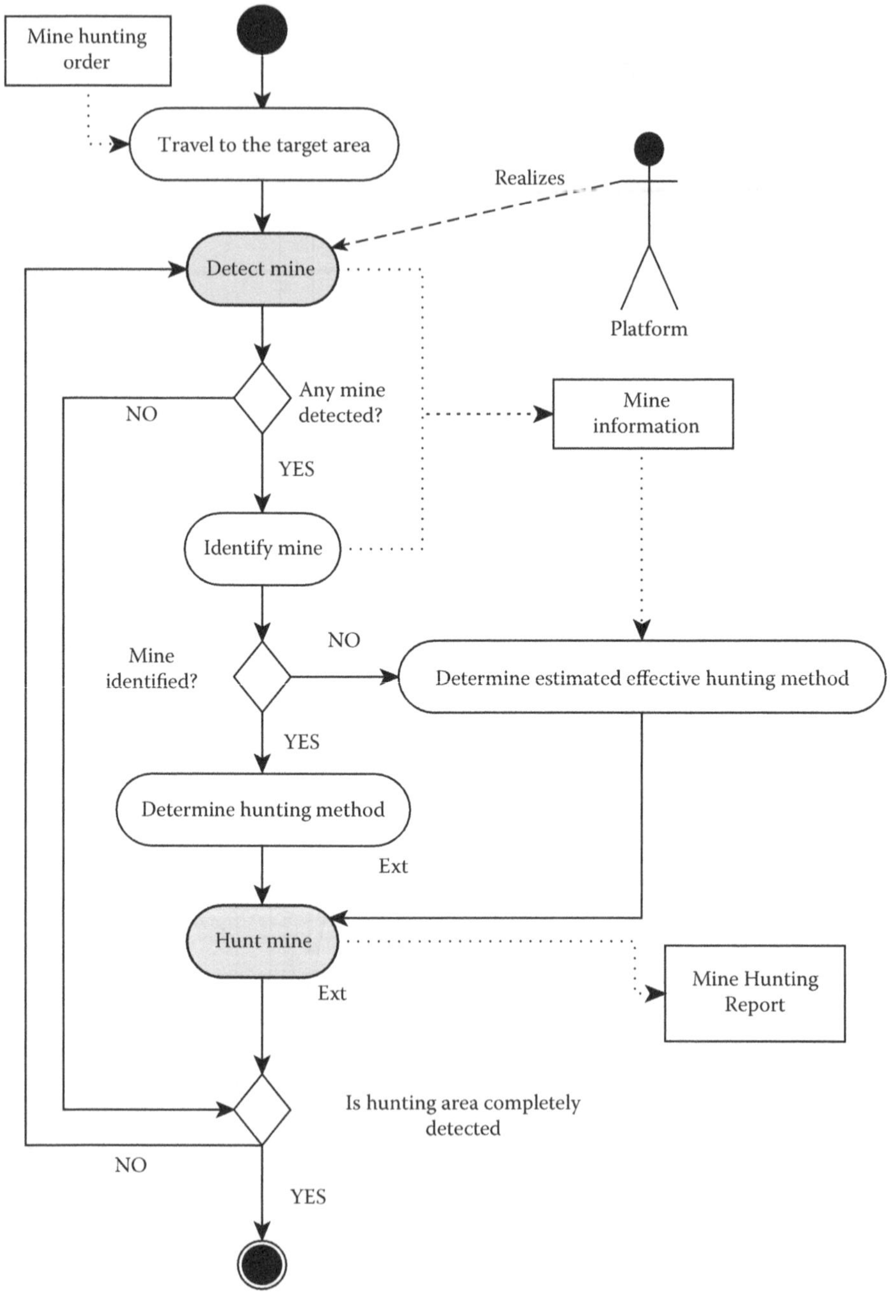

FIGURE 7.5
Example task flow diagram. (From Karagöz, N.A., A Framework for Developing Conceptual Models of the Mission Space for Simulation Systems, PhD thesis, Middle East Technical University, Department of Information Systems.)

arrival of the Mine Hunting Order input. All of these tasks except for "Detect mine" are realized by the Commander role as specified in the Mission Space diagram, therefore the "realizes" relation is shown only for the "Detect mine" relation on the diagram.

The shaded tasks denote the existence of task flow diagrams, which include the details of these tasks. The Mine Information output, which is partly produced by the Detect Mine task and then updated by the Identify Mine task, is used as an input to the "Determine estimated effective hunting method" task. The Hunt Mine task is an extension point with extensionId equals to five. The extensionId information is not shown on the diagram but recorded as an attribute of the task. The two extending missions that are shown in Figure 7.4 extend the Perform Mine Hunting mission depending on the selected hunting method. A Mine Hunting Report is produced as a result of the execution of the Hunt Mine task. The variations in the task flow diagram are represented with decision points, which may have any number of outgoing control flows. However, the guard conditions shown on each outgoing control flow must not contradict with each other.

7.2.1.3 KAMA Tool

In order to facilitate efficient utilization of the KAMA method and the notation, a tool was developed to support the conceptual modeler (Karagöz 2008). It consists of a graphical modeling editor and a conceptual model repository. Conceptual modelers can define and manage conceptual model elements, develop and navigate in conceptual model diagrams, share conceptual models via a common repository, perform search among the existing conceptual models and diagrams, so that they can look for reusable conceptual model elements and diagrams. The tool also provides basic configuration management functionality such that versioning of conceptual model elements and diagrams, keeping change history records and baselining of conceptual models. Predefined verification rules can be executed by the tool and anomalies are reported. Conceptual model diagrams can be examined through different perspectives by using fish-eye and hyperbolic views. Conceptual model elements and diagrams can be exported and then imported as XML (Extensible Markup Language) files, which enables sharing conceptual models with other modeling tools.

7.2.2 Federation Development and Execution Process (FEDEP)

FEDEP (Federation Development and Execution Process) (IEEE 2003) defines the processes and procedures that are intended for HLA (High Level Architecture) (IEEE 2000) users. It does not aim to define low-level management and systems engineering practices native to HLA user organizations but rather aims to define a higher-level framework into which such practices can be integrated and tailored for specific use. Single simulation system in a federation is called a federate and federation can be defined as a set of simulation systems. Therefore, the federation conceptual model includes the conceptual models of the simulation systems that make up the federation. The purpose of FEDEP architecture is to facilitate interoperability among simulation systems and promote reuse of simulation systems and their components.

FEDEP provides process definitions that encourage the use of conceptual models in the simulation system development life cycle. These definitions have originated at the Department of Defense and then standardized by IEEE as recommended practices. The designers of HLA identified the need for a flexible process according to which HLA applications will be developed. The main idea was avoiding unnecessary constraints on the construction and execution processes because these processes could vary significantly within or across user applications. However, it was then realized that it is possible to define a process at a more abstract level which should be followed by all HLA developer organizations. Figure 7.6 shows the top level view of this FEDEP process flow which includes seven steps. The focus will be on the first two steps in the scope of this chapter.

Step 1—Define federation objectives: The federation user, the sponsor, and the federation development team define and agree on a set of objectives and document what must be accomplished to achieve those objectives. This step includes two key activities: (a) identify user/sponsor needs and (b) develop objectives. The aim of the first activity is to develop a clear understanding of the problem to be addressed by the federation; therefore it is essential for developing a valid conceptual model. This understanding is generally recorded as a "statement of needs" document, which may vary widely in terms of scope and degree of formalization.

Despite this variance, it is generally accepted that this document should at a minimum include; high-level descriptions of critical systems of interest, expectations about required fidelity and required behaviors for simulated entities, key events that must be represented in the federation scenario and output data requirements. In addition, the needs statement should indicate the constraints related with the development of the federation such as funding, personnel, tools, facilities, due dates, nonfunctional requirements, etc.

The needs statement is then analyzed and refined into a more detailed set of specific objectives for the federation. The aim of this activity is developing more concrete, complete and measurable goals and also performing an early assessment of feasibility of the federation and risks related to development. This activity requires a close collaboration between the federation sponsor and the federation developer.

Step 2—Perform conceptual analysis: The second step of FEDEP includes development of the conceptual model. FEDEP defines the conceptual model as "an abstraction of the real world that serves as a frame of reference for federation development by documenting simulation-neutral views of important entities and their key actions and interactions." The federation conceptual

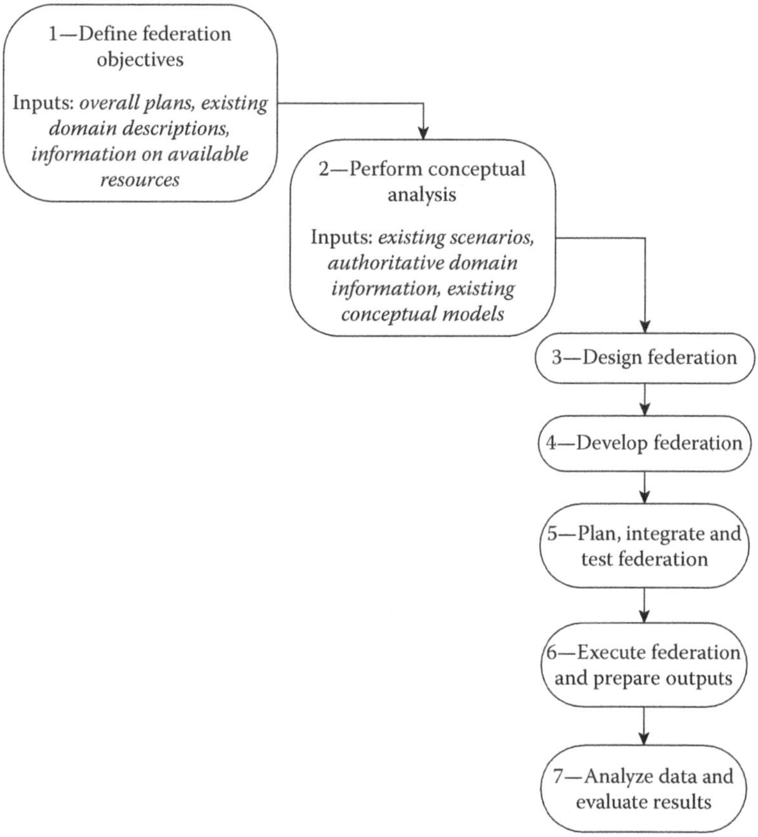

FIGURE 7.6
FEDEP high-level process flow. (Reproduced from IEEE Computer Society, IEEE 1516.3, Recommended Practice for High Level Architecture (HLA) Federation Development and Execution Process (FEDEP), 2003.)

model is defined as "the document that describes what the federation will represent, the assumptions limiting those representations, and other capabilities needed to satisfy the user's requirements. Federation conceptual models are bridges between the real world, requirements, and design."

In order to comply with these definitions, this step of the FEDEP process begins with developing federation scenarios that are based on the federation requirements. The relationship between the activities of this step, the consumed inputs and produced outputs are depicted in Figure 7.7. Federation scenarios define the boundaries of conceptual modeling activities. Authoritative information sources should be identified prior to scenario construction. A federation scenario includes "the types and numbers of major entities that must be represented by the federation, a functional description of the capabilities, behavior, and relationships between these major entities over time,

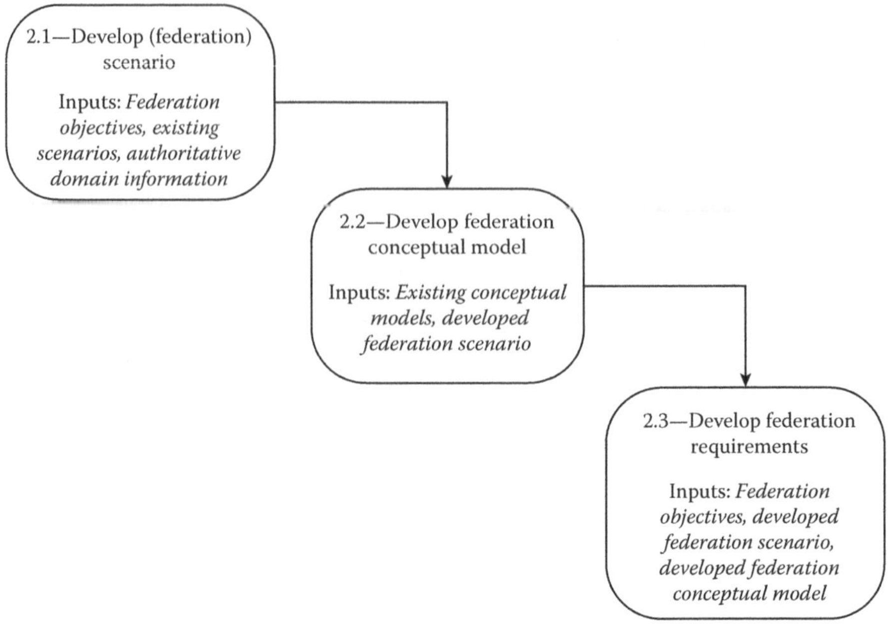

FIGURE 7.7
Perform conceptual analysis steps. (Reproduced from IEEE Computer Society, IEEE 1516.3, Recommended Practice for High Level Architecture (HLA) Federation Development and Execution Process (FEDEP), 2003.)

and a specification of relevant environmental conditions that impact or are impacted by entities" in the federation. Initial conditions (e.g., geographical positions for physical objects), termination conditions, and specific geographic regions should also be provided.

Following the scenario definition, a conceptual representation of the problem space is developed based on the interpretation of the user needs and federation objectives. The federation conceptual model provides an implementation-independent representation that serves as a vehicle for transforming federation objectives into functional and behavioral descriptions for system and software designers. The model also provides a crucial traceability link between the stated federation objectives and the eventual design implementation. This model can be used as a basis for the later federation development steps and can highlight problems early through a validation process that involves the user/sponsor.

The federation conceptual model development activities start with defining entities, actions and the assumptions and limitations regarding these. Defining entities include identifying static and dynamic relationships between entities, and identifying the behavioral and transformational (algorithmic) aspects of each entity. Static relationships may include association that shows a simple relationship among any entity, generalization that shows is-a relationship, and aggregation that shows part-whole relationship. Dynamic relationships may include temporally ordered sequences of entity interactions with associated trigger conditions. The characteristics of entities and interaction parameters used among entities may also be defined during this activity. FEDEP does not impose any specific notation for representing conceptual models, but states that it is important that this notation should be appropriate to develop conceptual models that provide insight into the real-world domain.

Since the objective of a conceptual model is to represent the real-world domain, it should be reviewed by the user/sponsor to ensure the adequacy of domain representation. As most of the researchers have pointed out, conceptual modeling is not a one-shot activity but an iterative one, therefore any changes caused by these reviews should be performed under control. As the conceptual model evolves, it is transformed from a general representation of the real-world domain to a more specific expression of the capabilities of the federation as constrained by the federates and available resources.

FEDEP is one of the first initiatives to emphasize the position of conceptual modeling in the simulation system development lifecycle, together with its boundaries, inputs and outputs. However, it does not provide sufficient detail on conceptual modeling to enable modelers to develop conceptual models. Not imposing a modeling notation enables FEDEP to be flexible and convenient for strategic level modeling, but on the other hand decreases its usability, especially at the tactical level. Modelers need more detailed guidance and illustrative conceptual models. They need to see the concrete relationship between conceptual model and simulation system requirements and design and data to show the return on investment resulting from the utilization of a conceptual model.

7.2.3 Conceptual Models of the Mission Space (CMMS)

The US Defense Modeling and Simulation Office (DMSO) has initiated a project to promote the use of conceptual models and has led an effort to provide an integrated framework and toolset for developing CMMS (Conceptual Models of the Mission Space) (DMSO 1997). The main goal of the project was to resolve the interoperability problems among many simulation system development projects developed in many different platforms. DMSO defined the CMMS as "the first abstractions of the real world that serve as a frame of reference for simulation system development by capturing the basic information about important entities involved in any mission and their key actions and interactions." CMMS provides the following:

- A disciplined procedure by which the simulation developer is systematically informed about the real-world problem to be synthesized
- An information standard the simulation domain expert employs to communicate with and obtains feedback from the military operations domain expert

- A real world, military operations basis for subsequent, simulation-specific analysis, design, and implementation, and eventually verification, validation, and accreditation/certification
- A singular means for identifying reuse opportunities in the eventual simulation implementation by establishing commonality in the real-world activities

The objectives of the CMMS are to "enhance interoperability and reuse of models and simulations by accessing descriptions of real-world operational entities, their actions, and interactions through the identification of authoritative sources of information for models and simulations and the integration of information from independent KA sources" as specified in DMSO (2007).

CMMS defines mission spaces for each mission area and aims to develop conceptual models for these mission spaces. The MSM (Mission Space Model) is a by-product of a simulation's front-end analysis; it is a simulation and implementation independent functional description. These functional descriptions represent a view of real-world operations, entities, actions, tasks, interactions, environmental factors, and relationships among all of them. Since CMMS is defined for all stakeholders, it serves as a bridge between the domain experts and the developers. Domain experts act as authoritative knowledge sources when validating the MSMs.

CMMS is a framework that includes tools for gathering and storing knowledge, reusing this knowledge and providing a common repository for information storage and tools for conversions among various model representation notations. CMMS is composed of four main components:

- Mission space models: consistent representations of real-world military operations
- Technical framework: standards for knowledge creation and integration, includes:
    - A common syntax and semantics for describing the mission space
    - A process definition for creating and maintaining conceptual models
    - Data interchange standards for integration and interoperability of MSMs
- Common repository: a DBMS for registration, storage, management, and release of conceptual models
- Supporting tools, utilities, and guidelines

According to CMMS process definition, the four basic steps in conceptual model development are as follows:

- Collect authoritative information about the simulation context.
- Identify modeling elements (entities and processes) to be represented in the simulation system.
- Develop modeling elements (representational abstraction).
- Define interactions and relations among modeling elements.

CMMS has been renamed as FDMS (Functional Descriptions of the Mission Space) but the project faded away by the end of 1999 and the term FDMS also have fallen out of use within a few years after that.

The CMMS project focused on resolving the interoperability issues caused by the usage of many different conceptual modeling notations such as IDEF1X (IEEE 1998), notations provided by general purpose case tools and legacy text formats by establishing a common data interchange format. The technical framework included a common data dictionary for representing conceptual models, common representation templates, and tool-specific style guides for managing various modeling tools used for conceptual modeling.

7.2.4 Defense Conceptual Modeling Framework (DCMF)

In 2003, The Swedish Defense Research Agency initiated a project to further study the conceptual modeling concepts and improve the CMMS. As they made progress in their research, they realized that they were moving further from the original CMMS concepts and renamed the project as DCMF (Defense Conceptual Modeling Framework) (Mojtahed et al. 2005). The major tasks were analyzing the CMMS in depth, studying the KA and elicitation phases, analyzing the language issues such as ontology, terminology, common syntax and semantics and developing new methods when required.

As part of the project, Lundgren et al. (2004) identified the problems and limitations related with CMMS and proposed solutions to these problems as listed below:

- Unsupported KA: A complete methodology for KA in this domain should be developed.
- Lack of clarity of modeling elements: Modeling elements should be logically grouped by using metalevels or abstraction levels.
- Need for alternative knowledge representations (KR): A new method should be developed for KR.
- Limitations of processes: An action-centric approach should be used to add dynamic knowledge without limiting the process.

Based on these findings, the objectives of DCMF were defined as "to capture authorized knowledge of military operations, to manage, model and structure the obtained knowledge in an unambiguous way; and to preserve and maintain the structured knowledge for future use and reuse." DCMF stipulates that the conceptual model should be "(a) well documented, (b) readable and usable for a person as well as a machine, (c) composable (includes units that can be composed to form more comprehensive conceptual models), (d) traceable the whole way back to the original sources, and finally (e) usable as a basis for simulation models." Although the DCMF documentation does not include a thorough definition, the simulation model is described as a more detailed model, which may include design-specific information built on the conceptual model.

The main purpose of the DCMF can be described as "to facilitate and support development, reuse and interoperability between simulation models." DCMF accomplishes this purpose by (a) providing a common language for all stakeholders and serving as a bridge between the military experts and the developers, (b) creating libraries of validated conceptual models with certified quality levels, and (c) using the KM3 language as an enabler for transforming the conceptual model into other formats.

The DCMF project outcomes included a process definition, a language definition, and a list of available tools and analysis methods related with conceptual modeling. One of the major improvements of DCMF over CMMS is DCMF's knowledge engineering focus on the conceptual analysis phase. The DCMF process consists of four main phases as shown in Figure 7.8.

Knowledge Acquisition (KA) is the learning phase that focuses on acquiring information and knowledge. DCMF defines this phase in 3 steps; the first step includes the determination of the focused context, the second step is the identification of the authorized knowledge sources, and the third step includes the actual acquisition of knowledge, which is sometimes called knowledge elicitation. DCMF suggests the use of structured and well-documented techniques for knowledge elicitation such as interviews, prototyping, questionnaires, etc. It is also noted that these kinds of analyses may include linguistic processes. A typical linguistic process includes phonetic, lexical, morphological, syntactic and semantic analyses of existing documents or voice records that are used as information sources.

Knowledge Representation (KR) phase aims to analyze the structure of the information and formalize the acquired information. The human readable and probably ambiguous information is transformed into a machine

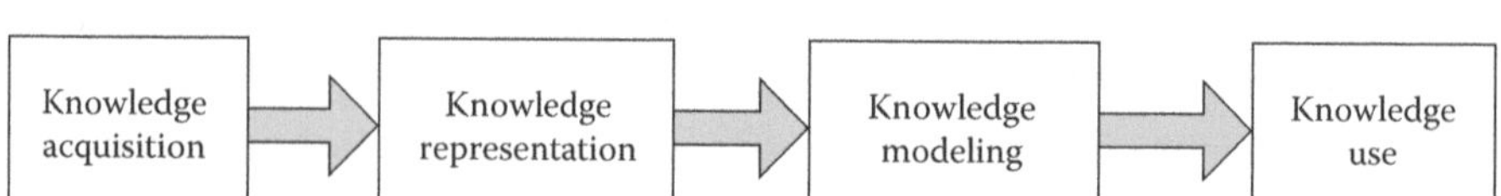

FIGURE 7.8
DCMF process: Main phases. (Reproduced from Mojtahed, V., Lozano, M.G., Svan, P., Andersson, B., and Kabilan, V., *Technical Report*, FOI-R—1754—SE, 2005.)

readable and unambiguous format. This can be done by using methods like SPO (Subject-Predicate-Object), 5Ws (Who-What-Where-When-Why), and KM3, which are explained in Mojtahed et al. (2005). These analyses will result in an ontology, which consists of the context of the domain, the definitions of terms and their relationships and interactions.

Knowledge Modeling (KM) phase focuses on the semantic analysis and modeling of the information. Although previous KR phase may produce usable artifacts, building a common general model at the right level of abstraction requires further study. Different models can be generated based on the same set of data. These models should be suitable for future use and reuse. In order to provide this facility, the DCMF proposes using knowledge components that represent smaller knowledge parts. This approach provides flexibility, increases the rate of reuse and composability of conceptual models. Knowledge modeling also involves the merging of these knowledge components or conceptual models; therefore it will be a good idea to store these artifacts in a knowledge repository.

The last phase of the DCMF process is Knowledge Use (KU), which deals with the actual use of the artifacts produced as a result of the previous phases. DCMF suggests using effective mechanisms that provide different visualizations of the knowledge for various users. These users may include the sponsor, consumer, producer and controller. The original intent of a knowledge component and any changes made to it should be recorded for an effective usage mechanism.

KM3 is at the same time a specification, a tool and a language. KM3 is a specification for the creation of generic and reusable conceptual models. It is a tool for structuring knowledge in the form of generic templates. It is a common language that enables different stakeholders in developing conceptual models. KM3 follows an activity-centric approach and represents activities as KM3 actions. KM3 specification includes both static and dynamic descriptions. The static descriptions are specified by the attributes of an object whereas the dynamic descriptions are specified by the inclusion of rules into the object descriptions. All changes to model elements are described by rule definitions, which specify the conditions under which an action starts and ends. A rule is composed of an activity role and an atomic formula. Atomic formulas can be combined conjunctively (OR-Connections) or disjunctively (AND-Connections) to create complex formulas.

7.2.5 Base Object Model (BOM)

BOM (Base Object Model) is a SISO (Simulation Interoperability and Standardization Organization) standard that intends to "provide a component framework for facilitating interoperability, reuse, and composability." The SISO BOM Product Development Group produced the "BOM Template Specification" (SISO 2006a) and the "The Guide for BOM Use and Implementation" (SISO 2006b) for describing the BOM related concepts. The objective of BOM is to encourage reuse, support composability, and help enable rapid development of simulation systems and simulation spaces.

BOM standard is compatible with the FEDEP definition and defines the conceptual model as "a description of what the [simulation or federation] will represent, the assumptions limiting those representations, and other capabilities needed to satisfy the user's requirements." BOMs are defined to "provide an end-state of a simulation conceptual model and can be used as a foundation for the design of executable software code and integration of interoperable simulations" in SISO (2006a). BOMs by definition are closer to the solution domain and the developer rather than the problem domain and the domain expert.

A BOM is composed of a group of interrelated elements, which are the model identification, conceptual model information, model mapping information and object model definition as shown in Figure 7.9. The components of this template can be represented using tabular format or UML diagrams. In addition, the BOM DIF (Data Interchange Format) enables the transfer of the BOM information between tools. It should be noted that, although the BOM specification uses the HLA OMT (Object Model Template) constructs, it does not restrict the use of a BOM to HLA-specific implementations.

The model identification component is used to associate important identifying information with the BOM. The conceptual model component includes different views to represent the conceptual model information. This information should be transformed into an object model, which is preferably a composition of HLA object classes, interaction classes, and data types. In order to enable this transformation, the required mapping information is provided as Model Mapping component. The Lexicon component is used to document the terms and ensure that they are consistently used in the correct form.

Below, we review the components of a BOM, with an emphasis on the conceptual model component.

7.2.5.1 Model Identification

One of the goals for using BOMs is to facilitate reuse; therefore each BOM has to contain a minimum but sufficient degree of descriptive information such as name, type, version, security classification, point of contact, etc. This information is mostly based on the IEEE Std. 1516.2-2000 (IEEE 2000) and is represented as a table. Every BOM is required to have a Model Identification table that includes the name, type, version, point of contact, and other required information as specified in the IEEE Standard 1516.2-2000 (IEEE 2000).

7.2.5.2 Conceptual Model Definition

The conceptual model of a BOM describes how the pattern of interplay within the conceptual model takes place, the various state machines that

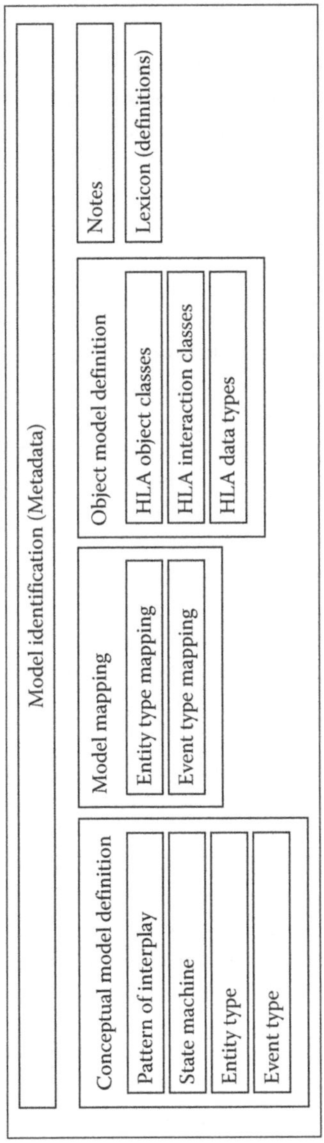

FIGURE 7.9
BOM composition. (Reproduced from Base Object Model (BOM) Template Specification, SISO-STD-003-2006, 2006a.)

may be represented, the entity type and event types defined in the conceptual model. This definition is closely matched with the FEDEP steps related with conceptual modeling and provides a description of what the simulation component, simulation or federation "will represent, the assumptions limiting those representations and other capabilities needed to satisfy the user's requirements" (IEEE 2003).

The pattern of interplay template component is used to identify the sequences of actions necessary for fulfilling the pattern of interplay that may be represented by a BOM. In addition to the main course of events, the variations and exceptions are also represented as pattern descriptions. A pattern of interplay may be composed of many pattern actions each of which includes the sender entity, the receiver entity and optionally the variations and exceptions.

The state machine template component is used to identify the behavior states of a conceptual entity that are required to support one or more patterns of interplay. BOM DIF defines the State Machine Table for describing one or more state machines. A state machine table includes the name of the state machine, the conceptual model entities that support the states defined, and the behavior states that are supported by a conceptual entity. A name and exit condition is defined for each state. Each exit condition identifies an exit action and the next state upon satisfying the exit action.

The entity type template component provides a mechanism for describing the types of entities. It is used to identify the conceptual entity types required to support the patterns of interplay and executing the various state machines. An entity type may play the role of a sender or receiver in a pattern of interplay or may be associated with a state machine. The entity type is identified by a name and associated characteristics. An example entity type may be a "waiter" having the "name" and "assigned tables" as characteristics.

The event type is used to "identify the type of conceptual events used to represent and carry out the actions variations and exceptions within a pattern of interplay." The two types of BOM events are BOM Triggers and BOM Messages, which represent undirected and directed events, respectively. In an undirected event the sender of the event is known but the receiver is not specified, so that any entity that has interest may receive the event. In a directed event both the sender and receiver entities are specified. A BOM trigger is an undirected event that may occur as a result of a change in the state of an entity and affects other entities that have interest in such observable changes. For a BOM trigger, the source entity and the trigger condition are known, but the target entities cannot be identified. A BOM Message is a directed event that identifies both of the source and target entities. A Message is an event type with a target entity, and a trigger is an event type with a trigger condition.

7.2.5.3 Model Mapping

The model mapping template component provides a mechanism for mapping between the elements of the conceptual model and the class structure elements of the Object Model Definition. The two types of mapping supported are Entity Type Mapping and Event Type Mapping.

The entity type mapping is used to map entity types and their associated characteristics to class structures. An entity type is mapped into an HLA object class or HLA interaction class, and characteristics of an entity type are mapped to HLA attributes or HLA parameters.

An event type is mapped into an HLA object class or HLA interaction class. Source characteristics, target characteristics, content characteristics, and trigger condition of an event type are mapped to HLA attributes or HLA parameters.

These mappings are means for transforming conceptual model elements into object model elements.

7.2.5.4 Object Model Definition

The object model definition defines the structure of an object and interaction class, and their associated attributes and parameters. HLA object classes include HLA attributes and HLA interaction classes include HLA parameters. This BOM component also includes the inheritance relationships between classes.

7.2.5.5 BOM Integration

As BOM Specification explains, BOMs are used to represent the conceptual models in accordance with the FEDEP steps. These steps include; selecting BOMS that support an aspect of a conceptual model among the existing ones, developing new BOMs if required, integrating these BOMs to create BOM assemblies and generating Federation Object Model (FOM)/Simulation Object Model (SOM) from these assemblies. BOM specification also admits that although use of HLA is not a mandatory subsequent step, it is likely that BOM assemblies are intended to support an HLA based federation. This feature of BOM makes it dependent on the simulation specifications, which contradicts with the simulation-independency of a conceptual model. However, BOM can be used as an effective tool to transform conceptual models of the mission space into simulation space object models.

The BOM Template Specification (SISO 2006a) defines the format and syntax for describing the elements of a template for representing BOMs. It specifies the syntax and the semantics of the elements of BOM. This specification also provides a DIF for representation of BOMs using XML. The "Guide for BOM Use and Implementation" (SISO 2006b) introduces methodologies for creating and implementing BOMs in the context of a larger simulation environment. The document provides guidance for BOM development, integration and use in supporting simulation system development. The guide also includes examples of UML diagrams that may be used to represent BOM tables.

7.2.6 Robinson's Framework

Robinson discussed the meaning of conceptual modeling and the requirements of a conceptual model (Robinson 2007a) and then defined a framework for conceptual modeling (Robinson 2007b) in a series of papers. The framework is also described in Chapter 4. He defines conceptual modeling as the process of abstracting a model from a real or proposed system. Besides this definition, Robinson describes the following four requirements of a conceptual model; validity, credibility, utility and feasibility, which should be harmonized with the need to develop the simplest model possible to solve the project goals (see Chapter 1).

The conceptual modeling process in Robinson's framework begins with developing an understanding of the problem situation. Depending on the domain experts' knowledge and expression capability on the problem situation, different methods may have to be followed. The domain experts generally believe that they have a sound understanding of the problem situation and they can express it effectively. However, after an analysis study, it is often observed that there are missing parts in their state of knowledge. Robinson emphasizes the difficulty arising from the fact that each client and domain expert may possess a different view of the problem. In order to overcome these obstacles, Robinson suggests that modelers use formal problem structuring methods.

Determining the modeling and general objectives is a critical task for developing the intended model. These objectives are apparently related with the aims of the organization. Although the modeling activity by itself provides useful insight for the organization, the real benefit is in the learning that can be gained from using the model. Robinson describes an objective as composed of three components; which are achievement, performance, and constraints. The success of a simulation study is tightly related with the fulfillment of the client, which can be ensured by defining appropriate objectives. However, neither the problem situation nor the objectives are static, and therefore subject to change. Besides the modeling objectives, Robinson points out the importance of general project objectives such as flexibility, ease-of-use and run-speed. These kinds of general project objectives will have impact on the conceptual model design.

Robinson defines the model outputs as the responses expected from the system. The responses are used to identify whether the modeling objectives have been achieved and to point out the reasons why the objectives have not been achieved. Responses are generally reported in the form of numerical data or graphical charts. The model inputs are called as experimental factors, because these are the model data that can be changed to achieve the modeling objectives. The experimental factors are also closely related with the modeling objectives, which implies that a change in the objectives requires a change in the experimental factors.

The next step is determining the model scope and the model's level of detail. The former defines the boundaries of the model whereas the latter describes the depth of the model. The scope of the model can be described in terms of the entities, activities, queues and resources. The level of detail for these components can be determined by the judgment and past experience of the modeler, analysis of preliminary data and prototyping. During this process various assumptions and simplifications may be made. These are recorded and classified as high, medium, or low depending on their impact on the model responses.

Robinson demonstrates this framework with a modeling application at Ford Motor Company engine assembly plant (Robinson 2007b). He proposes assessing the model by checking the validity, credibility, utility and feasibility of the model. Robinson also points out the importance of expressing the modeler's mental model as a communicative model and states the usefulness of diagrammatic representations for this purpose. He lists some of the possible diagrammatic notations, however, does not impose any of them for use with his framework. Robinson defines the conceptual modeling as art and states that the framework brings some discipline to that art. The artistic characteristic of conceptual modeling, combined with the different perspectives of the modelers and the domain experts make it impossible to define an absolutely right conceptual model. Therefore, Robinson suggests a conceptual modeling framework should provide a means for communicating and debating on conceptual models rather than aiming to develop a single best conceptual model.

7.3 A Comparison of Conceptual Modeling Frameworks

A conceptual modeling framework that includes a method definition, a notation and a supporting tool is essential for effective implementation of the conceptual analysis phase. The method definition should include the process steps in detail; the notation should provide an easy-to-use interface for both the conceptual modelers and the domain experts and the tool should support both the method and the notation. All of the abovementioned frameworks and methods point out the requirements of a conceptual model, the inputs needed and the likely outputs that are produced. A comparison of conceptual modeling frameworks is provided in Table 7.1 in terms of the framework parameters mentioned above.

FEDEP includes a high-level overview of the conceptual analysis phase by providing the boundaries of the phase, required inputs and the produced outputs; however detailed guidance for developing conceptual models is lacking. The CMMS approach was promising when the objectives are considered, however the project faded away without being able to provide a common syntax and semantics for conceptual modeling. BOM does not include a detailed process definition for developing conceptual models, but there

TABLE 7.1

Conceptual Modeling Approaches

Approach	Method/Process	Notation	Tool Support
FEDEP	Includes a process definition intended for HLA	No specific notation is imposed	No tool support
CMMS (FDMS)	Process definition does not include detailed guidance	Common lexicon is defined. Data Interchange Format is defined.	No tool support
DCMF	Includes a process definition	Includes KM3 notation for representing conceptual models	Existing UML modeling tools and ontology tools can be used
BOM	Process definition does not include detailed guidance	Includes a text-based syntax and semantics definition. UML may also be used.	BOMworks tool has been developed (BOMworks 2009)
Robinson	Includes a process definition	No specific notation is imposed, but diagrammatic notations are suggested	No specific tool is imposed. Existing graphical modeling tools can be used
KAMA	Includes a process definition	UML-based graphical notation is defined	A graphical modeling tool has been developed (Karagöz et al. 2005)

is active ongoing work on the tool support. DCMF complies with the three parameters of the framework definition with a focus on the KA activities of the conceptual modeling phase. Robinson does not define a specific notation for conceptual modeling but proposes using diagrammatic representation techniques and does not mention about the tool support. KAMA defines a notation specific to the conceptual modeling domain and includes a detailed process definition. The KAMA tool can be used for developing, sharing and verifying conceptual models.

All of these frameworks have some common limitations. It is difficult to provide a generic framework that is appropriate for all types of problem domains, because of their distinct requirements and objectives. Metamodel based notations may propose a solution to this problem by means of modifiable metamodels. However, in such a case the modelers should thoroughly analyze the tradeoff between a best-fit metamodel and a more general metamodel that allows more flexibility and reusability.

Conceptual models represented in diagrammatic notations are known to provide better understanding and communication, however as these diagrams get complicated these advantages are lost and cognitive issues arise (Kılıç et al. 2008). Diagrams with dynamically adjusted abstraction levels, or multidimensional viewing features may be utilized for overcoming these issues. The different perspectives of the conceptual modelers and the domain experts make it almost impossible to define the absolutely right conceptual model, which may also be considered as a cognitive issue.

Acknowledgments

The section on the KAMA framework is mostly reproduced from: Karagöz, N.A. 2008. A framework for developing conceptual models of the mission space for simulation systems, PhD thesis, Middle East Technical University, Department of Information Systems.

Some sections of this chapter are derived from the following resources:

- Mojtahed, V., M.G. Lozano, P. Svan, B. Andersson, and V. Kabilan. 2005. DCMF, Defense Conceptual Modeling Framework. *Technical Report*, FOI-R—1754—SE.
- IEEE Computer Society. 2003. IEEE 1516.3, Recommended practice for High Level Architecture (HLA) Federation Development and Execution Process (FEDEP).
- SISO (Simulation Interoperability and Standardization Organization). 2006a. Base Object Model (BOM) template specification, SISO-STD-003-2006.
- SISO (Simulation Interoperability and Standardization Organization). 2006b. Base Object Model (BOM) guidance specification, SISO-STD-003.1-2006.

References

Aysolmaz, B. 2007. Conceptual model of a synthetic environment simulation system developed using extended KAMA methodology. *Technical Report* 2006–2007: 2–17. Informatics Institute, Middle East Technical University.

Balci, O. 1994. Validation, verification, and testing techniques throughout the life cycle of a simulation study. *Annals of operations research* 53: 121–173.

Balci, O., and R.E. Nance. 1985. Formulated problem verification as an explicit requirement of model credibility. *Simulation* 45 (2): 76–86.

Borah, J. 2007. Informal simulation conceptual modeling: Insights from ongoing projects. In *Proceedings of the Simulation Interoperability Workshop*. www.sisostds.org

BOMworks. 2009. http://www.simventions.com/bomworks. Last visit on 20.03.2009.

Business Process Management Initiative (BPMI). 2004. Business process modeling notation, version 1.0. http://www.bpmn.org. Last visit on 12.01.2009

Defense Modeling and Simulation Office (DMSO). 1997. Conceptual Models of the Mission Space (CMMS) technical framework. USD/A&T-DMSO-CMMS-0002 Revision 0.2.1.

Globe J. 2007. Using SysML to create a simulation conceptual model of a basic ISR survivability test thread. In *Proceedings of the Spring Simulation Interoperability Workshop*. www.sisostds.org

Haddix, F. 1998. Mission space, federation, and other conceptual models, Paper 98S-SIW-162. In *Proceedings of the Spring Simulation Interoperability Workshop*. www.sisostds.org

IEEE. 1998. IEEE Std. 1320.2-1998 IEEE Standard for Conceptual Modeling Language Syntax and Semantics for IDEF1X.

IEEE Computer Society. 2003. IEEE 1516.3, Recommended Practice for High Level Architecture (HLA) Federation Development and Execution Process (FEDEP).

IEEE Computer Society. 2000. IEEE Std. 1516-2000 IEEE Standard for Modeling and Simulation (M&S) High Level Architecture (HLA): Framework and Rules.

Johnson, T.H. 1998. Mission space model development, reuse, and the conceptual models of the mission space toolset. In *Proceedings of the Spring Simulation Interoperability Workshop*. www.sisostds.org

Karagöz, N.A., U. Eryilmaz, A. Yildiz, O. Demirörs, and S. Bilgen. 2005. KAMA project research report. Center of Modeling and Simulation, Middle East Technical University.

Karagöz, N.A. 2008. A framework for developing conceptual models of the mission space for simulation systems, PhD thesis, Middle East Technical University, Department of Information Systems.

Karagöz, N.A., H.O. Zorba, M. Atun, and A. Can. 2008. Developing Conceptual Models of the Mission Space (CMMS): An experience report. In *Proceedings of the Fall Simulation Interoperability Workshop*. www.sisostds.org

Karagöz, N.A., O. Demirörs,. 2007. Developing Conceptual Models of the Mission Space (CMMS): A metamodel based approach. In *Proceedings of the Spring Simulation Interoperability Workshop*. www.sisostds.org

Kılıç, Ö., B. Say, and O. Demirörs. 2008. Cognitive aspects of error finding on a simulation conceptual modeling notation. In *Proceedings of the ISCIS Symposium*, Turkey, October.

Lacy, L.W., W. Randolph, B. Harris, S. Youngblood, J. Sheehan, R. Might, and M. Metz. 2001. Developing a consensus perspective on conceptual models for simulation systems. In *Proceedings of the Spring Simulation Interoperability Workshop*. www.sisostds.org

Lundgren, M., M.G. Lozano, and V. Mojtahed. 2004. CMMS under the magnifying glass: An approach to deal with substantive interoperability, 04F-SIW-0101. In *Proceedings of the Fall Simulation Interoperability Workshop*. www.sisostds.org

Mojtahed, V., M.G. Lozano, P. Svan, B. Andersson, and V. Kabilan. 2005. DCMF, Defense Conceptual Modeling Framework. *Technical Report*, FOI-R—1754—SE.

Pace, D.K. 1999a. Conceptual model descriptions, Paper 99S-SIW-025. In *Proceedings of the Spring Simulation Interoperability Workshop*. www.sisostds.org

Pace, D.K. 1999b. Development and documentation of a simulation conceptual model. In *Proceedings of the Fall Simulation Interoperability Workshop*. www.sisostds.org

Richter, H., and L. Marz. 2000. Toward a standard process: The use of UML for designing simulation models. In *Proceedings of the Winter Simulation Conference*, ed. J.A. Joines, R.R. Barton, K. Kang, P.A. Fishwick, 394–398. Piscataway, NJ: IEEE.

Robinson, S. 2004. *Simulation: The Practice of Model Development and Use*. Chichester: Wiley.

Robinson, S. 2007a. Conceptual modeling for simulation part I: Definition and requirements. *Journal of Operational Research Society*, 59 (3) 278–290.

Robinson, S. 2007b. Conceptual modeling for simulation part II: A framework for conceptual modeling. *Journal of Operational Research Society*, 59 (3) 291–304.

Ryan, J., and C. Heavey. 2006. Requirements gathering for simulation. In *Proceedings of the Third Operational Research Society Simulation Workshop*, 175–184. Birmingham: Operational Research Society.

Sargent, R.G. 1987. An overview of verification and validation of simulation models. In *Proceedings of the Winter Simulation Conference*.

Shannon, R. 1975. *Systems Simulation: The Art and Science*. New Jersey: Prentice-Hall.

Sheehan, J. 1998. Conceptual Models of the Mission Space (CMMS): Basic concepts, advanced techniques, and pragmatic examples. In *Proceedings of the Spring Simulation Interoperability Workshop*.

SISO (Simulation Interoperability and Standardization Organization). 2006a. Base Object Model (BOM) template specification, SISO-STD-003-2006.

SISO (Simulation Interoperability and Standardization Organization). 2006b. Base Object Model (BOM) guidance specification, SISO-STD-003.1-2006.

Sudnikovich, W.P., J.M. Pullen, M.S. Kleiner, and S.A. Carey. 2004. Extensible battle management language as a transformation enabler. *Simulation*. 80 (12) 669–680.

Tanriover, O., and S. Bilgen. 2007. An inspection approach for conceptual models for the mission space developed in domain specific notations of UML. In *Proceedings of the Fall Simulation Interoperability Workshop*. www.sisostds.org

Willemain, T.R. 1995. Model formulation: What experts think about and when, *Operations Research*, 43: 916–932.

Zeigler, B.P., H. Praehoffer, and T.G. Kim. 2000. *Theory of Modeling and Simulation*, 2nd ed. San Diego: Academic Press.

8

Conceptual Modeling in Practice: A Systematic Approach

David Haydon

CONTENTS

8.1 Introduction

This chapter presents a methodology for the design and implementation of a discrete-event simulation model. It is not the only way to implement such a model, nor is it necessarily the best—but it is in use and it works. (It is also used for other types of study, although the general approach is modified to suit the method of analysis to be used.)

The methodology has been developed over a number of years through trial and error. Ideas have been culled from a variety of sources. Some have been tried, found not to be useful and have been dropped. Other ideas have been found useful and have been kept. Others have been modified or parts of them used. The resulting methodology covers all aspects of the design and development of a simulation model—from requirements through design and development to testing. Just as this methodology has been constructed from pieces taken from various sources, it is suggested that the reader take those elements of this approach that are useful to them and incorporate them into the reader's own approach.

This approach is consistent with BS EN ISO9001—a quality assurance standard, although the details of the required procedures and documentation have been omitted for simplicity.

8.2 Software Project Life Cycle

The methodology presented here is based on the principle that a simulation model is a software application and that software engineering principles apply to its development. Simulation models tend to differ from most other software applications in that they generally have fewer data entities with simpler relationships between them. On the other hand, simulations tend to have much more complex decision logic than other software applications. So, although general software engineering principles can be applied they need to be modified to take account of the simpler data entities and more complex decision logic.

The traditional software engineering iterative waterfall software life cycle has been found to be a good model for simulation model development. The life cycle has the following stages:

- Requirements—what the model should do
- Design—how it should do it

- Implementation—developing the model to the design
- Verification—testing that the model conforms to the design
- Validation—testing that the model is fit for purpose
- Use

The term *iterative waterfall* is derived as follows. "Waterfall" refers to the process of completing one stage of the software life cycle, including review and approval, before starting the next stage. "Iterative" recognizes the fact that, in practice, later stages can have an impact on earlier stages. Iteration involves going back to an earlier stage and modifying decisions made at that stage and then following the effects of those changes through subsequent stages until the current stage and desired outcome is reached. For example, it is often the case that initial testing will show that some aspect of the model has a greater impact on the results than initially anticipated and that that aspect needs to be modeled in greater detail. This may mean that one or more of the requirements need to be modified (or additional requirements added). It will certainly require changes to the design. The software life cycle is illustrated in Figure 8.1.

At the end of each phase, the outputs of that phase should be reviewed by an independent, technically competent reviewer. The review should compare the outputs of the phase with the outputs of the previous phase to ensure that they are consistent. The review should be documented.

Software engineering has proved the value of separating the design and development into a number of independent phases. The use of this approach for the design and development of many simulation models has proved that it is also valuable for simulation. The rest of this chapter looks at each of the phases in more detail.

8.3 Requirements

This section considers the requirements phase of the software project life cycle. We discuss the purpose of the Requirements Document and outline its contents.

The purpose of the requirements phase is to document the requirements for the model. This may sound obvious but it is important to consider *why* the model is required. In the commercial world, models are developed to help address a specific problem and that will define the timescales and budget of the model development. It will also define the accuracy required from the project and hence the accuracy required from the model. The "Why?" is therefore to address some specific problem to the required accuracy and within the given timescales and budget. This also has implications for the

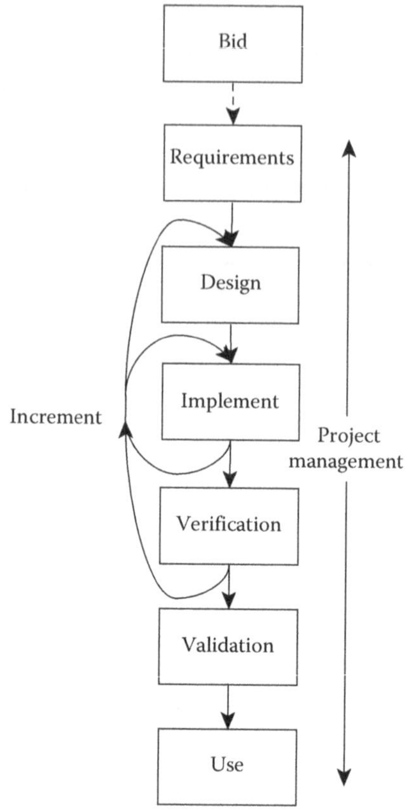

FIGURE 8.1
The iterative waterfall software life-cycle model.

analysis method to be used. In general, the method used should be simplest, quickest and cheapest method that will answer the "Why?" question. The selection of the analysis method should be driven by the requirements and is a design decision. It is not part of the requirements.

The Requirements Document provides the following:

- A clear statement of requirement against which to validate the model (test whether it is fit for purpose)
- A clear statement of what will and what will not be included in the model, i.e., the breadth and depth of the scope of the model

Although the Requirements Document is not usually a deliverable, it is useful to provide a copy to the customer to review and approve. This ensures that the customer is aware of what the model will cover and what it will not. Since the Requirements Document includes a description of the system to be modeled, the review by the customer gives them a chance to correct the analyst's understanding of the system and how it works. Since the Requirements Document is intended as a definitive statement of what the model will do it is important that the individual requirements be clearly identified—and ideally uniquely numbered for ease of reference. Requirements should be identified as mandatory, desirable or optional; defined as follows:

- Mandatory requirements must be met. The model will fail Validation if any of the mandatory requirements are not met.
- Desirable requirements should be met where doing so will not adversely affect timescales or budget.
- Optional requirements may be met if doing so has little or no effect on timescales or budget.

The words *shall, should,* and *may,* respectively, can be used in the formal statements of requirement to help differentiate between the requirement types.

The requirements phase does not include design decisions. All elements of the design must be left until the design phase.

8.3.1 Contents of the Requirements Document

The Requirements Document should include the following sections, which are described below:

- Purpose of the Development
- Stakeholders
- Study Objectives
- Overview of the System to be Modeled
- System Perspective
- General Requirements and Constraints
- Specific Requirements

8.3.2 Purpose of the Development

This section contains a top-level statement of the "Why?" question. As mentioned several times already, and worth repeating again, the "Why?" question drives the whole of the development and use of the model. The Requirements Document should only contain statements of requirements that contribute to the "Why?" If something does not contribute to the "Why?" then it has no value. In fact, it will be counterproductive since it will add to the complexity of the model and increase the time and cost of model development and testing—all for no benefit.

A simple statement of the purpose of the development helps to focus attention on the important requirements and also provides help in deciding what is important: "Does this contribute to the 'Why?' as stated in the purpose?"

It is worth emphasizing that the "Why?," as embodied in the statement of purpose in this section of the Requirements Document, is used throughout the development. At every stage of design and implementation, we should be asking, "Does this contribute to the 'Why?' Does it contribute to the purpose?"

Consider the following example from a past study:

> *The objective of the development is to be able to produce a realistic Recognised Theatre Logistics Picture (RTLP) to support the Future Logistics C2 Theme for a range of scenarios that involve a deployed Joint Force such that the RTLP can be used to make command and control (C2) logistics decisions.*

This statement leaves a great many requirements still to be determined, e.g.:

- Which scenarios
- Which C2 decisions
- What information will be required to make those decisions
- How should the information be generated and displayed

However, a simple clear statement of the overall purpose provides a framework against which the detailed requirements and design decisions can be set.

8.3.3 Stakeholders

This section should contain a list of the stakeholders, where a stakeholder is a person or organization with an interest in the development or use of the model. A stakeholder may be a person or organization with information about the system to be modeled.

8.3.4 Study Objectives

The study objectives are taken from the customer's Statement of Work and are the customer's stated objectives. Not all projects have a list of study objectives provided by the customer. The detailed objectives are sometimes unclear. In which case, a project start-up meeting may be required to discuss and clarify the study objectives (and document them in the minutes of the meeting). In either case, repeating (and expanding if necessary) the study objectives in the Requirements Document not only brings the relevant information together in one place for ease of reference but also means that the customer can confirm that the study objectives are correct (assuming that the document is reviewed by the customer).

The study objectives include timescales, available budget, and required accuracy.

8.3.5 Overview of the System to be Modeled

As the title of this section suggests, it contains a description of the real-world system that is to be modeled. By reviewing and approving the Requirements Document the customer confirms that the analyst's understanding of the system is correct and complete.

The overview includes a statement of the scope of the model: breadth defines which parts of the real-world system are to be included in the model and depth defines the level of detail. Having the scope reviewed and approved by the customer (as part of the Requirements Document) helps the customer understand what is being modeled.

8.3.6 System Perspective

This section details the relationships between the proposed model and other projects and/or models:

- Relationship to External Systems and Subsystems
- Relationship to Previous Projects
- Relationship to Current Projects
- Relationship to Successor Projects
- User Type Characteristics (required skills and experience of the expected Users of the model)
- Operational Scenario (how it is expected the model will be used)

External systems may include sources and/or sinks of data.

8.3.7 General Requirements and Constraints

General requirements and constraints are those that relate to the model as a whole. For example, use of COTS (Commercial Off The Shelf) software, standard hardware, predefined data formats, standard file formats (e.g., XML, bmp, JPEG), Identification Control (the use of model version numbers), Change Control (the control of modifications to the model).

8.3.8 Specific Requirements

If the system to be modeled can be split into a number of subsystems then the specific requirements should be spread over a number of sections—one per subsystem. The specific requirements detail the aspects of the real-world system that are to be included in the model. The detail of some specific requirements may be included as annexes. For example, details of particular algorithms to be used, data formats of external data sources.

8.3.9 Summary of the Requirements Phase

The requirements phase is intended to define and document what the model is required to do. It is not concerned with how the model does it or what type of model is to be used.

8.4 Design

This section considers the design phase of the software project life cycle. We discuss the purpose of the Design Document and outline its contents.

The purpose of the design phase is to document the design of the model. The design is "how" the requirements are to be met. The Design Document provides the following:

- A discussion of/justification for selecting simulation as the method of analysis
- A clear statement of the structure and content of the model
- Sufficient detail from which to implement the model
- Sufficient detail from which to develop the Test Plan

The Design Document is not normally provided to the customer. The customer is concerned with what the model does not how it does it.

8.4.1 Contents of the Design Document

The Design Document should include the following sections, which are described below:

- Purpose of the Development
- Stakeholders
- System Perspective
- Overview of the System to be Modeled
- Method of Analysis
- Simulation Structure

- Detailed Design
- Inputs and Outputs

8.4.2 Purpose of the Development

This section contains a restatement of the purpose of the development as stated in the Requirements Document. The "Why?" question is still relevant during the design phase. Design elements that do not contribute to the "Why?" add no value to the model and should be omitted. To repeat, the "Why?" question drives the whole development and use of the model.

8.4.3 Stakeholders

This is essentially a repeat of the stakeholder list contained in the Requirements Document. It is a useful reference to help identify source(s) of information that can be used to resolve design problems.

8.4.4 System Perspective

This is also a repeat of the equivalent section of the Requirements Document, except that there may be design details required for interfaces with other systems and/or models.

8.4.5 Overview of the System to be Modeled

This is usually copied word-for-word from the Requirements Document, usually with the following comment:

> *The system overview is contained in Ref..... It is repeated here for ease of reference but Ref.... remains the definitive statement.*

8.4.6 Method of Analysis

This section contains a discussion of and a justification for the selection of simulation as the method of analysis. A discussion of the advantages and disadvantages of the various mathematical, statistical, and other operational research techniques that could be used is beyond the scope of this chapter, as is the methodology for comparing and contrasting them in selecting the method of analysis. But the principle is simple. Each available technique is compared with the requirements. The simplest technique that meets the requirements should be selected. For example, if a spreadsheet model meets the requirements then a spreadsheet model should be used (unless there is a simpler technique available). (The term "spreadsheet model" is used somewhat loosely. It is possible to implement complex simulation models using a spreadsheet application—but the resulting model would still be a simulation model. Since the "simulation part" of the model is more complex than the "spreadsheet part," we regard such a model as a simulation model rather than a spreadsheet model.) (Note that if a simple spreadsheet model meets the requirements but would not be suitable for answering the "Why?" question, then some requirements have been missed.)

In general, the following requirements are needed for a simulation model to be the most appropriate technique:

- Interactions between different parts of the system—so that the system must be considered as a whole
- Dynamic effects—through feedback and/or time-dependent behavior—so that the system behavior depends on its past history and not just on its current state
- Randomness—typically requiring that system capacity be greater than the average throughput so as to allow for peak demands

In practice, the method of analysis is provisionally chosen before the start of the project—during the project bid/estimation phase (which is outside the scope of this chapter). This is so that the required resources and timescales can be estimated in order for the project to be authorised. This section will usually confirm and justify the provisional choice of technique.

8.4.7 Simulation Structure

Determining the appropriate simulation structure is critical to the success of a simulation study. To date, it has not been possible to extend this methodology to include a method for designing the optimum simulation structure. Every model is different. An approach that works well for one project and that yields a simple yet flexible structure will not work for another project. But it has been possible to derive an approach that provides the information necessary to determine a good structure. This section describes that approach.

The approach relies on the principle that discrete-event simulations work by considering that the system changes from one state to another at discrete times and that we can consider the system to be unchanging between those times. It follows that we are interested in what happens when the system changes from one state to another and what causes those changes. We define simulation events to be those unconditional activities that occur at specific times—typically end activities. When the event occurs, the system changes state. For example, if a tank is being filled, there may be an event when the tank becomes full. The system (or that part of it) would change state, say, from "being filled" to "full," and we would expect the flow to be stopped. Simulation activities are the conditional activities that take place when certain conditions are met. Activities can only start at the time of a simulation event. This is because if the conditions have not been met at a given time then the system state is such that the activity cannot start. The conditions will remain unfulfilled until there is a change in the state of the system. Since the system state only changes when there is an event, the conditions can only be met when there is an event. The design of the simulation is the process of selecting a set of simulation entities that can generate all of the required simulation events.

The approach to designing the simulation structure is to list all of the required events—including virtual events that do not exist in the real world but are required by the model. Then, for each event, list all of the entities that are involved with that event. Activity Cycle Diagrams (ACD) are a useful tool in identifying events, entities, and activities. If every event were associated with one and only one entity then the structure is complete. Every entity is included in the simulation structure.

In practice, most events will involve multiple entities. Designing the simulation structure is a matter of selecting a subset of entities that cover all of the events while minimizing the number of events associated with more than one of the selected entities. The selection is done by trial and error, guided by skill and experience. (We did say that it has not been possible to find a method for designing the optimum simulation structure!) The process can sometimes be simplified by identifying subsets of the system where a single entity can be selected and then removing that subset from consideration.

This approach has the advantage of providing a systematic approach to the problem of designing the simulation structure. It does not necessarily provide a simple solution to the problem but it does ensure that nothing is overlooked.

8.4.8 Detailed Design

The detailed design consists of listing all of the activities and events. For each activity, list the preconditions that must be met and the processing that takes place when those conditions are met. For each event, list the processing that takes place when the event occurs.

8.4.9 Inputs and Outputs

This section details the required inputs and outputs. In most simulations, the inputs consist of the attributes of the entities and the parameters of the activities. The outputs usually consist of utilizations, queuing times, etc.

8.4.10 Summary of the Design Phase

The design phase consists of selecting (or confirming) the method of analysis and determining the simulation structure. There are many approaches to designing the simulation structure but an approach based on consideration of the simulation events has been found effective. Once the simulation events and entities have been selected, the detailed design is produced by considering the conditions for activities to start and the processing involved at the start and end of those activities.

8.5 Implementation

Implementation is the process of turning the design into a working model. It depends on the simulation package in use and is outside the scope of this chapter.

8.6 Verification

Verification is the process of testing that the model as implemented conforms to the design. Model verification can be undertaken by using simple data sets for which the expected results can be easily calculated. These data sets are likely to exercise limited areas of the model. Data sets should be selected that, between them, cover all of the functionality of the model. The results of runs using these data sets should be compared with the expected results.

Verification of the complete functionality of the model can be performed by combining selected simple data sets and checking that the results are consistent with the results of the component runs. (Note that interactions between model elements would typically increase waiting times and reduce the combined throughputs.) The complexity of the test data can be gradually increased until the data sets are similar to those that will be used for the project.

The final verification tests are soak tests. A series of tests with inputs much higher than those that would be used in a study and a series of very long runs. Both series of tests are intended to ensure that the model behaves properly under extreme conditions.

The results of verification testing should be documented in the Verification and Validation Log.

8.7 Validation

Validation tests that the model is fit for purpose:

- Does the model meet the requirements in the Requirements Document?
- Can the model be used for the intended study, i.e., does it answer the "Why?" question?

Validation testing should use real data where this is available.

Validation with real-world data is not always possible since real-world data may not exist or may not available for many of the model functions. In which case, the model can be validated by using Subject Matter Experts (SME) to assess the results of the model for a range of scenarios.

Validation is often a mix of real data and SME judgment.

The results of validation testing should be documented in the Verification and Validation Log.

8.8 Example of the Methodology

As an example of the use of the methodology, consider a project to design a Message Handling Centre. Messages are received by the Centre, are processed, and in a small proportion of cases, result in messages being sent by the Centre to external systems and/or organizations for further action. There are a number of different message types with each type requiring different processing. The design project is responsible for the complete design of the Message Handling Centre:

- The size and physical layout of the building
- The procedures for handling the messages, including which parts of the processing can be handled by automatic systems and which require manual input
- The numbers, skills and qualifications, shift patterns, etc. of the Centre staff.

Due to the complexity of the message handling, it has been decided that the design project will require analytical support and that a simulation model is the most cost-effective way to provide that support.

8.8.1 Requirements

Following a Requirements Analysis, the following key requirements have been identified:

- To clarify the Message Handling Centre requirements and inform the design of the Centre
- To facilitate discussion of the proposed design, both within the design team and with the customer and external stakeholders
- To de-risk the design process
- To support the design of the business processes within the Centre
- To help plan resource management for the future in order to deal with peaks and troughs in demand
- To investigate the impact of resource levels on system performance

8.8.2 Simulation Structure

The methodology for designing the simulation structure starts by listing all of the simulation events and simulation entities within the scope of the system to be modeled. In the case of the Message Handling Centre, the simulation events are these:

- Message generation (of messages input to the Centre)
- Manual processing of messages by Centre staff
- Automatic processing of messages by Centre systems
- Processing of messages by systems external to the Centre
- Processing of time-expired messages
- Shift changes
- Dynamic reallocation of staff between teams to meet changes in workload for each team

There are several candidates for the time-dependent simulation entities (the entities that control the simulation timing):

- Messages
- Manual processes
- Automatic processes
- Staff
- Centre systems
- External systems
- Workstations

But messages are the only one of the above entities that are involved in all of the main activities. Also, the message dwell time (how long a message remains in the Centre) is one of the main outputs required. Thus the main simulation entity should be the message. (This leaves shift changes and dynamic reallocation to be handled separately.)

8.8.3 Simulation Activities

The simulation activities in which the simulation entities (the messages) take part are the processing tasks of the Message Handling Centre. How should these activities be linked to form the activity cycle(s) through which the simulation entities will move? The obvious way in which to link the activities is in the order in which they appear in the relevant process. We would then have an activity cycle for each process and the structure of the model would reflect the structure of the Centre processes. While this approach has the advantage of mirroring the Centre process structure, it does have a number of disadvantages:

- Messages in the Centre will queue for a team or person, or for a Centre or external system. Messages in the above simulation process structure will queue for a process task. Although tasks that use a shared resource can be linked by requiring a common resource, having multiple queues for each resource type complicates the management of the queue discipline.
- Changes to a process or the addition of another process would require changes to the model. (Such changes are likely to be required and would be better handled as data changes, if possible.)
- Dynamic reallocation of staff according to team workload depends on calculating the current demand on each team. With the above model structure, that demand would be spread over a number of queues and would be time consuming to calculate.
- Although the screen layout of the model would show the structure of the processes, it would give no indication of the physical layout of the Centre.

None of the above disadvantages are insurmountable. It is to be expected, however, that very high throughputs of messages will be required during runs of the model. It is therefore necessary that the model design be as efficient in runtime as possible. So, for example, it is likely that a design that allows a simple method of calculating team demand would be preferable to one that does not.

An alternative simulation structure would be one that focuses on the physical layout of the Centre. The activity cycles would be based on the activities performed by Centre staff and Centre and external systems. This would answer the above disadvantages as follows:

- Messages in the simulation could queue for a team or person, or for a Centre or external system. Queue discipline would be implemented in single queues for each resource type.
- Since the process structures are not modeled by the structure of the model, they would need to be modeled in data. Changes to a process or the addition of another process would be done by data changes.
- The current demand on each team could be calculated from the demand on the team and its members—indicated by the work queueing for the team and its members.
- The screen layout of the model could show the physical layout of the Centre.
- This alternative simulation structure has the disadvantage that the process structure needs to be defined in data and that it is not obvious where a message must be routed once it has completed a task. The data for that task/process must be referenced to determine to where the message should be routed. Either every activity must be followed by a router that routes the message to the next task or the model requires a central routing function that can handle every message and every process. The latter approach is more complex, but the former leads to duplication of code and possible inconsistencies in the model.

8.8.4 Simulation Design

The final simulation design focuses on the processing of the messages by the Centre staff and systems. The process maps for the processing tasks that are applied to each message type are defined in data. The process maps can therefore be modified and new message types added to the model without making changes to the model itself. This gives a highly flexible model but one that is difficult to implement and test. The final check on the suitability of the design is to review the key requirements and assess how well the model design meets those requirements.

- To clarify the Message Handling Centre requirements and inform the design of the Centre—*any well-designed model should meet this requirement.*
- To facilitate discussion of the proposed design, both within the design team and with the customer and external stakeholders—*the physical layout of the Centre is mirrored in the screen layout of the model and the process maps of the message processing can be easily changed.*

- To de-risk the design process—*any well-designed model should meet this requirement.*
- To support the design of the business processes within the Centre—*designing the business processes is simplified by being able to experiment with the processes without making changes to the model.*
- To help plan resource management for the future in order to deal with peaks and troughs in demand—*shift patterns and dynamic reallocation of staff are two ways of dealing with peaks and troughs in demand. Since the design has single work queues for each resource type, model implementation and testing will be simpler and runtimes will be shorter than if the model had to manage multiple queues for each resource type.*
- To investigate the impact of resource levels on system performance—*any well-designed model should meet this requirement.*

The resulting design is by no means the only design that could have been produced, and if the key requirements had been different, a different design may have been developed. Just as the structure of the model should mirror the structure of the real-world system, it should also mirror the study requirements.

8.9 Summary

In this chapter we have shown how software engineering principles can be applied to conceptual modeling for discrete-event simulation models. Those principles have been modified to reflect the qualitative differences between simulation models and other types of software applications.

The methodology starts with the requirements phase, which addresses the "Why?" question. Why do we need a model and what should it do? The "Why?" question applies to every phase of the model development. The output from the requirements phase is the Requirements Document. The Requirements Document has two uses: as an input to the design phase and as a clear statement of what the model is intended to do that can be reviewed and approved by the customer.

The requirements phase is followed by the design phase. The output from the design phase is the Design Document, which documents the design, including why it was decided to use a simulation model and a discussion of the design of the simulation structure. Design is followed by implementation, and then by verification and validation.

The methodology described in this chapter is not the only conceptual modeling methodology, nor is it necessarily the best. But it has been used in the development of many simulation models in a commercial environment and has proven useful.

Part III

Soft Systems Methodology for Conceptual Modeling

9

Making Sure You Tackle the Right Problem: Linking Hard and Soft Methods in Simulation Practice

Michael Pidd

CONTENTS

9.1 Introduction

This chapter argues that simulation analysts should carefully consider the context for their technical work before starting to build a simulation model. It is common for analysts to complain that, though their work was excellent, it was never used or implemented because of what they refer to, somewhat dismissively, as "organizational politics." Rather than dismiss such politics, which some people regard as part of any organization, it is better to use methods that help an analyst to understand how the power and interests of different stakeholders can affect the outcome of their work. That is, analysts need to develop skills that enable them to accommodate to organizational realities rather than to bemoan their existence. Not to do so is to greatly increase the risk of failure, however excellent the technical methods used in developing a simulation model.

Most simulation text books devote their space to three technical areas that readers should master if they are to become proficient as simulation practitioners.

1. Modeling: extracting the relevant parts of a system of interest and representing them, appropriately, within a simulation model. This is a skill that develops though practice and requires the analyst to take a systems viewpoint so as to tease out system components that have the most effect on performance. Since this is not a skill that is easily expressed in technical terms, this is one of the weakest parts in many books, exceptions being Robinson (1994, 2004) and Pidd (2003, 2009).
2. Statistical methods: most discrete-event simulations include stochastic behavior that is represented by sampling from appropriate probability distributions. Thus the modeler needs to know which distributions are appropriate, how they should be represented in the model, and how to analyze the resulting behavior of the simulation. The approaches needed rest on standard statistical theory and are relatively easy to write down in unambiguous terms, and thus these ideas are well described in books such as Law and Law (2006) and Lewis and Orav (1989).
3. Computing: though it is possible to build a simple simulation model with little or no computing knowledge or experience, it soon becomes apparent that it helps to know much more than which button to press. Knowing how the simulation program or package works, is a great help in developing simulations that are accurate and run fast enough for proper use. Hence there are books that devote much space to showing readers how to develop simulation programs using particular software; examples include Kelton et al. (2004) and Harrell et al. (2004).

The programs for simulation events such as the annual Winter Simulation Conference (http://www.wintersim.org/) also focus on the same topics, though also include reports of work in particular application domains such as manufacturing, health care, aerospace, criminal justice, or another domain.

When starting to work in any application domain, a simulation analyst needs to take her understanding of modeling, statistics and computing and bring them to bear in the domain. When starting work in a domain that is wholly new to them, all simulation analysts experience some confusion in which they are unsure what level of detail is required and what should be included or excluded from the model and the modeling. Obviously, this becomes easier with experience in the application domain, as the analyst learns which features are important. However, there is also a danger that an analyst with long experience in an application domain starts to take things for granted that, later, turn out to be important. Hence it is important that analysts carefully consider what elements should be included in their study—no matter how familiar they are with the domain.

This chapter presents, briefly, some of the main ideas in problem structuring and discusses how they can be useful in conceptual modeling. It introduces an informal approach to problem structuring and lists some of the formal approaches advocated in the literature and then continues with a more detailed exposition of soft systems methodology (SSM). Finally, it uses a real-life simulation study conducted for a UK police force to show how aspects of problem structuring methods (PSMs) can be useful in practice.

9.2 Problem Structuring

This suggests that there is a need for approaches to help someone new to a problem domain to get to grips with its important features and a similar need for approaches that help prevent the experienced from becoming overconfident. Clearly, there is no silver bullet that will guarantee success, but there are approaches that are intended to help an analyst tease out the important features of a problem situation. There are approaches that are intended to help analysts understand the main features of problems in which they are asked to work. These appeared in the Operational Research/Management Science (OR/MS) community in the UK and Europe and are sometimes known as "soft" OR or as PSMs (Rosenhead and Mingers 2001). The term *soft* seems to have become popular following the development of SSM by Peter Checkland and his colleagues at Lancaster University, an approach described later in this chapter. The use of the adjective *soft* is unfortunate, since it may carry the idea of trivial or simple and this is far from the case, and this is one reason why the many people prefer to write about PSMs rather than soft OR.

The term *problem structuring* carries two different meanings, which are summarized in Figure 9.1. As originally used, "problem structuring" referred the work done in problem solving to structure the issues before any detailed analysis is conducted. As an example of this, Pidd and Woolley (1980) report on the problem structuring approaches used by OR practitioners in part of the UK about 30 years ago. The idea of this type of problem structuring is to develop understanding, in particular, to understand the context for the simulation project, such as the ways in which different stakeholders "see"

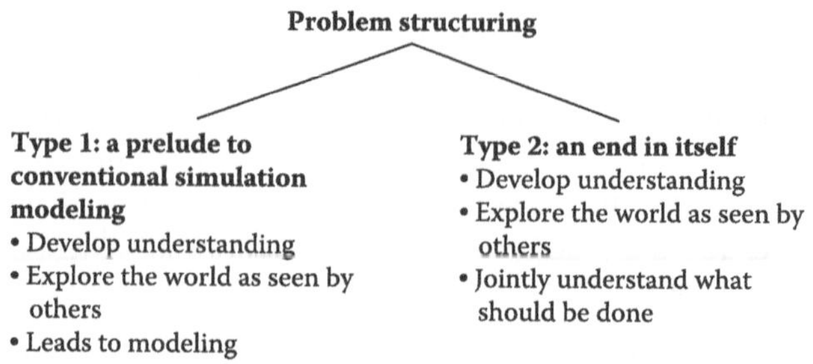

FIGURE 9.1
Two views of problem structuring.

the world and what they are hoping to achieve from the project. It continues through the project right up to implementation, supposing that this occurs, but is the main focus of the early stages of a typical simulation project. Pidd and Woolley (1980) report that though OR analysts were concerned to properly structure the problems they were tackling, there was no real evidence of them using formal methods to help with this.

The second use of the term relates to tackling "wicked problems" (Rittel and Webber 1973). These are characterized by clashing objectives, a shortage of data, and multiple stakeholders, who may have very different opinions from other stakeholders on what is desirable. Such wicked problems are, in essence, unsolvable, in the sense complete resolution or improvement. However, it is usually still possible to make progress in their resolution by structuring the interrelated issues in such a way that stakeholders can hold an intelligent debate about what might be done. The formal PSMs described in Rosenhead and Mingers (2001) are techniques and approaches that can be used to structure such debate and discussion. This type of problem structuring is a deliberate contrast with the idea of problem solving, since there is no assumption that problems can be solved in any permanent sense, rather the aim is enable stakeholders to make progress when faced with wicked problems. This second use of the term problem structuring is now more common than the first and can be seen as an attempt to introduce procedural rationality Simon (1972, 1976), into tackling wicked problems. That is, this form of problem structuring provides a systematic way to collect information, to debate options, and to find some acceptable way forward.

9.2.1 Complementarity

In recent years, it has become clear that the same methods developed for structuring wicked problems can also serve as a preliminary to formal modeling; that is, they can help with the first type of problem structuring. It might be argued that this amounts to overkill if the simulation project is very simple and straightforward. However, it is not at all unusual for what seems like a simple simulation project to become more and more complex as work proceeds. This can happen for many reasons as stakeholders become aware that work is underway and, not unreasonably, wish to have their voice heard. Hence, the argument of this chapter is that conducting formal problem structuring is valuable in almost all simulation projects.

The subject of this book is conceptual modeling, which is the process of understanding what might be included in a model and representing this in a way that is relatively independent of the simulation software to be used. Some writers (e.g., Robinson 2004, 2008) insist that the conceptual model must always be independent of the software being used, but that seems too stringent a requirement given the inclusive nature of much simulation software. Pidd (2009) defines a model as "an external and explicit representation of part of reality as seen by the people who wish to use that model to understand, to change, to manage, and to control that part of reality." Since it is only part of reality, a model will always be a simplification; some things will be included, others will be excluded. Before deciding what should be in and what should be out, it makes sense to consider the context of the proposed work and this is the role of PSMs in simulation.

When problem structuring approaches are used in combination with analytical approaches such as computer simulation, it is sensible to regard the two approaches as complementary (Pidd 2004). It is, though, important to realize that such complementary use is based on the mixing of paradigms and methodologies (Mingers and Gill 1997) and that care is needed when doing so. Detailed discussions of this complementary use can be found in Pidd (2004), which reports on a research network involving both academics and practitioners in the UK established to consider the difficulties and challenges. Kotiadis and Mingers (2006) discuss some of the challenges faced when attempting to link PSMs with "hard" OR, specifically with discrete-event simulation modeling in health care and is optimistic about such complementarity. Pidd (2009) compares and contrasts formal PSMs with more classical management science techniques, including simulation.

From this point on the term *problem structuring* applies to the use of systematic approaches to help diagnose a problem and understand the main issues as a prelude to detailed simulation modeling. The aim is to find ways to implement John Dewey's maxim (quoted in Lubart 1994): "A problem well put is half solved." It seems as if he had in mind that a poorly posed problem will be very hard, if not impossible, to solve—as expressed in the title of this chapter: making sure that you tackle the right problem. There can, of course, be no guarantee of this, but problem structuring approaches can help reduce the risk of working on the wrong problem. As with simulation modeling itself, users of PSMs grow more expert in their use as their experience develops. There is, though, no silver bullet, no magic formula that will guarantee the correct diagnosis of a problem in such a way that the right simulation model is built and that this is used appropriately.

9.2.2 Informal Problem Structuring: Critical Examination

Since many simulation practitioners are engineers, it makes sense to start by discussing an approach known as critical examination that has been used by engineers for many years. Though engineers are rarely regarded as poets, critical examination is based on six questions that are neatly summarized in a verse by Rudyard Kipling from the *Just So Stories* ("The Elephant's Child"):

I keep six honest working men
(They taught me all I knew);
Their names are What and Why and When
And How and Where and Who.

These make a very good starting point for considering the main aspects of a problem for which a simulation approach is being considered.

The first question in the verse revolves around what. Of course, there are many different questions that could be asked, which begin with what. The most obvious and one for which there is rarely a straightforward answer without working through all six questions is, "What's going on?" or "What's the problem we need to work on here?" It is perhaps better to ask, "What are the main issues that concern people?" In a manufacturing simulation, these might include some or all of cost reduction, uniform high quality, integrating work centers, or reducing stocks. In a simulation of a call center, they might include some or all of meeting performance targets for answering calls, establishing equipment needs, designing a call routing system, and determining a shift pattern. Note that these issues are rarely independent and may be in conflict with one another. At the early stage of a simulation project, it is important to simply identify these issues and to keep them in mind as part of the development of a conceptual model.

The second question starts with why. Perhaps the most common variants on this are to ask, "Why are these issues important?" "Why do particular people think these are important?" and "Why is this important now?" Of course, the latter two questions spill over into the who and when questions. It is not unusual for problems to be known, but not tackled. Sometimes there is good reason for this—there are just more important things to be done, or people have found workarounds that have been good enough. It is very common for answers to the why questions to become more subtle and complex as the work proceeds. Hence it is best to regard problem structuring as something that goes on throughout a project.

Experienced modelers know that they sometimes only have a real appreciation of the problem they are tackling when the work is complete. It was this realization that led Pidd and Woolley (1980) to conclude that this form of problem structuring is characterized by four features:

1. It is *inclusive:* the questioning and deliberation are not just concerned with the technical aspects of the work, but also considers how the model might be put to work and how stakeholders might be persuaded to act on any recommendations.
2. It is *continuous:* the questioning and deliberation are iterative or cyclic and continue throughout the project. In the terms introduced by Kolb (1983), it is a learning cycle during which participants learn the aspects that need to be included in the model and its use.
3. It has some *hierarchical features:* one problem tends to spawn another and decisions must be made on how detailed or specific the model is intended to be.
4. It is *informal:* which explains the title of this section. That is, people get on with it, cutting corners where appropriate and sometimes regretting this later.

With this in mind, the third informal question asks when and concerns the time dimension. Typical examples might be: "Is this a once-off problem or one that recurs?" or "Has this been a problem for some time but only recently become important enough for action?" or "When will the model be needed?" or "When will the changes to the systems need to be implemented and properly working?" The first two relate to the earlier why questions and the latter two give some idea of the resources that will be needed to do the work and of the level of detail that can be achieved in the model. If the model needs to be built and tested in a couple of weeks, it is unlikely to include much detail.

The fourth informal question asks how. The first common example asks: "How am I going to model this?" referring to the technical approach that may be needed. The second common example asks: "How did all of this start to emerge?" Clearly this and the other five "honest working men" are close relatives or friends, and in this form it relates closely to the who and when questions. But it also relates to the what question in facing up to how things are done at the moment or how people might envisage things to operate in the future. This depends both on the analyst's reflection and deliberation and also on the opinions of the people who are interviewed at this stage of the work.

Fifth, we can ask the where questions. Often these are less important when taken at face value, for the location of the system of interest may be obvious. However, even this should not be taken for granted. Location can be very important now that instantaneous electronic communication around the world is available at low cost. Tasks that once had to be located in one particular place may now be located elsewhere in the world. Examples include the transfer of medical images and resulting diagnosis on a different continent from the place that the patient is located, the location of telephone help-desks, the processing of routine documents and the 24/7 development of computer software. The why question might also become "Where is this problem occurring?" and this suggests the need for careful understanding of the system of interest.

Finally, and often the most important, are the who questions. Since most organizations are inherently political, the people, their motivations, and their actions become very important. Put simply, in many situations, some people have much more power to get things done than others. Equally, some people have much more power to stop things being done and this may be just as important. In a privately owned business it may be obvious that the owner calls the shots. In a public body there are often many stakeholders whose views must be considered and their views may conflict. Hence, irritating though it can be to people of a technical bent, a careful identification of the main players can be crucial in getting things done—even for something as basic as data acquisition.

As the preceding argument demonstrates, informal problem structuring is not difficult to understand. This presentation of critical examination should not, though, be used to as a reason wander aimlessly around asking aggressive questions of other people. The idea is that the analyst keeps these questions in her head and, in interacting with other people and using previous experience, teases out answers that will inform the modeling work and its implementation. There are times, however, when something more than this informal approach is needed, when a more formalized methodology is needed to manage a complex situation.

9.3 Formal Problem Structuring Methods

It is impossible, in the space available, to give more than a flavor of commonly used, formal PSMs. A good survey is found in Rosenhead and Mingers (2001) and detailed accounts can be found in works produced by the developers and advocates of the various approaches. The *Journal of the Operational Research Society* (Shaw, Franco, and Westcombe 2007) produced a special issue devoted to recent developments and it seems likely that others will appear in the future. In this chapter, the aim is to present a very brief survey and then illustrate the ideas by focusing on one approach, SSM, in more detail. Other problem structuring approaches have been used in a complementary fashion with discrete simulation modeling; for example, see Sachdeva, Williams, and Quigley (2007). Likewise, PSMs have been used with system dynamics models and Howick (2003) is an example of a paper linking this form of simulation with cognitive mapping, another commonly used problem structuring approach. For a discussion of the use of PSMs in simulating airline operations that involved multiple stakeholders, see Den Hengst, de Vreede, and Maghnouji (2007).

It is likely that formal PSMs are of most use in situations where strategic issues loom large, rather than in tackling low-level, operational problems. The various formal methods assume that stakeholders may legitimately disagree with one another, that they may behave politically and that there may be disagreement about ends (why and what should we be doing?) as well as about means (how can we increase throughput by 15%?). With this in mind, Rosenhead (1989, p. 12, Table 2) suggests that the formal methods share six distinctive characteristics:

1. Non-optimizing: seeks alternative solutions that are acceptable on separate dimensions without trade-offs
2. Reduced data demands: achieved by greater integration of hard and soft data with social judgments
3. Simplicity and transparency: aimed at clarifying the terms of conflict
4. Conceptualizes people as active subjects
5. Facilitates planning from the bottom-up
6. Accepts uncertainty and aims to keep options open for later resolution

The idea of formal PSMs seems to have arisen in the UK OR community in the 1970s and the methods have developed since then and are routinely taught on educational programs in Europe. Curiously, their penetration has been much lower in the US and some in the OR/MS community view them with great suspicion. Other communities, for example those involved in software engineering, have also developed approaches such as Dialog Mapping (Conklin 2002) and Design Rationale (Lee and Kai 1991), with many of the same characteristics. Rosenhead and Mingers (2001) discuss the approaches most commonly used in OR/MS and matches a descriptive chapter on each approach with another discussing an implementation. Their list of methods is as follows:

- SODA (cognitive mapping)
- Soft systems methodology
- The strategic choice approach
- Robustness analysis
- Drama theory and confrontation analysis
- Related methods: viable systems modeling, system dynamics, and decision analysis

It is impossible to do justice to all these approaches in a single chapter that relates their use to conceptual modeling. Hence, here we focus solely on SSM, which Mingers and Taylor (1992) reports as one of the most commonly used PSMs.

9.4 Soft Systems Methodology

Despite its unfortunate title, SSM is widely used. An early postal survey (Mingers and Taylor 1992) investigated the use of SSM and reports that the majority of SSM users did so with a view to easing a problem situation or to develop understanding. Users also claimed that a main benefit of SSM was that it provided a strong structure within which they could achieve these aims. These findings seem to support the view that SSM provides a formalized approach to gaining understanding within an organization, paying due regard to cultural issues.

Checkland (1981, 1999) describes the development of SSM and its main features. Checkland and Scholes (1999) provide a more practical view of the ideas, Wilson (1990) provides a systematic discussion of how the ideas might be operationalized, which is an issue also faced in Checkland and Poulter (2006). The description of SSM presented here is based on that in Pidd (2003, chapter 5). Paul and Lehany (1996) discusses some general issues in linking SSM to discrete simulation modeling, Baldwin, Eldabi, and Paul (2004) present a general methodology based on SSM for understanding stakeholders in health-care simulations and Lehany and Paul (1996) discuss a specific health-care application in which SSM and simulation are used in a complementary fashion. As mentioned earlier, Kotiadis and Mingers (2006) discuss the issues to be faced when linking PSMs to discrete-event simulation and Kotiadis (2007) specifically suggests ways of achieving a symbiosis between SSM and simulation.

9.4.1 The Overall Approach of SSM

The original book on SSM (Checkland 1991) presents a nine-step approach to its use. It seems that Checkland rather regretted this mechanistic presentation, for a rather more fluid description is provided in later works. Figure 9.2, which reflects Checkland's preferred depiction of SSM, shows the approach as a learning cycle with a number of features.

SSM aims to help people to understand and, possibly, to design human activity systems. In exploring what this means it is important to realize that the idea of a system is employed in SSM somewhat differently from its everyday use. Rather than assuming that systems, as such, exist, they are taken

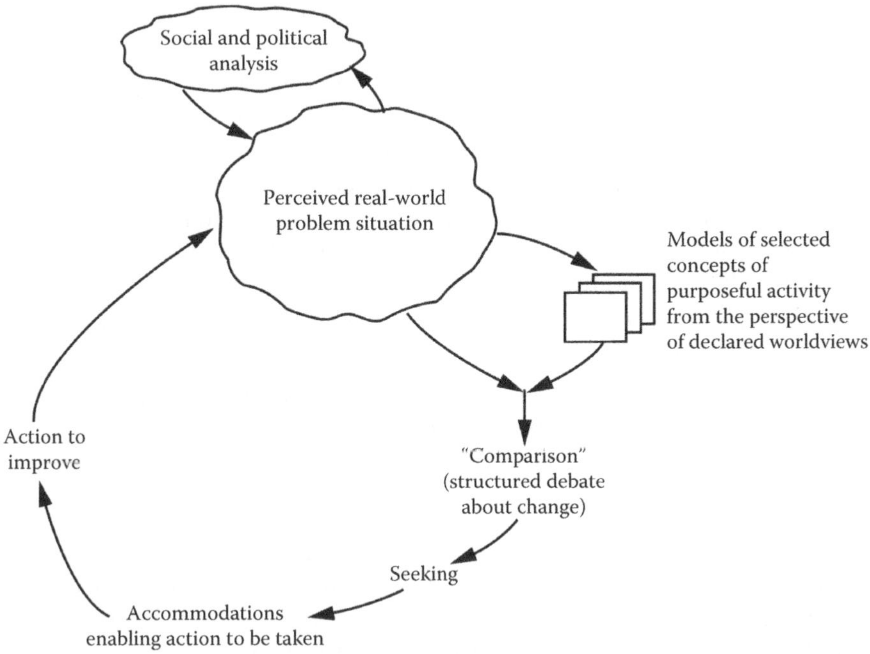

FIGURE 9.2
An overview of soft systems methodology. (Adapted from Checkland, P.B. and Holwell, S., *Systems Modelling: Theory and Practice,* John Wiley & Sons Ltd., Chichester, UK, 2004. Used with permission.)

as useful conceptualizations or convenient fictions. In these terms, human activity systems have the following characteristics:

- *Boundaries:* Some things are inside the system, others are not and constitute the environment of the system. Note, though that the boundary may not be obvious. For example, in a call center, is the location from which someone calls to be part of the model?
- *Components:* There is more than a single element within the boundary. A boundary that contains nothing is not a system and nor is a boundary that contains a single element.
- *Internal organization:* The elements are organized in some way or other and are not just chaotic aggregations.
- *Behavior:* The system is recognized as such because it displays behavior that stems from the interaction of its components; that is, this behavior is not just from those individual components.
- *Openness:* The system boundary is permeable in both directions and there is communication and interaction across the boundary. The

cross-boundary exchanges constitute the external relations of the system.

- *Human activity:* What people do, and how they do it, are prime concerns of SSM. It follows from this that human activity systems are dynamic as a result of human action.
- *Human intent:* People are not just machines that do things. What they do has meaning and significance for the individuals and groups concerned. Why people do things are often at least as important as what they do and how they do it.
- *Limited life:* Human activity systems are not eternal, and their life may be quite short.
- *Self-regulation*—A process of regulation, control, or governance, which maintains it through time, is a characteristic of an open system. These systems may be in equilibrium, but this stability is not the same as stasis.

This view of a human activity system, for which Checkland prefers the term *holon*, is somewhat wider than the classic engineering view of a system as something designed to achieve a purpose in that it incorporates the idea of human activity and human intent, recognizing that these are crucial to success. The stacked rectangles in Figure 9.2, labeled as "models of selected concepts of purposeful activity from the perspective of declared worldviews," do not imply that such models, or human activity systems, actually exist or even could exist. These are conceptualizations that serve to illustrate how things might ideally exist, and the idea is to understand what action might be taken, by those involved, to improve things.

The large cloud represents a perceived, real-world problem situation. In many SSM studies, this is the starting point of the work and this is likely to be the case if the SSM is used as a prelude to detailed modeling, possibly using simulation. The term *real-world problem situation* is carefully chosen. The word *perceived* is used because a study always begins with a recognition that something needs to be done; that is, some situation is unsatisfactory now or a system needs to be designed or reconfigured for the future. Since there are often different stakeholders (including the client and analyst), the perceptions of those people matter and different stakeholders may perceive things rather differently. However, SSM is not primarily intended for philosophical use, but for the world of action and in which something must be done. Hence, this is a real-world problem that needs to be tackled.

In this chapter, the main focus is the use of PSMs in conceptual modeling. That is, Type I problem structuring (Figure 9.1), which is a prelude to more formal mathematical or computer representations of a system of interest. Hence, it focuses on the role SSM in understanding how stakeholders view the issues of concern in a simulation study as part of conceptual modeling. Therefore, there is no attempt to complete the learning loop of Figure 9.2 within the chapter, for which the reader is referred to the numerous detailed accounts of SSM, including those listed at the end of this chapter.

9.4.2 Understanding a Perceived, Real-World Problem Situation

The main focus of this chapter is the use of SSM to understand a real-world problem so as to work toward improvement. The word *problem* is itself somewhat problematic (Pidd 2009, chapter 3) and so Checkland instead refers to a problem situation; that is, the set of interacting policies, people, equipment, actions, and intent that may or may not be causing difficulties now or in the future. One of the aims of SSM is to tease out these aspects so as to understand which are most important in seeking improvement. This is the finding out stage of SSM and seems transferable to most application areas and could serve as a useful starting point for many simulation studies to reduce the risk of fruitless endeavor later. In this finding out stage, above the cloud in Figure 9.2 is the need for social and political analysis to inform the developing understanding of this problem situation. In essence this is a formalization of the six questions involved in the critical examination of informal problem structuring.

The social analysis can be considered in two parts; firstly, a conscious attempt to identify the people occupying various roles in an typical modeling project, as follows. First, there is the "would-be problem solver": the person who has decided, been told or has requested to investigate the situation—most likely, you, the analyst. Second, the "client": the person for whom the investigation is being conducted. Finally, the "problem owners": which would include the various stakeholders with a range of interests. Note that any or all of these roles may overlap—for example, someone may be using SSM to help support their own work. The second part of SSM social analysis is to investigate the problem situation as a social system. The idea is to build on the knowledge of the significant roles that people occupy, to investigate the norms and values that are expressed. Roles are taken to be the social positions people occupy, which might be institutional (teacher) or behavioral (clown). Norms are the expected, or normal, behavior in this context. Values are the local standards used to judge people's norms. The idea of this analysis is that the analyst should try to understand how people play out their roles.

The political analysis requires the examination of the problem situation as a political system, in an attempt to understand how different interests reach some accommodation. This is an explicit recognition that power play occurs in organizations and needs to be accounted for. Needless to say, any analysis of power and its use needs to be undertaken carefully and, possibly, covertly. Even when sitting in a bar, there is little point asking people what their power ploys are!

9.4.3 Power-Interest Grids

One widely recommended way to consider power is to use of a power-interest grid, which is a 2x2 classification on dimensions of power and interest. In simple terms, power is the ability to get things done or to stop things happening and this is very different from only having an interest in what is happening. Checkland does not discuss their use, but they seem a useful addition to the official accounts of SSM. Various formulations of power-interest grid can be found on the Internet and Eden and Ackerman (1998), discussing problem structuring using cognitive mapping, uses a sporting analogy and labels the quadrants as players, subjects, context setters and crowd, as shown in Figure 9.3.

The players have the most power to affect the outcome and, one hopes, the most interest in doing so. Thus, stakeholders with high power and high interest need to be managed very closely and examples might include senior managers in whose domain the work is being conducted. On the other hand, the crowd has both limited interest and limited power to affect the outcome: examples might include the owner of the land on which a manufacturing plant is based. The crowd, unlike a passionate football crowd, only gets excited when actions are taken that might threaten them. If the team is owned by a remote group interested only in the financial results, it may be reasonable to regard them as context setters having much power but little actual interest. As shown in Figure 9.3, the context setters should be kept informed of developments so as to ensure that they are on-side. Finally, the subjects have a great interest in what is happening but little direct power and, at the very least, their views must be noted since they could walk away and this might, eventually, lead to failure. From an ethical standpoint, too, it is important in some situations that the interests of subjects be protected

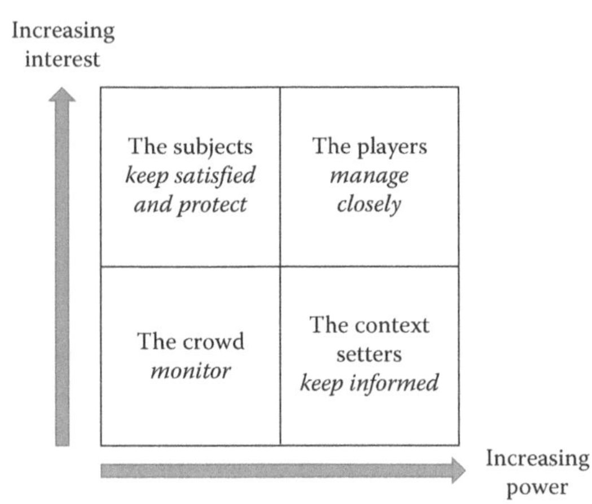

FIGURE 9.3
A power-interest grid.

during and after the project. It should be clear that the players are crucial, since they have high power and interest. However other stakeholders must not be ignored and a stakeholder analysis is always profitable and need not take long.

Following investigation of the problem situation, Figure 9.2 shows that an SSM study requires the construction of models of purposeful activity from the declared worldviews. Two aspects of this merit discussion here. First, it is important to realize what is meant by a model in SSM, since this is not the same as a simulation model. A model in SSM is something that captures the essential activity needed in an idealized implementation of the system of interest. These are usually developed from root definitions, which is a concept discussed later in this chapter. Second, note the reference to declared worldviews. The aim of the social and political analysis is to understand the different worldviews of the people and groups involved in the problem situation. SSM takes for granted that there may be different worldviews—that is, people may legitimately disagree about the ends and means of a study. The different viewpoints are teased out and represented in root definitions.

9.4.4 Root Definitions

The idea of a root definition is to provide a minimal definition of a system, viewed partly in input:output terms, to enable discussion between stakeholders about what is required. The idea is that the root definition is in some sense neutral, in that a particular structure is required that separates the definition and its supporting worldview from the stakeholder(s) to whom it belongs.

A root definition consists of six elements as follows (summarized in Pidd 2010):

- *Customers:* These are the immediate beneficiaries or victims of what the system does. A customer can be an individual, several people, a group or groups. This is very close to the total quality management (TQM) notion that the customer is the next person to receive the work in progress. The customers help define the main external relations of the system being conceptualized.
- *Actors:* In any human activity system there are people who carry out one or more of the activities in the system, these are the actors. They form part of the internal relations of the system. There may be several actors or several groups and their relationships also form part of the internal relations of the system.
- *Transformation process:* This is the core of the human activity system in which some definite input is converted into some output and then passed on to the customers. The actors take part in this transformation process and, ideally, a root definition should focus on a single

transformation. The transformation process is an activity and its description therefore requires the use of verbs.

- *Weltanschauung:* This is the, often taken for granted, outlook or worldview that makes sense of the root definition being developed. It is important to specify this because any system definitions can only make sense with some defined context. Thus a root definition needs only a single Weltanschauung.
- *Ownership:* This is the individual or group responsible for the proposed system in the sense that they have the power to modify it or even to close it down. This can overlap with the actors of the system or the customers.
- *Environmental constraints:* All human activity systems operate within some constraints imposed by their external environment. These might be, for example, legal, physical, or ethical. They form part of the external relations of the system and need to be distinguished from its ownership.

The mnemonic CATWOE, is often used to summarize these six elements, taking the initial letters of the above six terms.

9.5 Using Root Definitions

To illustrate the use of root definitions, consider a simulation study (Gunal, Onggo, and Pidd 2007) conducted for a police force that began with a request for help in improving the performance of its Contact and Response Centers (CaRCs). CaRCs are the primary point of contact between members of the public and the police force. People needing help or wishing to report and incident phone an emergency number and are connected to the nearest CaRC in which a call-taker talks to them and types a database entry, which is passed for response. The calls are graded by their severity so as to enable an appropriate response. The response is requested from local police units by radio operators who are also housed in the CaRC. The initial issue that presented itself was the poor performance of the CaRCs in answering the phone. The police force had agreed targets for answering calls but was nowhere near meeting them. Some callers had to wait a long time and some even complained about receiving an engaged tone. Neither was impressive for an emergency service that was provided for fearful or endangered citizens who may need help.

This study will be used to show how root definitions can illuminate stakeholder analysis and shed light on people's different concerns. Example root definitions will be developed to interpret how different groups might see the CaRCs themselves and also to tease out the expectations that different stakeholders may have of the study itself. The main stakeholders in this study were these:

- The admin branch of the police force who had asked for help from the simulation team. The admin branch is best regarded as the crowd in terms of their power and interest, since though they set up the study, the outcome does not directly affect them, and they have limited ability to change things, except through other people.
- The police authority, which is a governance structure that, in the UK, has responsibility for ensuring that the police force is accountable to the government and to the population. In terms of the power-interest grid, the police authority is best regarded as a context setter, since it is accountable for expenditure and performance yet has no detailed interest in the working of the CaRCs.
- Members of the public clearly have a great interest in the performance of the CaRCs, but have no real direct power to do anything about them and they are best regarded as subjects whose interests need to be protected.
- The operators who worked in the CaRCs, answering calls, and deciding what resources were needed to resolve a situation and these are also subjects, since they have a great interest in working conditions but little direct power to affect the outcome.
- The senior officers who are responsible for the operation and performance of the CaRCs, who are best regarded as the players, since they do have power to change things as well as having very major interests in those changes.

A root definition (CATWOE) could be constructed for each stakeholder group and we now consider a few examples to illustrate the use of root definitions. Note, however, that these root definitions are not representative of the actual stakeholders of the actual simulation study but are used to illustrate the structure and point of root definitions.

9.5.1 Root Definitions for the CaRCs

Consider, for example, members of the public who might call a CaRC, seeking help from the police. How might such people see a CaRC in terms of a root definition?

- *Customers:* clearly, most members of the public would see themselves as the main beneficiaries of a properly run CaRC.
- *Actors:* it seems likely that any member of the public who thought about this would regard the staff and officers of the CaRCs as the principal actors.

- *Transformation:* a member of the public is likely to see a CaRC as existing to receive calls that are transformed into appropriate and timely action.
- *Weltanschauung:* the previous three elements only make sense within a worldview that sees responsive policing as important for public safety and security.
- *Ownership:* since the CaRCs are funded through the police budget, it is clear that the owner is the police force itself.
- *Environmental constraints:* the CaRCs must operate within defined budgets, using available technology and responding in such a way as to provide an appropriate level of service.

Thus, seen in these terms, the CaRCs are a system that takes calls from the public and provides an appropriate and timely response for the benefit of the public who see such a response as necessary. The CaRC is run by the police force using staff and officers who operate within defined budgets using available technology.

As a slight contrast, discussions with the senior officers who manage the CaRCs may lead to a root definition something like the following.

- *Customers:* it is possible that the managers of the CaRCs might see the police force itself as the customer, since the CaRCs are part of responsive policing, in which appropriate resources should be deployed to incidents in a timely manner. This does not mean that these officers would ignore the needs of the public, but they may have different customers in mind.
- *Actors:* it seems likely that managers would regard the staff and officers of the CaRCs as the principal actors.
- *Transformation:* as mentioned in the discussion of customers, the transformation might be to turn information from the public into responsive policing.
- *Weltanschauung:* in the light of the previous elements, a worldview that makes sense is that the police force must engage in responsive policing.
- *Ownership:* since the CaRCs are funded through the police budget, it is clear that the owner is the police force itself.
- *Environmental constraints:* the CaRCs must operate within defined budgets, using available technology and responding in such a way as to provide an appropriate level of service.

Thus, in these terms, the CaRCs are needed to support responsive policing and are organized so as to provide a good responsive service, operated by staff and officers within budget and technology constraints and owned by the police force.

9.5.2 Root Definitions for the Simulation Study

As well as using root definitions to capture how different stakeholders might see the CaRCs, the same approach can be used to think through the simulation study itself. Consider, for example, the admin branch of the police force who commissioned the work. Perhaps they are concerned to ensure that the CaRCs meet performance targets as part of an effort to show that this is an excellent police force. With this in mind, a possible CATWOE for a simulation study might be as follows

- *Customers:* since the admin branch commissioned the study, they would probably see themselves as the customers.
- *Actors:* members of the admin branch are likely to see the simulation modelers as the main actors, whom they assist.
- *Transformation:* in these terms, the transformation is to move from being unsure why performance is poor to knowing what could be done to improve it.
- *Weltanschauung:* the previous three elements only make sense within a worldview that believes that a simulation model will provide useful performance information.
- *Ownership:* since the simulation modeling is commissioned by the admin branch it is clear that they are the main owners as well as being customers. It is, though, true that the modelers could also close down the project.
- *Environmental constraints:* the simulation modeling must be completed within agreed budgets and timescales, possibly using agreed software.

Seen in these terms, the simulation project is one commissioned by the admin branch so that they may develop ways to improve the performance of the CaRCs within agreed budgets and timescales in the belief that a simulation model will enable them to do this.

What about the staff and officers who work as operators in the CaRC? How might they see the simulation project. For simplicity we will assume that they share the public's view of the CaRCs, but not the admin branch's view of the modeling project. Hence, a possible CATWOE for their view of the modeling project might be as follows.

- *Customers:* the admin branch.
- *Actors:* admin branch and simulation modelers.

- *Transformation:* to move from a situation in which staff and officers in the CaRCs use their expertise to manage the CaRCs to one in which a more technocratic approach is used.
- *Weltanschauung:* the people who don't run CaRCs always think they know best how to improve their performance.
- *Ownership:* admin branch, and certainly not the staff and officers who work in the CaRCs.
- *Environmental constraints:* the project will have to be completed with whatever cooperation and time they can give in their over busy working lives.

Seen in these terms, the modeling project is one commissioned by the admin branch and only possible with the help of CaRC staff and officers, which may change the way they work in ways recommended by people who have never worked in a CaRC. If this is their view, they may be rather negative about the simulation project for fear that their working conditions will worsen or they may need skills that they do not possess. This emphasizes the view that the operators, as subjects, cannot be ignored and must be kept in the loop.

9.6 Using Root Definitions to Support Conceptual Modeling

It would be possible to develop root definitions for the other stakeholders both for the CaRCs themselves and for the simulation project so as to interpret their different views. In this way, it is possible to tease out different worldviews and assumptions about the operation of the CaRCs and of the modeling project. It ought be clear that gaining this understanding may be crucial in gaining the cooperation that will be needed if the work is to proceed with any chance of success. It may, of course, be argued that any experienced analyst will intuitively think through such issues—but, the perfect never have anything to learn.

Old hands in the simulation community will remember when developing a model always involved writing code, whether in a general purpose language or a simulation language. This was a tedious and error-prone process that forced the modeler to think very hard before writing code and provided an incentive for the development of tightly defined models for specific tasks. Contemporary simulation tools rightly free us from this drudgery, but their ease of use brings a temptation to dive straight to the keyboard and mouse and, later, to go on enhancing models. Like most temptations, this can result in initial pleasure but subsequent regret. Developing a conceptual model before diving for the computer can help reduce the appeal of this temptation, which is probably why Robinson (2004) defines a conceptual model as a non-software-specific description of the simulation model that is to be developed, describing the objectives, inputs, outputs, content, assumptions and simplifications of the model. This is a rather broad definition that might be better thought of as the conceptualization of a simulation model or a simulation project, rather than a conceptual model. Leaving aside this semantic difference, however, it should by now be clear that such a conceptual model depends heavily on the degree to which the modeler has an understanding of the appropriate simplifications required in the simulation model and an appreciation of the project context within which the simulation model will be developed and used.

This chapter has argued that formal PSMs can be used to gain an understanding of the context within which a simulation model will be built and may be used. It should also be clear that this is true of informal methods such as critical examination as well as of formal approaches such as SSM. The aim is to see the issues from the viewpoint of stakeholders with interest in the project or the power to do something about it. Stakeholder analysis using power-interest grids provides some insight and allows an analyst to decide how best to devote their efforts in listening to people's views and trying to satisfy them. Root definitions, as employed in SSM, allow the analyst to interpret how stakeholders see the issues involved in the study. Thus, PSMs can assist an analyst in developing this understanding and appreciation.

Acknowledgments

This chapter is based on an advanced tutorial delivered at the 2007 Winter Simulation Conference and published as Making sure you tackle the right problem: Linking hard and soft methods in simulation practice. In *Proceedings of the 2007 Winter Simulation Conference,* ed. S.G. Henderson, B. Biller, M.-H. Hsieh, J. Shortle, J.D. Tew, and R.R. Barton, 195–204. 9–12 December, Washington, DC.

References

Baldwin, L.P. T. Eldabi, and R.J. Paul. 2004. Simulation in healthcare management: A soft approach (MAPIU). *Simulation modeling practice and theory* 12: 541–557.

Checkland, P.B. 1981. *Systems Thinking, Systems Practice.* Chichester: John Wiley & Sons, Ltd.

Checkland, P.B. 1999. *Systems Thinking, Systems Practice: Includes a 30-Year Retrospective.* Chichester: John Wiley & Sons, Ltd.

Checkland, P.B., and J. Poulter. 2006. Learning for action: A short definitive account of soft systems methodology, and its use practitioners, teachers and students. Chichester: John Wiley & Sons Ltd.

Checkland, P.B., and J. Scholes. 1999. *Systems Methodology in Action,* 2nd edition. Chichester: John Wiley & Sons Ltd.

Checkland, P.B., and S. Holwell. 2004. "Classic" OR and "soft" OR: An asymmetric complementarity. In M. Pidd, *Systems Modeling: Theory and Practice.* Chichester: John Wiley & Sons Ltd.

Conklin, J. 2002. Dialog mapping. http://cognexus.org/index.htm. (Accessed, March 2010).

Den Hengst, M., G-J. de Vreede, and R. Maghnouji. 2007. Using soft OR principles for collaborative simulation: a case study in the Dutch airline industry. *Journal of the Operational Research* 58: 669–682.

Eden, C., and Ackermann F. 1998. *Making Strategy: The Journey of Strategic Management.* London: Sage Publications.

Gunal, M.M., S. Onggo, and M. Pidd. 2007. Improving police response using simulation. *Journal of the operational research society* 59: 171–181.

Harrell, C.R., B.K. Ghosh, and R.O. Bowden. 2004. *Simulation Using ProModel,* 2nd edition. New York: McGraw-Hill Professional.

Howick, S. 2003. Using system dynamics to analyze disruption and delay in complex projects for litigation: Can the modeling purposes be met? *Journal of the operational research society* 54: 222–229.

Kelton, W.D., R.P. Sadowski, and D.T. Sturrock. 2004. *Simulation with Arena.* New York: McGraw-Hill.

Kolb, D.A. 1983. Problem management: Learning from experience. In *The Executive Mind,* ed. S. Srivasta. San Francisco, CA: Jossey-Bass.

Kotiadis, K. 2007. Using soft systems methodology to determine the simulation study objectives. *Journal of simulation* 1: 215–222.

Kotiadis, K., and J. Mingers. 2006 Combining PSMs with hard OR methods: The philosophical and practical challenges. *Journal of the operational research society* 57: 856–867.

Law, A.M. 2006. *Simulation Modeling and Analysis,* 4th edition. New York: McGraw-Hill.

Lee, J., and K.C. Lai. 1991. What's in design rationale? *Human-computer interaction* 6 (3&4): 251–280.

Lehany, B., and R.J. Paul. 1996. The use of SSM in the development of a simulation of out-patients at Watford General Hospital. *Journal of the operational research society* 47: 864–870.

Lewis, P.A.W., and E.J. Orav. 1989. *Simulation Methodology for Statisticians, Operations and Analysts and Engineers, Volume 1.* Pacific Grove, CA: Wadsworth & Brooks/Cole.

Lubart, T.I. 1994. Creativity. In R.J. Steinberg (ed.), *Thinking and Problem Solving,* 2nd edition. London: Academic Press.

Mingers, J., and A. Gill. 1997. *Multi Methodology.* Chichester: John Wiley & Sons Ltd.

Mingers, J., and S. Taylor. 1992. The use of soft systems methodology in practice. *Journal of the Operational Research Society* 43(4): 321–332.

Paul, R., and B. Lehany. 1996. Soft modeling approaches to simulation model specifications. In *Proceedings of the 1996 Winter Simulation Conference,* 8–11 December, Hotel Del Coronado, Coronado, CA. Baltimore: Association for Computing Machinery.

Pidd, M. 2004. *Computer Simulation in Management Science,* 5th edition. Chichester: John Wiley & Sons Ltd.

Pidd, M. 2010. *Tools for Thinking: Modeling in Management Science,* 3rd edition. Chichester: John Wiley & Sons Ltd.

Pidd, M., ed. 2004. *Systems Modeling: Theory and Practice.* Chichester: John Wiley & Sons Ltd.

Pidd, M., and R.N. Woolley. 1980. A pilot study of problem structuring. *Journal of the operational research society* 31: 1063–1069.

Rittel, H.W.J., and M.M. Webber. 1973. Dilemmas in a general theory of planning. *Policy Sci* 4: 155–169.

Robinson, S. 1994. *Successful simulation: A practical approach to simulation projects.* London: McGraw-Hill.

Robinson, S. 2004. *Simulation: The Practice of Model Development and Use.* Chichester, UK: John Wiley & Sons Ltd.

Robinson, S. 2008. Conceptual modelling for simulation part I: Definition and requirements. *Journal of the operational research society* 59(3): 278–290.

Rosenhead, J.V. 1989. Rational analysis for a problematic world: Problem structuring methods for complexity, uncertainty and conflict. Chichester: John Wiley & Sons Ltd.

Rosenhead, J.V., and J. Mingers (eds.). 2001. *Rational Analysis for a Problematic World Revisited.* Chichester: John Wiley.

Sachdeva, R., T. Williams, and J. Quigley. 2007. Mixing methodologies to enhance the implementation of healthcare operational research. *Journal of the operational research society* 58: 159–167.

Shaw, D., A. Franco, and M. Westcombe (eds.). 2007. Special issue: Problem structuring methods II. *Journal of the operational research society* 58: 545–700.

Simon, H.A. 1972. Theories of bounded rationality. In H.A. Simon (1982), *Models of Bounded Rationality: Behavioural Economics and Business Organization.* Cambridge, MA: MIT Press.

Simon, H.A. 1976. From substantive to procedural rationality. In H.A. Simon (1982), *Models of Bounded Rationality: Behavioral Economics and Business Organization.* Cambridge, MA: MIT Press.

Wilson, B. 1990. *Systems: Concepts, Methodologies, and Applications,* 2nd edition. Chichester: John Wiley & Sons Ltd.

10

Using Soft Systems Methodology in Conceptual Modeling: A Case Study in Intermediate Health Care

Kathy Kotiadis

CONTENTS

10.1 Introduction

This chapter explores how soft systems methodology (SSM), a problem structuring method, can be used to develop an understanding of the problem situation and determine the simulation study objectives based on the experience gained in a real life simulation study in health care. Developing an understanding of the problem situation and determining the modeling objectives are two of the four phases of Robinson's (2004) conceptual modeling. Robinson (2004) divides conceptual modeling into the following phases:

- I. Developing an understanding of the problem situation
- II. Determining the modeling objectives
- III. Designing the conceptual model: inputs outputs and model content
- IV. Collecting and analyzing the data required to develop the model

Robinson's (2004) phases can also be used to describe the output of the conceptual modeling processes. There are two main processes involved in conceptual modeling: knowledge acquisition and abstraction (Kotiadis and Robinson 2008). This chapter explores how SSM contributes to these processes in order to get an understanding of the problematic situation and determine the study objectives.

This chapter is divided into four main sections. The introduction with its remaining subsections form the first section. The first section explores the conceptual modeling processes that the case study contributes toward and reflects on the appropriateness of SSM to conceptual modeling by looking at what others have said when they used SSM in their simulation study. The second section explores the case study, which is broken down into subsections. These subsections include a description of the problem and the motivation for using SSM, a brief description of SSM in general and how it was applied to this case study in terms of knowledge elicitation and abstraction. However special attention is paid on how SSM was conducted and extended to determine the simulation study objectives and the section concludes with a set of guidelines. The third section provides a discussion about the opportunity to further adapt SSM to determine the study objectives and the benefits of the proposed approach. The final section provides a summary of the chapter.

10.1.1 The Conceptual Modeling Processes

More recently Robinson (2008) defined the conceptual model to be "a non-software specific description of the computer simulation model (that will be, is or has been developed), describing the objectives, inputs, outputs, content,

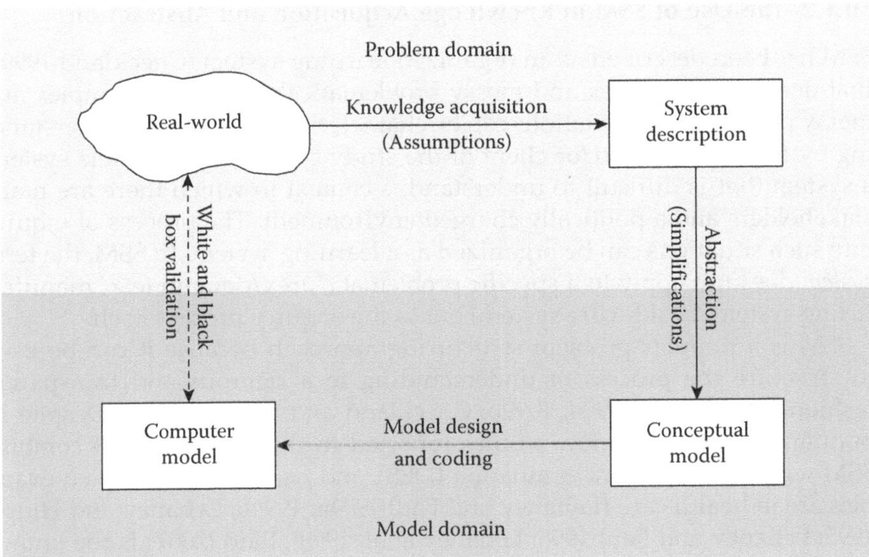

FIGURE 10.1
Artifacts of conceptual modeling. (From Kotiadis, K. and Robinson, S., *Proceedings of the 2008 Winter Simulation Conference*, Institute of Electrical and Electronic Engineers, Inc., Miami, FL, 2008. With permission.)

assumptions and simplifications of the model." (p. 283). This definition of a conceptual model includes assumptions and simplifications. Kotiadis and Robinson (2008) discuss how these two artifacts of conceptual modeling (assumptions and simplifications) have been linked with the conceptual modeling processes of knowledge acquisition and abstraction (Figure 10.1). Knowledge acquisition is about acquiring knowledge and information about the real world from subject matter experts (SMEs). However, the real world is often not understood or fully known or knowable (if one is modeling a system that does not exist) by the SMEs and *assumptions* must be made about the real world (Robinson 2008). The second process, model abstraction, is about the modeler obtaining the conceptual model by agreeing with the problem owners on what parts of the system description to model and at what level of detail. This process involves reducing the level of detail and/or the scope in the conceptual model in comparison to the system description. This process of partially representing the system description in the conceptual model involves simplification. Robinson (2008, p. 283) explains that "*simplifications* are incorporated in the model to enable more rapid model development and use, and to improve transparency" of the model. The process of simplification should focus on maintaining sufficient accuracy for addressing the problem situation/modeling objectives (Kotiadis and Robinson 2008).

In the next subsection we will look at the appropriateness of SSM with respect to the processes that have just been described.

10.1.2 The Use of SSM in Knowledge Acquisition and Abstraction

SSM has been described as an organized learning system (Checkland 1999a) that deals with complex and messy problematical situations. Complex and messy problematical situations can be characterized by a lack of understanding by the modeler and/or client of the study context, a nonvisible system, a system that is difficult to understand, a context in which there are many stakeholders and a politically charged environment. The process of inquiry into such situations can be organized as a learning system. In SSM, the term *system* does not apply to a specific problematic area/domain (e.g., manufacturing system, health-care system) but to the enquiry process itself.

SSM is a popular problem structuring approach because it can be used to structure the process of understanding in a rigorous and transparent fashion (Checkland 1999a, 1999b; Checkland and Scholes 1999). Despite its popularity there are a few studies reported in the literature that combine SSM with discrete-event simulation (DES), and nearly all the known examples are in health care (Lehaney and Paul 1994a, 1994b; Lehaney and Hlupic 1995; Lehaney and Paul 1996; Lehaney et al. 1999). Pidd (2007) is the known exception who provides a general in depth discussion on the use of Problem Structuring Methods in simulation, which includes SSM.

The oldest papers discussing the link between SSM and DES modeling are by Lehaney and Paul (1994a, 1994b), one of which is of particular interest as it demonstrates how an SSM model (the purposeful activity model (PAM)) can be developed as an activity cycle diagram using a case study of an outpatient facility (Lehaney and Paul 1994a). A second paper by Lehaney and Paul (1994b), which is complimentary to the first one, uses the SSM model to question health-care participants in the development of the simulation model. However, neither of these papers are focused on how the simulation study objectives are determined.

Lehaney and Hlupic (1995) review the use of DES for resource planning in the health sector and suggest the use of SSM as an approach for improving the process and research outcomes. Lehaney and Paul (1996) examine the use of SSM in the development of a simulation of outpatient services. The paper explains how the discussion of a PAM was used to determine the system that should be modeled out of a number of systems that could be potentially modeled. The authors argue that this multimethodology allows the participation of the staff in the modeling process and they conclude that the participation paved the way for the acceptance of the conceptual model and gave rise to the final simulation being credible. However, the paper's main contribution to conceptual modeling is that it suggests that SSM can be used to "tease out" of participants what actually should be modeled. Lehaney and Paul, however, do not specifically map the process for others to replicate.

In a subsequent paper Lehaney et al. (1999) report on an intervention that utilized simulation within a soft systems framework, and they call it *soft-simulation*. The project uses SSM to map out the activities of a simulation study through a case study of an outpatient dermatology clinic. Again it cannot be readily used by other stakeholders to determine a conceptual model as it is focused on the modeler and the activities that he or she should undertake in a simulation study.

Baldwin et al. (2004) propose an approach to thinking, influenced by SSM, to enhance stakeholder understanding and communication in health-care simulation studies during the problem structuring phase. They acknowledge the benefits of using SSM thinking for DES and contribute to that by linking some of their concepts. However, they focus on project management issues rather than conceptual modeling.

The studies mentioned above advocate the benefits of using SSM in simulation studies. Although these studies do not specifically focus on conceptual modeling it is evident, although not always explicit, that SSM has contributed to it to some extent. Because, these studies do not specifically focus on the role of SSM in knowledge acquisition and model abstraction there is an opportunity to reflect on SSM's contribution to these processes. These conceptual modeling processes are about understanding the problematic situation in a rigorous and transparent fashion, and abstracting from this situation the conceptual model. Although one could argue that SSM's main contribution is to problem structuring (knowledge acquisition), it does also have a role to play in model abstraction. In fact this chapter shows how SSM can help determine the simulation objectives, which is something that has not been seen in any other simulation study (Kotiadis 2007).

10.2 The Case Study: Intermediate Health Care

In this study, SSM was originally used for a similar reason to the one mentioned by Lehaney and Paul (1996): to understand which dimension of the problem and system to model. In our case the health system was a complex integrated health system for older people, called Intermediate Care (IC) in a locality in Kent. However, SSM was also found to be useful in other aspects that had not previously been anticipated, such as determining the conceptual model, particularly for conceptual modeling phases I, II, and most of phase III (Robinson 2004).

The study was part of a larger IC evaluation project commissioned for a locality in Kent, England in 2000 by the Elderly Strategic Planning Group and the Joint Planning Board for Care of the Elderly in East Kent (Carpenter et al. 2003). At that time IC services were a relatively new concept and their introduction can be attributed to the growing population of older people that in many cases were found to be inappropriately using the expensive and scarce secondary care resources (hospital-based resources). The Department of Health response to the National Bed Enquiry (Department of Health 2000) stated that it intended a major expansion of community health and social care services (IC) that in contrast to acute hospital services would focus on rapid assessment, stabilization, and treatment.

It was decided that DES modeling should be deployed as it had proved itself useful in other health-care studies in evaluating resources. However, we were having difficulty deciding how to model the IC system. The actual IC system at the beginning of the study was in its development phase and it was not particularly understood as a whole by any one person in the system. Knowledge acquisition in order to gain an understanding the system is generally considered the initial step of any simulation study and the initial output of conceptual modeling. In this study it was not immediately apparent how to carry out this step because of the size of the system, the newness of IC, and the difficulties in observing the system as the changes in the system were slow and its services were geographically dispersed. More importantly, no one had an overall understanding of how the system worked. This meant a problem structuring approach like SSM could be used to get an understanding of how the system worked and how it could/should work.

10.2.1 A Brief Look at SSM in General

For those that are not familiar with SSM, the following paragraphs present a brief overview of the methodology. To gain a more detailed understanding of SSM the reader should refer to Checkland (1999a, 1999b) and Checkland and Scholes (1999). In this case Checkland's (1999a) *four main activities version* of the SSM methodology was deployed, which consists of the following stages:

1. Finding out about a problem situation, including culturally/politically
2. Formulating some relevant PAMs
3. Debating the situation, using the models, seeking from that debate both of these:
   a. Changes that could improve the situation and are regarded as both desirable and (culturally) feasible
   b. Accommodations between conflicting interests that will enable action-to-improve to be taken
4. Taking action in the situation to bring about improvement

The processes of knowledge acquisition and model abstraction largely map on to SSM's first and second stage. Therefore this chapter focuses only on these steps but readers interested in the other stages of SSM or the methodology as a whole should consult Checkland (1999a, b).

The following paragraphs concentrate on introducing the SSM's tools that contribute to knowledge acquisition and model abstraction. There are two main tools used to assist the modeler in knowledge acquisition (finding out about the problem situation): drawing rich pictures and analyses one, two, and three. Pidd (2007) provides an in-depth discussion on the latter. The CATWOE, root definition, and performance measures (three Es) are tools that contribute toward the process of constructing the PAM, which is an SSM model. The PAM contributes toward the process of abstraction. The remainder of this subsection provides some reflections on these tools.

10.2.1.1 Rich Pictures

Rich pictures involve a holistic drawing of the situation of interest. The pictures do not have a specific format or language but aim to encompass the key elements of a situation such as processes, issues and stakeholders. Because they are easy to understand they can be drawn in a participative way with stakeholders. For example they can be drawn by the modeler during a semi-structured interview with a stakeholder or they can be drawn by a stakeholder or modeler in a participative way with a group of stakeholders. Therefore the advantage of using this tool to find out about the problematic situation is that it is non technical and enables a wide participation of stakeholders.

In terms of using this tool in knowledge acquisition for a simulation study, it is likely at the end of drawing a rich picture the modeler may not have a clear understanding of what should be modeled from this problematic situation, but will have a good grasp of the overall situation. This is particularly important when the modeler(s) or stakeholders are not familiar with overall situation but only a part of it. Also any initial assumptions about the situation can be brought forward through discussion and dealt with.

This SSM tool has not been particularly reported on in simulation studies. This could be attributed to the fact that almost all studies have been in health care. Drawing rich pictures is best achieved in a participative environment (including the modeler(s) and stakeholders) with a reasonable amount of time at hand to undertake the process, which is difficult to arrange with health-care professionals. Another reason is that this tool would be used at the start of the modeler(s)/stakeholders collaboration and the output of this tool tends to look like a child like drawing on flip chart paper. The modeler may feel that this tool does not fit with the image often sold to the client of working toward a computer model. On the other hand, modelers that use visual interactive simulation software have the opportunity to use the animation to discuss the problematic situation of interest but are limited to what can be drawn using the package.

10.2.1.2 Analyses One, Two, and Three

In addition to the use of rich pictures, Checkland (1999b) advocates analyses one, two, and three, otherwise respectively known as role analysis, social system analysis, and political system analysis.

Role analysis, or analysis one, is an analysis of the intervention system that involves exploring three main roles: the role of the client (who has caused the study to take place) the role of the "would-be problem solver" (who wants to do something about the situation) and the role of the problem owner. All or some of these roles may overlap.

Social system analysis, or analysis two, is based on the notion that a social system is a continually changing interaction of three elements: roles, norms and values. Roles are social positions of importance to the problem situation that are institutionally defined or behaviorally defined. The expected behaviors are otherwise known as norms. In addition, performance in a role will be judged according to values.

Political system analysis, or analysis three, is about understanding how power is expressed in a particular problematic situation. For example, power may be in the form of personal charisma or even membership to a particular committee.

Understanding the roles within a problem situation, typical behavior of the stakeholders and the allocation of power can mean that the modeler can manage the stakeholders during the conceptual modeling process (and the rest of the simulation intervention) and arrive at a conceptual model that is agreeable to all, desirable and feasible. All three analyses compliment each other and do not necessarily require to be undertaken in any particular order. Also the nature of some of the questions being asked should involve a certain amount of sensitivity and that can mean that some or all of this analysis may need to be done covertly (Pidd 2007). A possible solution to this is taking advantage of the rich picture drawing session to observe stakeholders and ask leading questions as part of analyses one, two, and three. This covert analysis is feasible as stakeholders can be made to feature within the drawings.

10.2.1.3 The Purposeful Activity Model

A PAM is an SSM model. The initial stage to developing a PAM is to define the system of interest using a structured approach involving a set of SSM tools. The SSM tools used as part of the process to develop the PAM also contribute toward knowledge acquisition. However, in simulation conceptual modeling the PAM largely contributes to the process of abstraction.

Some SSM authors refer to the PAM as a "conceptual model," but this has a different meaning to a simulation conceptual model. We will use conceptual model only with reference to simulation and PAM with reference to the SSM model. Checkland (1999a) provides both extensive guidance and examples of how to use the SSM tools (transformation process, CATWOE, root definition, measures of performance) to arrive at the PAM. These tools offer guidance on how to format a set of definitions to help develop the PAM. The definition of the system, called the root definition, can be loosely compared to a company's mission statement. However, central to the root definition is a need to demonstrate the transformation process (T) of some input to output.

Essentially the process undertaken to develop the root definition is an exercise in focusing the mind on the experimental frame prior to constructing the PAM.

Checkland (1999a) suggests that prior to constructing the PAM one should also define the measures of performance. He explains that it is necessary to "define the criteria by which the performance of the system as a whole will be judged" (p. A25) and suggests using the criterion of Efficacy (check that the output is produced), the criterion of Efficiency (check that the minimum resources are used to obtain the output), and the criterion of Effectiveness (check at a higher level that this transformation is worth doing because it makes a contribution to some higher level or long-term aim). Efficacy, Efficiency, and Effectiveness form the main measures of performance and are often referred to as the 3 Es. Checkland puts forward a further two criteria, Ethicality (is the transformation morally correct?) and Elegance (is this an aesthetically pleasing transformation?), that may be useful in some cases.

Checkland (1999a) suggests that once the definitions have been expressed (CATWOE, Root definition, measures of performance), it should be easy to construct the PAM. The process of constructing the PAM involves a certain amount of simplification as it is a record of the *necessary* activities to support the system's transformation of input to output (see example in Figure 10.2). Checkland (1999a, b) recommends listing seven activities plus or minus two. The PAM is a simplification of the system description and has the potential to be used instead of the communicative conceptual model (e.g., an Activity Cycle Diagram) (Lehaney and Paul 1994a).

Checkland and Scholes (1999) explain that a PAM provides an idealistic view of the elements in a system and does not represent reality as participants are asked to think outside the current bounds of what is there. This enables the participants to compare reality with the idealistic view with the aim of reaching consensus on any feasible changes. The following section explains how the PAM contributes to the process of abstraction.

10.2.2 Applying SSM to the IC Health System

In this section, we will initially explore SSM's contribution to knowledge acquisition and then to model abstraction in this particular case study.

10.2.2.1 Knowledge Acquisition Using Analyses One, Two, and Three

In this study analysis of the roles, social and political system started early on and continued throughout most of the study. There were several people involved in this system and as the study was over a period of about three years, there were a lot of changes that redefined each analysis. For example, some employees within the system were promoted and therefore acquired more power and so could commission extensive changes to the system. Understanding the political situation meant knowing what type of system would be feasible. From

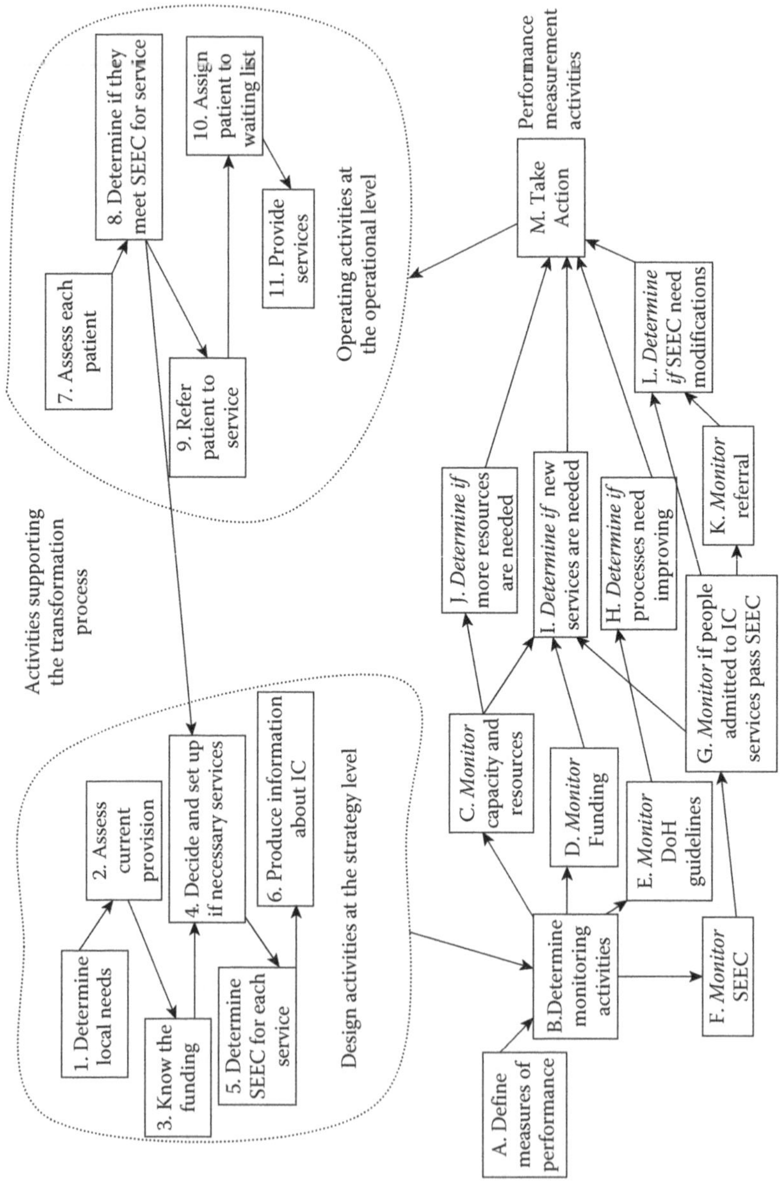

FIGURE 10.2
The purposeful activity model of the IC system. (From Kotiadis, K., *Journal of Simulation*, 1, 215–222, 2007. With permission.)

the analysis of the roles we were able to understand what action some key stakeholders were prepared to undertake within the system. This also meant knowing who to persuade to organize meetings with other stakeholders to obtain information. Undertaking the social analysis enabled a better understanding of the stakeholders within the system and enabled the modeler to align herself to the culture of this health and social care system and through interaction gain access to information and insights. Also some behavior was directly or indirectly included in the simulation model.

10.2.2.2 Knowledge Acquisition Using Rich Pictures

During the analyses above it became apparent that it would not be possible to observe the system as a whole because of its size, the slow pace of change in the system, and its geographic dispersion. Therefore, apart from observing individual services within the system and talking individually to stakeholders, it was very useful to meet concurrently with a number of stakeholders with knowledge of individual sections of the system as well as the key stakeholders with power over the entire system. In this particular case study these meetings were mostly arranged on the back of preexisting meetings, which meant that there was only limited time available and the stakeholders on many occasions were preoccupied by the business from the other meeting. In addition these meetings were scheduled weeks apart from each other so the process was slow.

We did not have the time to build up rich pictures with the dedication that one would observe in a stand alone SSM study but some ready made pictures of the processes within the system were brought along for discussion. It should be noted that these were not referred to as rich pictures but as diagrams. The stakeholders were told that we were using these to help us understand their system. In fact these diagrams were not typical rich pictures as they did not directly include issues or stakeholders. However these did emerge and get discussed within these meetings.

In hindsight it would have been better use of time for the modeler and stakeholders if a workshop was arranged for dedicated knowledge acquisition through rich picture building on the system as a whole. This opportunity could have been identified in the political or role analysis. However this was not a failure in terms of the analyses, but a failure in recognizing how these key stakeholders could be best used to the advantage of the study. The information acquired using these tools (drawing rich pictures and analyses one, two, and three) should compliment each other and support the process of knowledge acquisition.

Despite the missed opportunity to have a dedicated knowledge acquisition workshop, as no one had an overall understanding of how the system worked, the short meetings helped structure this ill-defined problem situation. Within these meetings the modeler realized that the IC system was not working as a whole because the individual services had not yet integrated their operational functions. SSM was able to deal with this because it enabled action research to take place. More specifically, the process of action research comprises of enquiry, diagnosis, action planning, action/intervention, evaluation and learning (Hart and Bond 1995). The stakeholders, who were aware of this lack of system integration, were interested in action research, which means they were willing to take action to improve the system during the study and not just as a result of the findings of the study. Therefore, it was sensible to aim at building a simulation model of a future integrated system rather than of the current situation and use SSM to determine what was considered by the stakeholders to be a desirable and feasible future system.

10.2.2.3 Abstraction Using CATWOE, Root Definition, Es, and PAM

In SSM it is common practice to develop several PAMs within an intervention. In this case, however, a primary task approach was adopted, in which only one relevant system (rather than many subsystems) that could potentially map on to an organizational boundary was concentrated upon. The focus was on building one overall PAM because the main focus at that time was to get an overall understanding of the IC system that was agreeable to all involved in order to build the simulation model. The study participants were asked a number of questions about the system individually or in group meetings, but for reasons such as having to explain to such a dispersed and diverse audience the SSM tools and how to use them within relatively short meetings, the participants were not directly involved in structuring the definition or constructing the PAM. However a group of key stakeholders that examined the definitions and PAM agreed that this was the most agreeable and feasible view of how the system could work.

A root definition of the relevant system, the IC system, was put together from information gained in knowledge acquisition in order to build the PAM. The mnemonic CATWOE, one of Checkland's best-known SSM tools and central to the process of deriving a root definition, was used to define the Customer, Actors, Transformation Process, Weltanschauung (the worldview), Ownership, and Environmental Constraints. The definitions in terms of the IC system according to each of Checkland's acronym guidelines are as follows:

- Customers—the victims or the beneficiaries of the transformation process—are the people over 65 who require rehabilitation or convalescence.
- Actors—those who will do the transformation process—are the IC employees, i.e., nurses, therapists, etc.
- Transformation Process—the conversion of input to output—is designing and operating a system of strategic and operational level activities to support IC in the locality.

- Weltanschauung—the worldview that makes the Transformation Process meaningful in its context—is a belief that these strategic and operational level activities are important in providing effective care for the older people.
- Ownership—those who could stop the transformation process—are the local health and social care authorities.
- Environmental Constraints—elements outside the system that are taken as given—include local IC funding, Department of Health guidelines, etc.

Using Checkland's (1999a) guidance on how to cast a root definition, but also taking into account the CATWOE definitions above, the following root definition was developed for the IC system:

> Root Definition (RD) = A local health and social care–owned system operated by IC staff that supports IC in our locality *by* designing and operating a system of IC strategic and operational activities *in order to* provide effective care for the older people, *while* recognizing the constraints of local IC funding and Department of Health guidelines.

In terms of this research the three measures of performance, or the 3 Es, are the following:

- Efficacy (E1): to check that the IC function is supported through IC strategic and operational activities
- Efficiency (E2): to check that the minimum IC resources are used to support the strategic and operational activities
- Effectiveness (E3): to check that the strategic and operational activities enable older people to be rehabilitated in the most appropriate service for their needs

The criteria, Ethicality and Elegance (Section 10.2.1), were not defined in this study, as they were not considered to add to the evaluation of our system.

The measures of performance were broken down into a number of activities and incorporated in the monitoring activities part of the PAM of the IC system (activities A–M in Figure 10.2). The reader should note that this is not common practice in stand alone SSM studies.

The Root Definition, CATWOE, and the 3 Es guided the construction of the activity model that aims to show the transformation process T (activities 1–11 in Figure 10.2). The process of building the activity model "consists of assembling the verbs describing the activities that would have to be there in the system named in the RD and structuring them according to logical dependencies" (Checkland, 2001, p. 77). Checkland (1999a, p. A26, Figure. A6) provides a set of guidelines in constructing the PAM. Information supplied by the stakeholders or observed during the first SSM stage (finding out about the problem situation), was used to determine the activities essential to the SSM transformation process.

The core PAM of the IC system can be seen in top part of Figure 10.2 (activities 1–11) and the right-hand section of this (activities 7–11) is closest to the computer model. Therefore the PAM is a simplification of the system description, but also with further abstraction provides the model description. More specifically, the PAM includes all the main IC operational activities (simplification/reduction in the level of detail in the conceptual model from that of the system description), but also describes in a focused way what actually takes place in the computer model (reduction in the scope of the conceptual model from that of the system description). Therefore, through this level of abstraction part the conceptual model can be derived from this simple representation of the IC system.

10.2.2.4 Determining the Simulation Study Objectives

We now focus on how the simulation modeling objectives were derived from the PAM. Central to this process is breaking down the three performance criteria (Efficacy, Efficiency, and Effectiveness) that were defined for the system into a number of activities that lead to an evaluation of this system. This process is not part of the standard SSM in constructing the PAM, but an extension aimed at producing a comprehensive line of questioning that supports the evaluation. This SSM extension forms the construction of the performance measurement model (PMM). Similar to the process of constructing the standard PAM, the information for these activities was obtained through the interviews, group meetings and the literature. Although the stakeholders did not participate in the construction of the PMM, they deemed it to be feasible and desirable.

In order to construct the PMM, the performance criteria were further broken down into monitoring activities (e.g., in the bottom part of Figure 10.2 the activities starting with "monitor") and corresponding activities to determine the action needed (e.g., in the bottom part of Figure 10.2 the activities starting with "determine if"). Subsequently they were circled, arranged and linked using arrows in a logical order and given a letter rather than a number to distinguish them from the transformation activities in the PAM. The next step was to decide which of the performance activities in the PMM could contribute data or be explored in a simulation model, largely represented by the operational level activities in Figure 10.2. In fact the PMM activities apart from helping the modeler determine the simulation study objectives can also be used to determine the inputs and outputs of the model.

Nearly all of the PMM activities influenced the development of the simulation study objectives, except for D, E, and H (Figure 10.2) as they could not be directly explored in the simulation model that focused on the operational level activities. Also activities A, B, and M do not form part of the simulation study objectives as they relate to the process of building the PMM. The remaining PMM activities were logically grouped into the following questions that formed the simulation study objectives:

- Are IC services working to their capacity (C and J)?
- Are the IC patients admitted to the most appropriate IC service (G, K, and L)?
- Are there any service gaps (F, G, and I)?

If SSM had not been deployed, no doubt the first objective regarding capacity would have been derived, since it is a typical simulation study question (Davies and Roderick 1998, Jun et al. 1999). In this study, capacity is examined by including all the places/beds available for each IC service in the whole system simulation model and monitoring queues.

The second and third questions, however, are more original and can be attributed to the use of SSM. To answer the second question the model emulates the decision making process using a rule base that determines the service each patient should be sent to, based on a large number of patient characteristics (attributes). At the end of a run one can see if a particular patient or group of patients entered the service that they had actually entered in real life. The model is able to answer the third question by determining whether there is an appropriate service for each level of IC need by examining if there are gaps in the services mix. For example, it can be used to examine the effects of adding a new service or removing an existing service (Kotiadis 2006).

10.3 Using SSM to Determine the Simulation Objectives

These simulation study objectives guided the development of the simulation model, which met the needs of the stakeholders and led to an implementation of the study findings. Implementation of a simulation study has been described as a four stage model consisting of the following stages (Robinson and Pidd 2008):

1. The study achieves its objectives and/or shows benefit.
2. The results of the study are accepted.
3. The results of the study are implemented.
4. Implementation proved the results of the study to be correct.

Stage 3 and 4 are what are referred to in the literature as *implementation*. A number of authors (Wilson 1981, Lowery 1994, Jun et al. 1999, Fone et al. 2003) who report a lack of implementation also provide advice to analysts with one notably common theme: to involve and gain the commitment of the users (health-care administrators and clinicians). Therefore if user participation will lead to implementation, then it is important to know how to involve the users in the model development. The most important part of model development is conceptual modeling because it is about deciding what to model and how to model it (Robinson 2004). In this chapter we have seen how SSM has been used to determine what to model in a case study in health care. The simulation study objectives are key to deciding what to model. We will now discuss some issues on determining the simulation objectives when using SSM for others adopting this approach to keep in mind. The first issue to be discussed is whether SSM can be further modified or adapted to determine the simulation objectives and the second issue whether there are any benefits in using SSM in determining the simulation objectives.

10.3.1 Can SSM be Adapted?

In this study, the SSM processes were adapted and therefore it is likely that others might wish to further adapt or even modify the existing SSM tools and approach to better map on to the specific needs of simulation studies. Checkland (1999a) in his 30-year retrospective of the use of SSM provides a lengthy discussion on what constitutes a claim to using SSM, but regardless of that emphasizes that SSM should be moldable to the situation. We initially reflect on an aspect of SSM that could be modified, although it was not modified it in this study, and then we reflect on an aspect of SSM that was adapted, but could be further adapted or even modified in other studies.

In SSM the PAM does not necessarily represent the current activities of a particular system of interest, which could be considered to conflict with the development of a DES conceptual model. In DES in general, the modeler builds a model of the current system that is verified and validated against it in an iterative manner until it is represented with sufficient accuracy. Exploring what could be there commences *if* and only *if* the model is considered to be a valid representation of the real system. In this study it was beneficial to actually move outside the current system constructs as it enabled creativity and led to alternative simulation study objectives. However, it may suit other simulation modelers to focus just on activities that are actually taking place in constructing the PAM, which would still enable modelers to benefit from deploying the SSM structured approach in defining the system. If this latter approach is adopted, the modeler could revisit the PAM after validation and verification of the simulation model and modify it to determine the simulation scenarios.

Another extension relating to the construction of the PAM is that the activities are grouped to strategic or operational level activities. This, to this author's knowledge, is not a step in stand alone SSM studies but could be a useful extension when used in simulation studies as there is a clear opportunity to map out the operational activities when done with stakeholders that may not easily distinguish the difference between the two. Another benefit of including the strategic level activities is that these provide the system owners perspective, which can lead to a PMM and subsequently objectives more aligned to their needs.

In this study, greater emphasis was placed on the performance criteria than is usually placed in other SSM studies (Checkland and Scholes 1999, Wilson 2001, Winter 2006). The core PAM (strategic and operational) activities were also linked with an extended model of performance criteria that are referred to as the PMM, which is to this author's knowledge again unreported as a step in the SSM literature. This stage was largely internalized and emerged after a series of discussions and reflections when there was a reasonable correspondence between the two, i.e., the PMM activities would satisfy the needs of the strategic and operational activities. Based mainly on the experience gained from this study, it is proposed that the following generic guidelines can be used by others to construct the PMM and arrive at the simulation study objectives:

1. Find out how the performance criteria developed relate to the real-life situation. Reflect on how each activity, supporting the transformation process in the PAM, can be evaluated.
2. Break down the performance criteria into specific monitoring activities, which are activities that involve observing and recording information. Where possible these activities should be in the format "monitor…."
3. Consider what action might be taken based on each of the monitoring activities or their combinations. Where possible record this action in the format "determine if…."
4. Where possible try and list the monitoring activities first and then link them according to logical dependencies to the "determine if" activities. Similar to the core PAM, circle each activity in the PMM and if helpful assign each a letter of the alphabet (rather than a number used in the core PAM).
5. Consider each of the performance measurement activities and determine which can be evaluated in a simulation model. These selected performance measurement activities can form simulation study objectives, but if necessary group these activities and relabel them to form simulation study objectives.

However, the PMM and process to derive the PMM can be further adapted or modified in order to better support the particular needs of a simulation study.

10.3.2 What are the Benefits of using SSM?

The more obvious benefits of using SSM to determine the simulation study objectives in the IC case were: (a) aiding the process of knowledge elicitation for this complex system that was difficult to understand, (b) enabling abstraction to take place leading to the model content and objectives, and (c) making the process more transparent and comprehensive and engendering creativity. Since the first benefit has been adequately covered in the case study and the second benefit in the previous section, this discussion focuses on the third benefit, regarding transparency.

There are two reasons for needing transparency in determining the simulation study objectives. First, there is a need for transparency so that simulation novices can learn how to determine a conceptual model. Conceptual modeling is currently treated as an art, which means that some aspects of conceptual modeling are difficult if not impossible to teach and most of the effort is spent demonstrating to students how to construct communicative models such as an activity cycle diagram. Unfortunately, this assumes that we have already gone through Robinson's (2004) conceptual modeling stages. The approach using SSM, described in this chapter, is reasonably transparent and provides both steps and tools to assist the less experienced modeler to appreciate the thought process in reaching the simulation study objectives.

The second reason why transparency is important is that it helps to establish trust between the modeler and clients. Pidd (1999) points out that although transparency through simplicity is important, a model should not be limited by the technical abilities of the clients. Therefore the trust established in the SSM process can be carried to the DES model without needing to simplify the DES model in order for the client to have a complete understanding of it.

Another benefit in using SSM is that it can surface objectives that would otherwise not be obvious. Lehaney et al. (1999) report that SSM saves time by surfacing issues that might have otherwise been left dormant within a simulation study. This is also the case in the study described here. The use of SSM ensured that the simulation objectives were appropriate and that the correct problem was addressed. In fact, based on this experience, it is reasonable to suggest that SSM leads to creativity in the conceptual modeling process.

Therefore the final and less obvious benefit from using SSM is that it can enable creativity to take place. Creativity is considered to encompass "seeing a problem in an unusual way, seeing a relationship in a situation that other people fail to see, ability to define a problem well, or the ability to ask the right questions" (Büyükdamgaci 2003, p. 329), which makes it an important element in problem definition. However being creative is difficult because by nature the brain is "hard wired" by its inherent abilities and predispositions (personality type) as well as the individual's past experience to function in a particular way (Büyükdamgaci 2003). This could mean defining the problem in a similar way to ones that have previously been experienced, called "functional fixation" (Duncker 2003). In this study the analyst had no prior simulation modeling experience in health care and very little modeling experience in general reducing the risk of functional fixation, but was at risk of irrational attitudes experienced under high stress levels, namely "defensive avoidance" and "hypervigilance" (Janis and Mann 1977). The former is about avoiding the problem by ignoring it and the later is about giving in to panic behavior and making decisions based on insufficient information. Fortunately, in this study, panic and stress led to the use of SSM!

10.4 Summary

This chapter set out to explain how SSM, a problem structuring method, can be used to develop an understanding of the problem situation and determine the simulation study objectives based on the experience gained in a real life simulation study in health care. Developing an understanding of the problem situation is the output of the conceptual modeling process of knowledge elicitation and determining the simulation study objectives is part of the conceptual modeling process of abstraction that leads to the computer model. Figure 10.3 depicts the relationship in general of the SSM tools and some of the artifacts of conceptual modeling discussed in this chapter. The following paragraph provides an explanation of Figure 10.3.

The SSM tools that can be of use to the simulation modeler are (a) rich picture drawing, (b) analyses one, two, and three, (c) CATWOE and root definition(s), (d) the performance measures (3 Es), and (e) the PAM. The SSM tools a, b, and c can help structure the process of knowledge acquisition and the output of these tools can be produced with the stakeholders and provides and agreeable view of the problematic situation. In Figure 10.3 there arrows going in both directions to represent the output of the process being deposited in the stakeholders' minds as well as the SSM tools. The PMM is another tool, not listed as an SSM tool, as it is an extension to the usual SSM approach (3 Es) that provides the opportunity to abstract the simulation study objectives and to some extent the inputs and outputs. In this chapter guidelines are provided on how to go about constructing the PMM. In addition to the PMM extension, the PAM is also constructed in a particular way; the PAM lists activities that are broken down to strategic and operational level activities. The computer model content can be abstracted at a high level from the operational level activities. However the objectives, inputs and outputs, derived from the PMM, also inform the construction of the computer model.

Using SSM in conceptual modeling provides structure and transparency to the process of knowledge acquisition and abstraction and paves the way for stakeholder participation and ultimately acceptance of the simulation study finding and implementation of the recommendations.

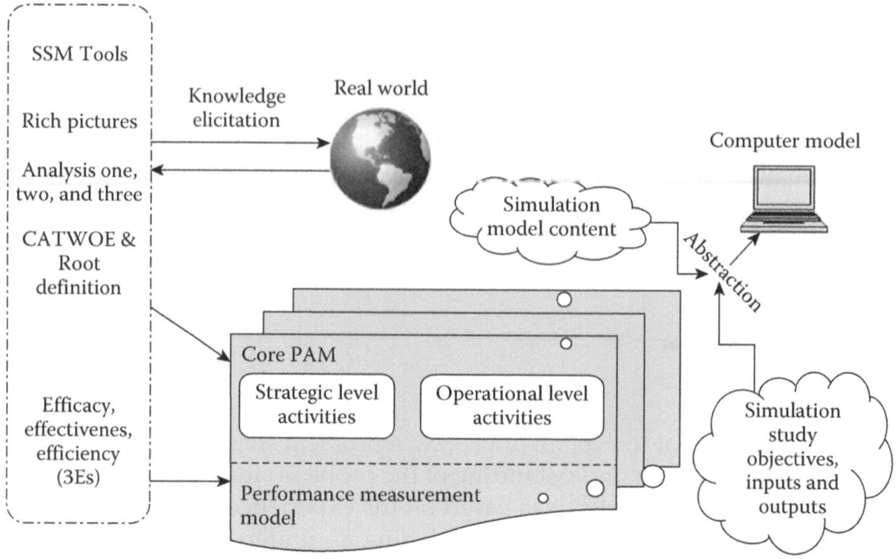

FIGURE 10.3
Using SSM in conceptual modeling.

Acknowledgments

This chapter is reproduced, with major editing, from: Kotiadis, K. 2007. Using soft systems methodology to determine the simulation study objectives. *Journal of simulation* 1: 215–222.© 2007 Operational Research Society Ltd. Reproduced with permission of Palgrave Macmillan.

Some sections of this chapter are based on the following:

Kotiadis, K., and S. Robinson S. 2008. Conceptual modeling: Knowledge acquisition and model abstraction. In *Proceedings of the 2008 Winter Simulation Conference,* ed. S.J. Mason, R. Hill, L. Moench, O. Rose, T. Jefferson, and J.W. Fowler, 951–958. Miami, FL: Institute of Electrical and Electronic Engineers, Inc.

References

Baldwin, L.P., T. Eldabi, and R.J. Paul. 2004. Simulation in healthcare management: A soft approach (MAPIU). *Simulation modelling practice and theory* 12 (7–8): 541–557.

Büyükdamgaci, G. 2003. Process of organizational problem definition: How to evaluate and how to improve. *Omega* 31: 327–338.

Carpenter, G.I., K. Kotiadis, and M. Mackenzie. 2003. *An Evaluation of IC Using Computer Simulation: The Summary Report.* Canterbury: University Of Kent, Centre For Health Service Studies.

Checkland, P. 1999a. *Soft Systems Methodology: A 30-Year Retrospective.* Chichester: Wiley.

Checkland, P. 1999b. *Systems Thinking, Systems Practice: Includes A 30-Year Retrospective.* Chichester: Wiley.

Checkland, P. 2001. Soft systems methodology. In *Rational Analysis for a Problematic World Revisited,* 61–89, ed. J. Rosenhead, and J. Mingers. Chichester: Wiley.

Checkland, P., and J. Scholes. 1999. *Soft Systems Methodology in Action: Includes a 30 Year Retrospective.* Chichester: John Wiley and Sons.

Davies, R., and P. Roderick. 1998. Planning resources for renal services throughout UK using simulation. *European journal of operational research* 105: 285–295.

Department of Health. 2000. *Shaping the Future Nhs: Long Term Planning for Hospitals and Related Services: Consultation Document on the Findings of the National Beds Inquiry.* London: Department of Health.

Dunker, K. 1945. On problem solving. *Psychological monographs* 58 (5): 270.

Fone, D., et al. 2003. Systematic review of the use and value of computer simulation modelling in population health and health care delivery. *Journal of Public Health Medicine* 25: 325–335.

Hart, E., and M. Bond. 1995. *Action Research for Health and Social Care: A Guide to Practice.* Buckingham: Open University Press.

Janis, I.L., and L. Mann. 1977. *Decision Making.* New York: Free Press.

Jun, J.B., S.H. Jacobson, And J.R. Swisher. 1999. Application of discrete-event simulation in health care clinics: A survey. *Journal of Operational Research* 50: 109–123.

Kotiadis, K. 2006. Extracting a conceptual model for a complex integrated system in health care. In *Proceedings of the 3rd Simulation Workshop (SW06),* ed. J. Garnett, S. Brailsford, S. Robinson, and S. Taylor, *Proceedings of the OR Society's Two-Day Workshop (SW06),* 235–245. Birmingham: Operational Research Society.

Kotiadis, K., and S. Robinson S. 2008. Conceptual modelling: Knowledge acquisition and model abstraction. In *Proceedings of the 2008 Winter Simulation Conference,* ed. S.J. Mason, R. Hill, L. Moench, O. Rose, T. Jefferson, and J.W. Fowler, 951–958. Miami, FL: Institute of Electrical and Electronic Engineers, Inc.

Lehaney, B., S.A. Clarke, and R.J. Paul. 1999. A case of an intervention in an outpatients department. *Journal of the operational research society* 50: 877–891.

Lehaney, B., and V. Hlupic. 1995. Simulation modelling for resource allocation and planning in the health sector. *Journal of the royal society for the promotion of health* 115: 382–385.

Lehaney, B., and R.J. Paul. 1994a. Developing sufficient conditions for an activity cycle diagram from the necessary conditions in a conceptual model. *Systemist* 16 (4): 262–268.

Lehaney, B., and R.J. Paul. 1994b. Using soft systems methodology to develop a simulation of outpatient services. *Journal of the royal society for the promotion of health* 114 (5): 248–251.

Lehaney, B., and R.J. Paul. 1996. The use of soft systems methodology in the development of a simulation of out-patients services at Watford General Hospital. *Journal of the operational research society* 47: 864–870.

Lowery, J.C. 1994. Barriers to implementing simulation in health care. In *Proceedings of the 1994 Winter Simulation Conference*, ed. J.D. Tew, S. Mannivannan, D.A. Sadowski, and A.F. Seila, December 11–14, ACM, 868–875. Lake Buena Vista, FL.

Pidd, M. 1999. Just modeling through: A rough guide to modeling. *Interfaces* 29(2):118–132.

Pidd, M. 2007 Making sure you tackle the right problem: Linking hard and soft methods in simulation practice In *Proceedings of the 2007 Winter Simulation Conference*, ed. S.G. Henderson, B. Biller, M.-H. Hsieh, J.D.T. Shortle, and R.R. Barton, 195–204. Piscataway, New Jersey: Institute of Electrical and Electronic Engineers, Inc.

Robinson, S. 2004. *Simulation: The Practice of Model Development and Use.* Chichester: John Wiley and Sons.

Robinson, S. 2008. Conceptual modelling for simulation part I: Definition and requirements. *Journal of the operational research society* 59 (3): 278–290.

Robinson, S., and M. Pidd. 1998. Provider and customer expectations of successful simulation projects. *Journal of the operational research society* 49: 200–209.

Wilson, B. 2001. *Soft Systems Methodology: Conceptual Model Building and Its Contribution.* Chichester: Wiley.

Winter, M. 2006. Problem structuring in project management: An application of soft systems methodology (SSM). *Journal of the Operational Research Society*, Advanced online publication 15 March 2006: 1–11.

Part IV

Software Engineering for Conceptual Modeling

11

An Evaluation of SysML to Support Simulation Modeling

Paul Liston, Kamil Erkan Kabak, Peter Dungan, James Byrne, Paul Young, and Cathal Heavey

CONTENTS

11.1 Introduction

First published in September 2007, the Systems Modeling Language (SysML) is a recent language for systems modeling that has a growing community of users and advocates in the field of systems engineering. This chapter gives an overview of SysML, identifies why it is of interest to the simulation community, and evaluates the feasibility of using this standard to support the conceptual modeling step in the discrete-event simulation (DES) process.

It has been recognized for many years that conceptual modeling is an extremely important phase of the simulation process. For example, Oren (1981) noted that conceptual modeling affects all subsequent phases of a simulation project and comprehensive conceptual models are required for robust and successful simulation models. Costly development time can be greatly reduced by clearly defining the goals and content of a model during the precoding phase of a simulation study. Despite this acknowledged importance, relatively little research has previously been carried out on the topic of conceptual modeling, as highlighted earlier in Chapter 1 of this book.

Over the years a number of frameworks for conceptual modeling have been suggested with the aim of bringing standardization to what is perceived by many to be more of an art than a science (Kotiadis 2007). One of the earliest frameworks was put forward by Shannon (1975), which consists of four steps. The first step is to specify the model's purpose, the second is to specify the model's components, the third is to specify the parameters and variables associated with the components, and the fourth is to specify the relationships between the components, parameters and variables. While these steps are still valid today, alternative frameworks have been presented in the years since then that refine the steps and/or focus on different aspects of the conceptual model. Examples include Nance (1994), Pace (1999), and van der Zee and Van der Vorst (2005), among others. The most recent modeling framework, presented by Robinson (2008), draws more attention to the goal of the model by encouraging the modeler to explicitly identify the model outputs and inputs prior to considering content. The steps of this framework are as follows:

- Understand the problem situation
- Determine the modeling and general project objectives
- Identify the model outputs (responses)
- Identify the model inputs (experimental factors)
- Determine the model content (scope and level of detail), identifying any assumptions and simplifications

In each case, the proposed modeling frameworks provide a set of steps that should be followed in order to compile a useful conceptual model. However, even with these frameworks in place a means of gathering and communicating the above listed information is still required. Typically a document-centric approach is taken for this whereby word processing is used to document and communicate the relevant information. While this has been and largely continues to be the standard approach adopted, there is now a move toward a model-centric approach. A number of researchers have documented the benefits of using a model-based approach, using process modeling tools and techniques, to support the initial stages of a simulation project (Van Rensburg and Zwemstra 1995, Nethe and Stahlmann 1999, Jeong 2000, and Perera and Liyanage 2000). These techniques include the following:

- Petri Nets (Vojnar 1997, Ou-Yang and Shieh 1999, Balduzzi et al. 2001, Koriem 2000, Shih and Leung 1997, Evans 1988)
- Activity Cycle Diagrams (ACDs) (Richter and Marz 2000, Shi 1997)
- Discrete Event Specification System (DEVS) (Rosenblit et al. 1990, Thomasma and Ulgen 1988)

- UML Activity diagrams (Barjis and Shishkov 2001, Niere and Zundorf 1999)
- UML Statecharts (Richter and Marz 2000, Hu and Shatz 2004)
- Process flow diagrams (Robinson 2004)
- IDEF3 and IDEF0 (Van Rensburg and Zwemstra 1995, Al-Ahmari and Ridgway 1999, Jeong 2000, Perera and Liyanage 2000)

ABCmod (Birta and Arbez 2007) and Simulation Activity Diagrams (Ryan and Heavey 2007) are model-based approaches that have been developed specifically for conceptual modeling. These techniques are discussed in detail in Chapter 6 and Chapter 12, respectively.

A further standard that has been shown to be applicable to conceptual modeling is Business Process Modeling Notation (BPMN) (Onggo 2009). This is a graphical modeling approach that is used for specifying business processes. The notation has been designed to coordinate the sequence of processes and the messages that flow between the different process participants in a related set of activities (http://www.bpmn.org/). While this may address much of the information concerned in a conceptual model, there are further details that BPMN is not equipped to deal with. For instance while comparing BPMN with UML, Perry (2006) notes that BPMN is unable to model the structural view or the requirements of the process. The structural aspect of a system can be of importance in a conceptual model when it places a constraint on the system (e.g., logistical implications of relative location of processing stations). When the requirements and model purpose information cannot be held within the modeling environment then supplementary documentation is required, thereby diminishing the advantage of a model-centric approach.

The parallels between simulation modeling and software development are discussed by Nance and Arthur (2006) and Arthur and Nance (2007) who investigate the use of software requirements engineering (SRE) techniques in simulation modeling. A notable contrast in the level of standardization in the two fields is identified with the authors observing that simulation methodologies differ widely in formality and rigor. One tool that has promoted standardization in the software development world is the Unified Modeling Language (UML). This standard is used by software developers to define and communicate information, particularly in the requirements gathering and design stages of the software development process. Based on UML, a new graphical modeling standard called the SysML has been developed to support systems engineering. This chapter explores how this new standard could be used to support the conceptual modeling phase of a simulation project. First, an overview of the SysML standard is given with a brief history and an introduction to the primary constructs. The use of SysML within the simulation context is then discussed with reference to other research in the field and case study work that has been undertaken by the authors. Insights into the strengths and weaknesses of a SysML-based conceptual modeling approach are presented with discussion around the direction of future research.

11.2 The Systems Modeling Language (SysML)

The Systems Modeling Language (SysML) is "a general-purpose graphical modelling language that supports the analysis, specification, design, verification, and validation of complex systems. These systems may include hardware, software, data, personnel, procedures, facilities, and other elements of man-made and natural systems" (Friedenthal et al. 2008b).

SysML is essentially a UML profile that represents a subset of UML 2 with extensions (Friedenthal et al. 2008a). UML is a language for specifying, visualizing and constructing the artefacts of a software intensive system, which was designed to model object-oriented software systems, and has been used successfully in this field for over a decade (Booch 1999).

Although UML has also been used to represent non-software systems, it is not ideally suited to this purpose and requires non-standardized use of model elements that can ultimately lead to confusion and incorrect interpretation of diagrams. To adapt UML for non-software systems, the developers of SysML attempted to remove the software bias, and added semantics for model requirements and parametric constraints. The resulting SysML standard is a general purpose modeling language capable of specifying complex systems that include non-software components.

11.2.1 A Brief History

In January 2001, the International Council on Systems Engineering (INCOSE) Model Systems Design Workgroup decided to adapt the UML for Systems engineering applications. Subsequently, INCOSE collaborated with the Object Management Group (OMG), which maintains the UML Specification, and together they developed the set of requirements for the new language. These requirements were issued by OMG as part of the *UML for Systems Engineering Request for Proposal* in March 2003.

In response to the OMG request for proposal, a work group called *SysML Partners* was formed in May 2003, which is a consortium of industry leaders and tool vendors. They initiated an open source specification project to develop the SysML standard according to the outlined requirements and the first SysML draft was distributed in October 2003. By the summer of 2005, disputes forced the work group to split into two *(SysML Partners* and *SysML Submission Team)* and in November of that year two competing SysML specifications were submitted to OMG. However in early 2006 the two teams worked together with OMG to merge the competing SysML specifications into OMG SysML™ Version 1.0. This was published as an official standard in September 2007. Reflecting on the development process, Weilkiens (2006) notes that the separation of the work group "had a positive effect on the quality of the SysML specification" as it encouraged creativity and critical review.

To maintain momentum in the development of SysML, the *SysML Revision Task Force* was promptly set up to examine the specification and recommend suitable revisions. These recommendations have been incorporated into the most recent version (at time of writing) of the specification, OMG SysML™ v1.1, which was published on December 3, 2008, and is available from the OMG Web site, http://www.omgsysml.org/.

11.2.2 The SysML Diagrams and Concepts

SysML is a visual modeling language that provides semantics and notation for representing complex systems (OMG 2008). Although a number of software vendors were involved in the development of the specification, SysML has remained methodology and tool independent.

As noted earlier, SysML is a subset of UML with extensions, which means that certain elements of UML that were deemed unnecessary have been removed while other elements have been modified or added to enable the representation of non-software systems, see Figure 11.1. Specifically, SysML reuses seven of UML 2's 13 diagrams and adds two diagrams (Requirements and Parametric diagrams), giving a total of nine diagram types. The removal of some UML elements is questioned by a number authors (e.g., Holt and Perry 2008) who highlight that software is still included in the list of system types that SysML can be applied to. However, UML 2 is widely criticized for being overinflated (e.g., Kobryn 1999) and there is strong justification, in terms of learning effort and consistency of use, for keeping the number of SysML elements to a minimum.

The two primary diagram categories are structural and behavioral, with structural diagrams specifying the parts of the system and their properties and the behavioral diagrams specifying the functional capabilities of the system.

The system structure is represented by block definition diagrams and internal block diagrams, which are based on the UML class diagram and UML composite structure diagram, respectively.

- The *block definition diagram* describes the system hierarchy and system/component classifications through the representation of structural elements called blocks. Any block that exhibits behavior must have an associated state machine diagram.
- The *internal block diagram* describes the internal structure of a system in terms of its parts, ports, and connectors.

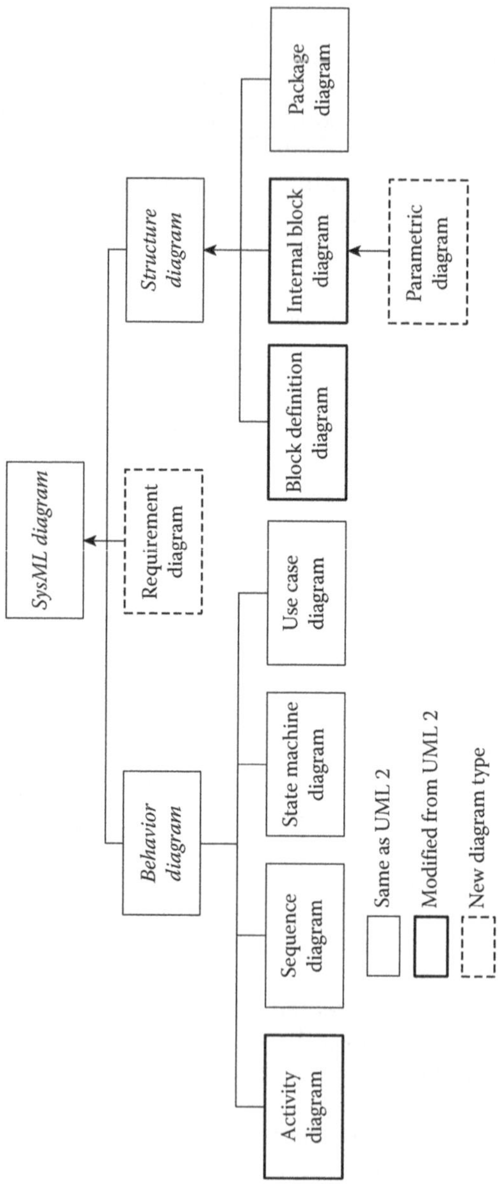

FIGURE 11.1
SysML diagram taxonomy. (From the SysML v1.1 Specification, fig. 4.4 or A.1. Object Management Group, Inc. (C) OMG, 2009. Reprinted with permission.)

- The *parametric diagram* is a restricted form of the internal block diagram and represents constraints on property values.
- The *package diagram* represents the organization of a model in terms of packages that contain model elements.

The behavior diagrams include the use-case diagram, activity diagram, sequence diagram, and state machine diagram.

- The *activity diagram* represents the flow of data and control between activities and shows how actions transform inputs into outputs.
- The *sequence diagram* represents the interaction between collaborating parts of a system in terms of a sequence of exchanged messages.
- The *state machine diagram* describes the state transitions and actions that a system or its parts performs when triggered by events.
- The *use-case diagram* provides a high-level description of the system functionality in terms of how a system is used by external entities (i.e., actors).

The *requirements diagram* is neither structural nor behavioral. It supports requirements traceability by representing text-based requirements and their relationships with other requirements and model elements.

SysML also supports crosscutting, which is a means of linking between diagrams to represent associations between different model elements. The benefit of crosscutting is only truly realized in software tools that support navigation between and consistency checks across the various diagrams of a SysML model. Currently available software tools for creating SysML models are discussed later in Section 11.2.4.

As a complement to the diagrams outlined above, tabular representations are also supported in SysML to capture model information such as allocation tables, which support requirements allocation, functional allocation, and structural allocation. This capability facilitates automated verification and validation and helps to identify missing data and gaps in the model. The next section discusses some of the strengths and weaknesses of SysML as reported to date.

11.2.3 Reported Strengths and Weaknesses of SysML

Being largely based on UML, SysML inherits many of its strengths along with some of its weaknesses. UML has been tried and tested for many years and has become accepted in the software industry thus providing SysML with a strong foundation to build on. The developers of SysML have also had the opportunity to cut some of the "semantic bloat" for which UML has been criticized. Although some still feel that unnecessary redundancy still exists in the standard, Weilkiens (2006) notes that in certain circumstances redundancy can bring the benefit of a less complex model since model elements and relationships can be omitted from some diagrams.

Willard (2007) states that the main benefit of SysML is that it provides a standard and comprehensive system specification paradigm. He notes that the consistency that this brings in terms of model syntax and semantics, together with unambiguous graphical symbols, can greatly improve communication. He proceeds to list the following side benefits that occur as a consequence to this main benefit:

- It can avoid the need to normalize system definitions, i.e., if all aspects of systems are defined in a common language, the need to translate definitions in cross sector projects is avoided.
- It can increase the potential and likelihood of reuse.
- It can mitigate vendor/tool interoperability problems.
- It can simplify distributed team environments.

Looking at the field of simulation each of these stated benefits has relevance and is desirable. For instance, the type of comprehensive standardization described by Willard (2007) is noted to be lacking in simulation particularly in the area of conceptual modeling for DES where, as discussed previously, standardization is nonexistent. The use of SysML could help overcome interoperability issues in relation to simulation modeling tools that are currently unsupported in this respect. Additionally, it is acknowledged that considerable amounts of insightful information are unearthed during the conceptual modeling phase of a simulation study. The current difficulty is that this information is often lost after the project concludes. If the information is held in reusable sections of a graphical model, it will be available not only for future simulation projects but also any other type of process improvement initiative and continual management. This potential use of SysML is discussed further in Section 11.3.

An identified weakness of SysML is that it gives too much freedom to the modeler. It is therefore possible for important information to be represented in an obscure manner in a SysML diagram, which, as Herzog and Pandikow (2005) point out, could be "easily overlooked by a human reader." UML models are true reflections of the systems they represent since the concepts that are used to develop UML models are also used to develop the software systems they represent. On the other hand, SysML models are just abstractions of the systems they represent. Any abstraction is open to interpretation; the freedom offered to SysML users means that these abstractions can be developed in various ways thus creating even further opportunity for confusion and miscommunication, particularly in larger scale systems.

A further weakness of SysML is the associated learning effort. For example, when asked by The Aerospace Corporation to evaluate SysML as a new general-purpose modeling language (under a project named Quicklook), the Tactical Science Solution (TSS) Team at George Mason University recorded in their final report (Alexander et al. 2007) that it took 1.5 man months to train the project teams to an acceptable competency in SysML and a SysML tool. They also note that users who had no background in UML took 50% longer to be trained to a "level of competency in which they were able to produce acceptable results" than users with a UML background. The team spent 300 hours training in total. They concluded that

> The TSS Team's training and engineering results show that a design team can learn and use SysML in a reasonable amount of time (five standard workweeks) without significant training or experience. In this way, Project Quicklook has dispelled the notion that organizations cannot use model-based systems engineering with SysML because the start-up resource cost is too high.

While this may be an acceptable training period for people such as systems engineers who will regularly use the standard, could it be considered a significant commitment in other cases where the nature of the users would be different? For instance in a simulation modeling context, it may be feasible for simulation modelers to learn the standard as they could continually use it. However, if the modeler is eliciting information from a manufacturing engineer during a project, it may not be feasible for the engineer to learn the standard in order to communicate with the simulation modeler by commenting on, adding to or modifying SysML diagrams. Noting this, some effort has already been made by SysML tool vendors to support communication with stakeholders who are unfamiliar with SysML, through alternative user interfaces. This, however, entails additional development effort on the part of the modeler; this topic is discussed further in the next section.

One of SysML's greatest strengths is the level of interest that it has received. The number of industrial partners who have contributed to its development illustrates practitioner recognition of the need for SysML and an eagerness to make standardized graphical modeling notation freely available. These partners include significant industry leaders such as IBM, Lockheed Martin Corporation, BAE Systems, the Boeing Company, Deere & Company, and NASA. As noted earlier, tool vendors also showed their support for the language by contributing to the development process. Herzog and Pandikow (2005) highlight that the number of tool vendors involved in drafting the SysML specification shows that "there exists not only a pull from the market, but also a push from the vendor community." This type of across-the-board support significantly strengthens the likelihood of widespread adoption of SysML.

11.2.4 SysML Tools

To date there are six commercial and two open source tools available for developing SysML models.

- *Artisan Studio* by Artisan Software Tools is a UML tool that has been developed to fully support the SysML profile (http://www.artisansoftwaretools.com/).
- *Tau G2* by IBM is a standards-based, model-driven development solution for complex systems (http://www-01.ibm.com/software/awdtools/tau/).
- *Rhapsody* also by IBM is a UML/SysML-based model-driven development for real-time or embedded systems (http://www-01.ibm.com/software/awdtools/rhapsody/).
- *MagicDraw* by No Magic is described as a business process, architecture, software and system modeling tool, having a specific plugin to support SysML modeling (http://www.magicdraw.com/).
- *Enterprise Architect* by Sparx Systems is a UML analysis and design tool with a module for developing SysML models (http://www.sparxsystems.com/).
- EmbeddedPlus Engineering offers a SysML toolkit as a third party plugin for *IBM Rational* (http://www.embeddedplus.com/).
- *TOPCASED-SysML* is a SysML editor that has been developed by the open source community (http://gforge.enseeiht.fr/projects/topcased-sysml/)
- *Papyrus for SysML* is an open source UML tool based on the Eclipse environment and includes all of the stereotypes defined in the SysML specification (http://www.papyrusuml.org/).

These are all the tools that fully support SysML at the time of writing but the standard is still relatively new and more tools are bound to emerge. For instance, Visual Paradigm is a UML design tool that has begun to include some SysML capability but to date has only implemented the Requirements Diagram. Additionally there is a SysML template available for Microsoft Visio, however this allows for SysML diagrams to be drawn rather than SysML models to be created. The distinction here is that SysML diagrams in Visio will lack the integration and interactivity that a truly model-centric approach will benefit from.

As can be seen above the majority of tools for developing SysML models are computer aided software engineering (CASE) tools in which the UML capabilities have been enhanced to accommodate the SysML specification. This has significant benefits in terms of providing features that have been tried and tested in a UML context for longer than the SysML standard has even existed. The drawback of the history of these tools is that they are primarily designed for the software market rather than for the broader user base that SysML is intended for and therefore may not meet the expectations of all users.

The type of features that these tools offer include integration with SRE tools to aid population of Requirement Diagrams and repository-based architectures to support model sharing and multiuser collaboration over local networks. These tools offer crosscutting functionality, which allows relationships between elements on different diagrams to be defined thereby tying the model together. These relationships allow for information that has been defined in one diagram to be automatically added to another and help maintain consistency throughout the model. This is the type of integration that is difficult if not impossible to achieve when using the SysML template in Microsoft Visio.

In terms of improving the communication of a SysML diagram, many tools allow users to upload and use images that are more representative of the model elements. They also allow users to toggle on and off model block compartments to hide and show information (e.g., block parameters) as required and prevent information overload. A number of tools allow for the execution of model diagrams. This functionality allows the user to step through the sequence of activities and see how various activities are triggered. This is useful for initial validation of the model and for subsequently communicating the details, and demonstrating the functionality, of the system to others. To further aid communication, particularly to stakeholders who are unfamiliar with SysML or not from a technical background, a number of tools allow for alternative graphical and often interactive views to be used. These non-SysML views are either developed within the tool itself or the tool is designed to interact with external GUI prototyping tools, as in the cases of the graphical panels in Rhapsody and the integration of Artisan Studio with Altia Design, respectively.

Recognizing that document-based reports are still often required, many SysML tools allow for templates to be automatically populated with information about model requirements, content and relationships. Information about the model can also be exported in XMI format (XMI [XML Metadata Interchange] is an OMG standard) to allow for communication between different SysML tools. The next section discusses current and potential use of SysML in the simulation modeling domain.

11.3 SysML and Simulation

Research and discussion around the use of SysML in the simulation domain has been presented by McGinnis et al. (2006), Kwon and McGinnis (2007), Huang et al. (2007), and Huang et al. (2008). McGinnis et al. (2006) examine engineering tools that support factory design. In particular, they focus on the tools that provide sufficient support for automatic generation of simulation models that address factory design decisions. They define a factory CAD or F-CAD system as an integration of a number of commercial off-the-shelf (COTS) tools: CAD tools, simulation tools, data management tools, collaboration tools, and SysML tools. They provide an example, based on a wafer fab, in which they use SysML for building a plant metamodel and for describing basic fab entities with logical representations. They conclude that it is also possible to introduce control rules by SysML to generate automatically the associated simulation models.

Kwon and McGinnis (2007) deal with the use of SysML for structural modeling of a factory and present a more generalized conceptual SysML framework based on the study of McGinnis et al. (2006). The framework is composed of four layers supported by a number of COTS tools.

- The first layer is an abstraction layer that has a collection of SysML diagrams that identify domain reference classes.
- The second layer is a domain reference layer that contains geometric models and a group of classes such as operations, attributes, ports, and constraints in order to facilitate the reuse of simulation models.
- The third layer is an instance layer that includes all the simulation data for a particular system.
- The last layer includes a generator that converts the language independent instance model from the instance layer to a specific simulation model.

Huang et al. (2007) demonstrate a method of translating a SysML model of a real system into different instance SysML models suited to different analysis techniques (simulation or queuing network analysis). They illustrate how a simulation model can be automatically generated from these models using XMI, a parser application, a database and a simulation modeling tool. Huang et al. (2008) discuss the possibility of on-demand simulation model generation based on the techniques described in the previous papers discussed here.

There is a strong case for improving/automating the process of developing a simulation model from SysML diagrams especially if the use of SysML becomes as widespread as anticipated. However before generating a simulation model in this manner the SysML model needs to be created; this in itself is not a trivial task. As noted earlier, it has been reported that it takes approximately 1.5 man months to reach an acceptable standard of proficiency in SysML. One approach to developing SysML models is to translate existing domain-specific models into SysML. A graph transformation approach has been proposed for undertaking this task (Paredis and Johnson 2008, Johnson et al. 2008) that uses the Triple Graph Grammars (TGG) approach (Schürr 1994) to accomplish bidirectional transformation between the domain-specific language and SysML. This approach has been demonstrated with the Modelica language (Mattsson et al. 1998) and Paredis and Johnson (2008) report that further work is being conducted to illustrate the approach with Matlab Simulink and eM-Plant. This approach of course is only useful if the system is already described in a formal model. The next section describes the authors' experience of building a SysML model from the perspective of gathering information and defining a conceptual model for a simulation study.

11.3.1 Conceptual Modeling with SysML: An Example

As part of a SysML evaluation we retrospectively looked at a process analysis conducted in an electronics manufacturing plant and built a SysML model of the information. The information presented here was gathered while undertaking the initial steps of Robinson's conceptual modeling framework, i.e., understand the problem situation and determine the modeling and general project objectives. The original report (essentially a first iteration conceptual model) can be classed as document-centric: a 24-page document primarily containing text with some supplementary sketches to illustrate certain aspects of the system. This data were collected through a series of on-site meetings/interviews over a three-day period.

The first meeting was with a senior manager who gave an overview of how the production facility operated and outlined the primary business objectives and constraints. Next, one of the production engineers gave an extensive line tour describing the processing steps and product flow. This engineer then identified three line managers who could provide more detailed information on the day-to-day running of the production lines, the problems that occur and the remedial action that is taken. In addition to these shop floor meetings, a further meeting was held with a member of the team responsible for calculating capacity requirements and staffing levels. This was a critical decision process for the company as the products experienced highly seasonal demand and staff training took up to five weeks due to the complexity of the products.

In addition to recording notes on each meeting, CAD files of the plant layout, sample documents and spreadsheets relating to the process steps and process flow diagrams were also collected. Interestingly, process flow diagrams were only available for four of the seven major process steps and within these four process flow diagrams there were three different formats. The key assembly process steps, around which the line was built, were surprisingly among those without corresponding process flow diagrams. The collected information was compiled into a structured document describing the flow of material from when the customer order is placed and components are requested from a third party warehouse, through the production process and out the shipping dock at the end of the line.

Using this data, a SysML model was built in Artisan Studio Uno. A high-level description of the production process was developed using an Activity

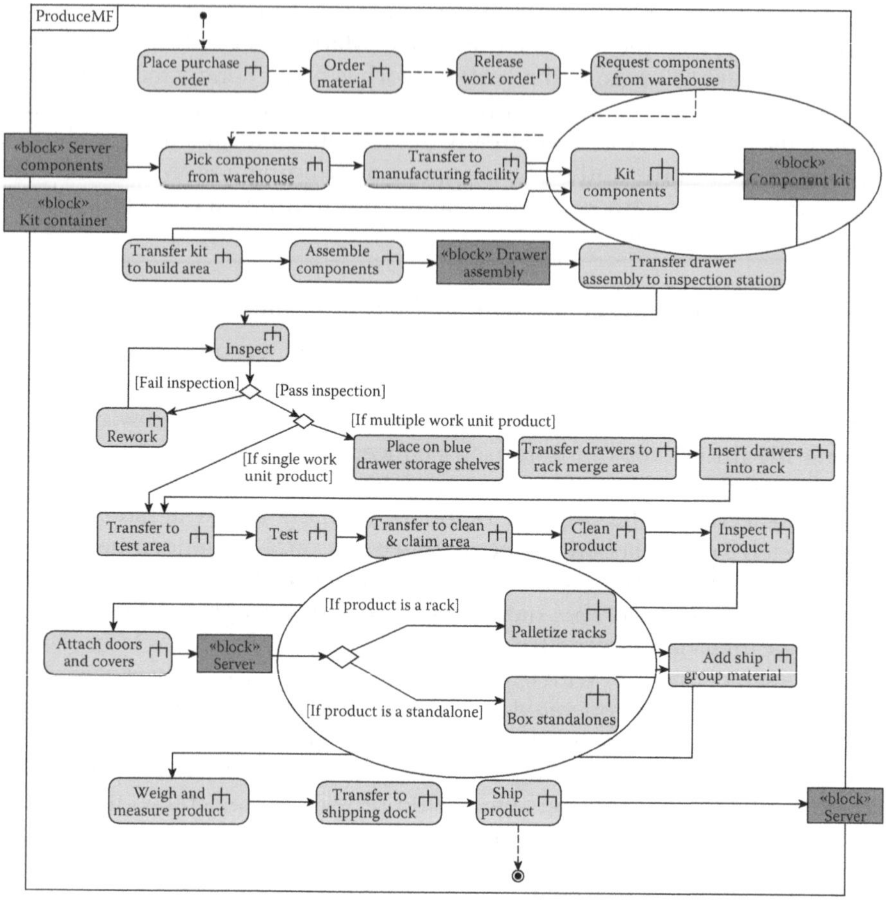

FIGURE 11.2
Activity diagram of overall production process.

Diagram (Figure 11.2). Although largely illegible at the scale shown here, one can see that the general format is not dissimilar to that generally used in process mapping. Decision nodes for instance are represented by a familiar diamond shaped symbol and the alternative routes are labelled with the associated decision criteria (see the lower zoomed section of Figure 11.2). A particularly useful attribute of SysML is the ability to segregate information and avoid overly complex diagrams. The activity diagram in Figure 11.2 describes the series of activities that occur from when a customer places a purchase order until the product is shipped. While this diagram does not describe in detail every step of what is a complex production process, it does allow the reader to very quickly get an understanding of what is involved in fulfilling a sales order. Additional more detailed information is made available to the reader in further activity diagrams. Here we give the example of the "Kit Components" activity, which yields the "Component Kit" (see the upper zoomed section of Figure 11.2) that is used in the subsequent assembly activity. The "rake" symbol on the upper right-hand corner of the "Kit Components" activity indicates that there is a more elaborate diagram associated with this model element. The diagram associated with the "Kit Components" activity is shown here in Figure 11.3.

SysML activity diagrams can also show how physical objects in the system interact with the represented activity. For instance in Figure 11.3, it can be seen that the "Kit Components" activity takes in components (high value and

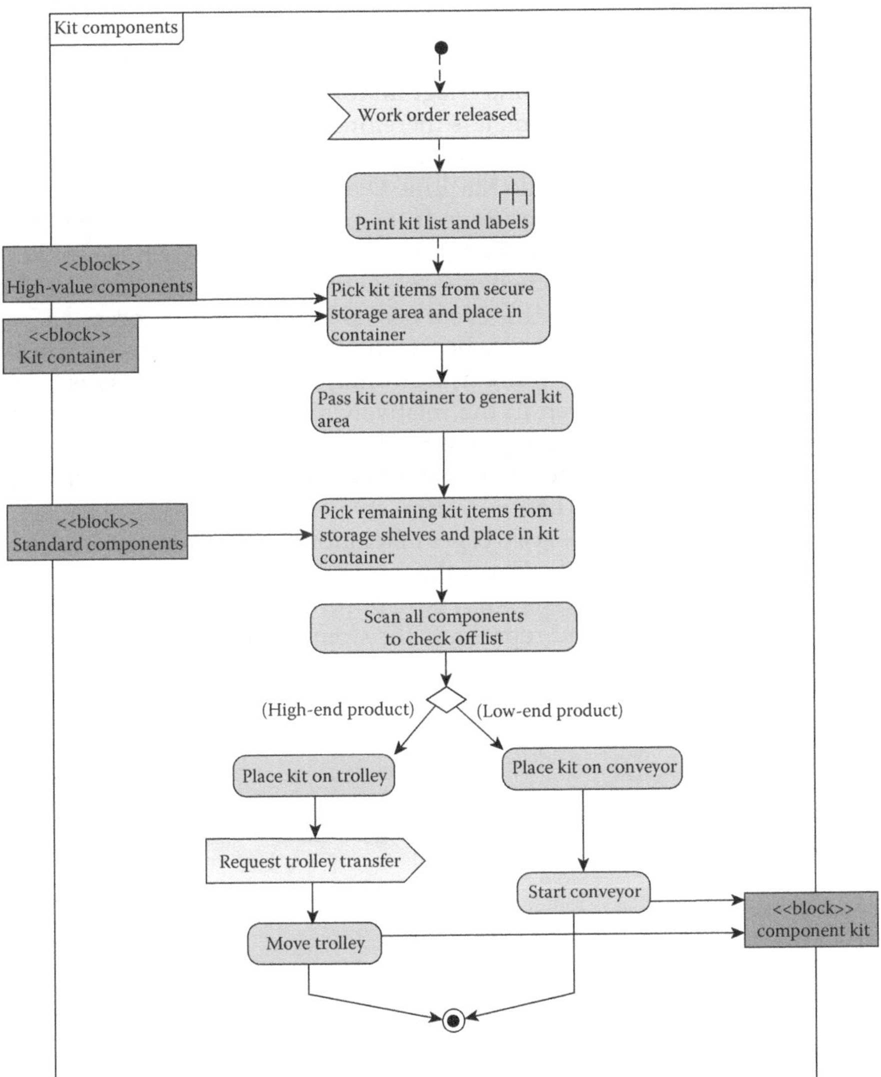

FIGURE 11.3
Activity diagram of the kitting process.

standard) and a kit container and outputs a component kit. Even with this straightforward kitting activity, the advantages of compartmentalizing data in order to prevent information overload can be seen. A further advantage of this modular approach is that once an activity has been described it is available to be referenced from any diagram or reused in another model. Indeed this idea of information reuse and cross referencing is an integral aspect of SysML modeling and helps ensure consistency across model diagrams.

An example of this type of integration is shown here with the "StartConveyor" step in Figure 11.3 although as this is essentially a document-based description of a model-centric approach the integration is not apparent. This "StartConveyor" step is in fact an event that had been previously defined in a State Machine Diagram for the conveyor (see Figure 11.4) and was added to this diagram. It is therefore the same piece of information shown in two different diagrams and therefore if it is changed on one it also changes on the other. The State Machine Diagram in Figure 11.4 shows how this event causes the conveyor to go from an "idle" state to an "operating" state. This diagram also shows that this can only happen when the conveyor is in an active state (i.e., switched on). The Activity Diagram in Figure 11.3 on the other hand shows when this event occurs in the Kitting Process.

The level of freedom offered when using SysML does bring certain difficulties to the modeling exercise as it is often unclear which way is best to represent information. Even in the relatively simple case of describing the generic types of component required by the manufacturing activity to fulfil a sales order it is possible to create initially confusing diagrams such as the Block Diagram in Figure 11.5.

This block diagram shows two types of relationship: the connector with the white triangle represents the Generalization relationship, which can be read as "is a type of" (e.g., a Chassis is a type of bulky component) and the connector with the black diamond represents the Composition relationship, which

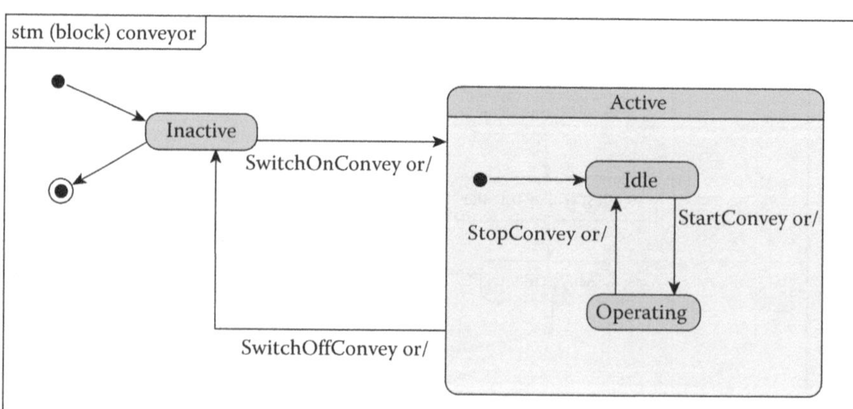

FIGURE 11.4
State machine diagram for the conveyor.

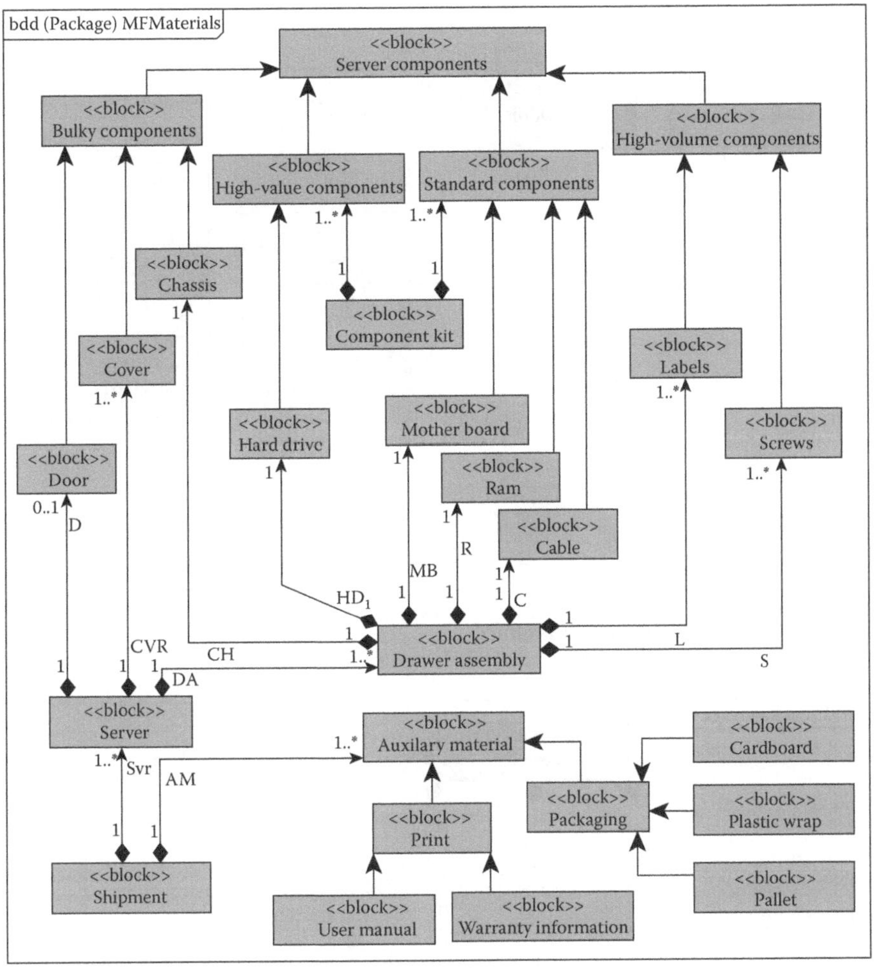

FIGURE 11.5
Block diagram for components.

can be read as "is made up of" (e.g., the Server product is made up of one or more drawer assemblies, one or more covers and possibly a door). Once the reader is aware of the meaning of the connectors the diagram becomes less confusing, however, considering that this diagram only shows generic component types there is potential for it to grow out of control when specific part numbers are included.

However, once relationships have been entered into a SysML model it is possible to generate alternative views of the same information. For example, the majority of the content of the Internal Block Diagram shown in Figure 11.6 and the Block Diagram in Figure 11.7 was generated automatically based on the composition relationships defined in Figure 11.5 (the

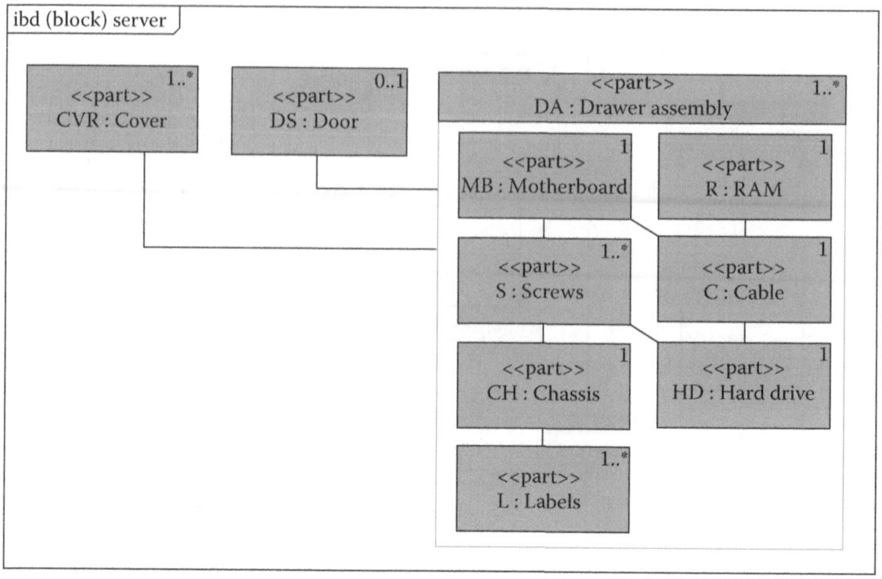

FIGURE 11.6
Internal block diagram for server product.

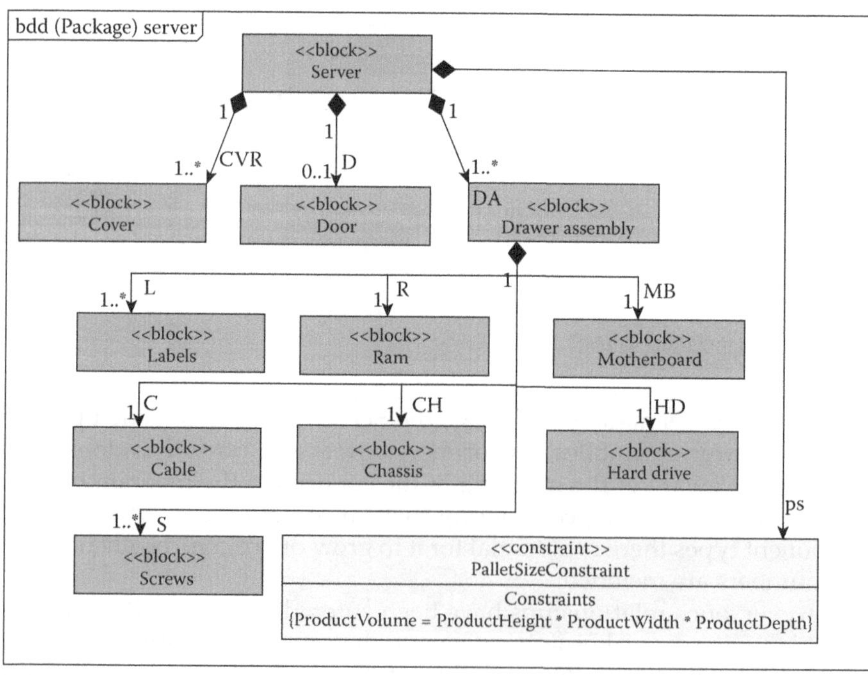

FIGURE 11.7
Block diagram for server product.

physical connections in Figure 11.6 were added manually and the constraint information in Figure 11.7 was based on a Parametric Diagram in the model) and both of these show more clearly what parts a Server product is made up of. The important point here is that a SysML model is more than a collection of pictures it is also the underlying logic that is represented in the pictures. This logic is valuable knowledge that once entered can be reused and centrally maintained.

The diagrams presented so far relate primarily to material and material flows. When examining a system during a simulation project, the flow of information through the system is also of importance as this often contains much of the system logic that must be incorporated into the simulation model. Figure 11.8 shows a sequence diagram that illustrates the connection between the company receiving a sales order from a customer for a new product and then placing a purchase order with a vendor for components. The diagram shows that the information is passed through a number of integrated software systems during this process. While these software systems may not need to be explicitly represented in a simulation model of the system, the logic of determining the production requirement and comparing this with the available material will need to be captured.

A feature of this model-centric approach that is of particular benefit in the simulation context is the ability to clone a diagram and edit it. This new diagram can be used to describe proposed design alternatives or experimental settings, and can be readily compared to the original. Furthermore, SysML modeling is suited to the concepts of simplifications and assumptions as used in conceptual modeling for simulation. Take for example the case shown earlier of the Kitting Process. If it is decided that the objectives of the model require this process to be modeled in detail, then the lower-level information in Figure 11.3 is used, if not, then it can be represented as a single activity as in Figure 11.2. The power of having this information in SysML is that the detailed information is retained and available for future projects. By scoping diagrams to packages (akin to organizing files into folders in Explorer) it is possible to signify which information forms part of the current conceptual model and which falls outside the scope of the current project. The issue with this approach is that if the SysML model of a production plant for instance is to be maintained as a central information repository that can be drawn upon for future simulation projects then there will be various conceptual models with overlapping information and alternative perspectives on the system and organizing packages in this manner will quickly become complex. The distinguishing difference between using SysML for conceptual modeling and for systems engineering is that conceptual modeling requires both an understanding of the real system and an understanding of the simulated system (with the assumptions and simplifications that make it different from the real system) whereas systems engineering deals only with the real system. Conceptual modeling therefore has additional perspectives and interpretations to deal with that make

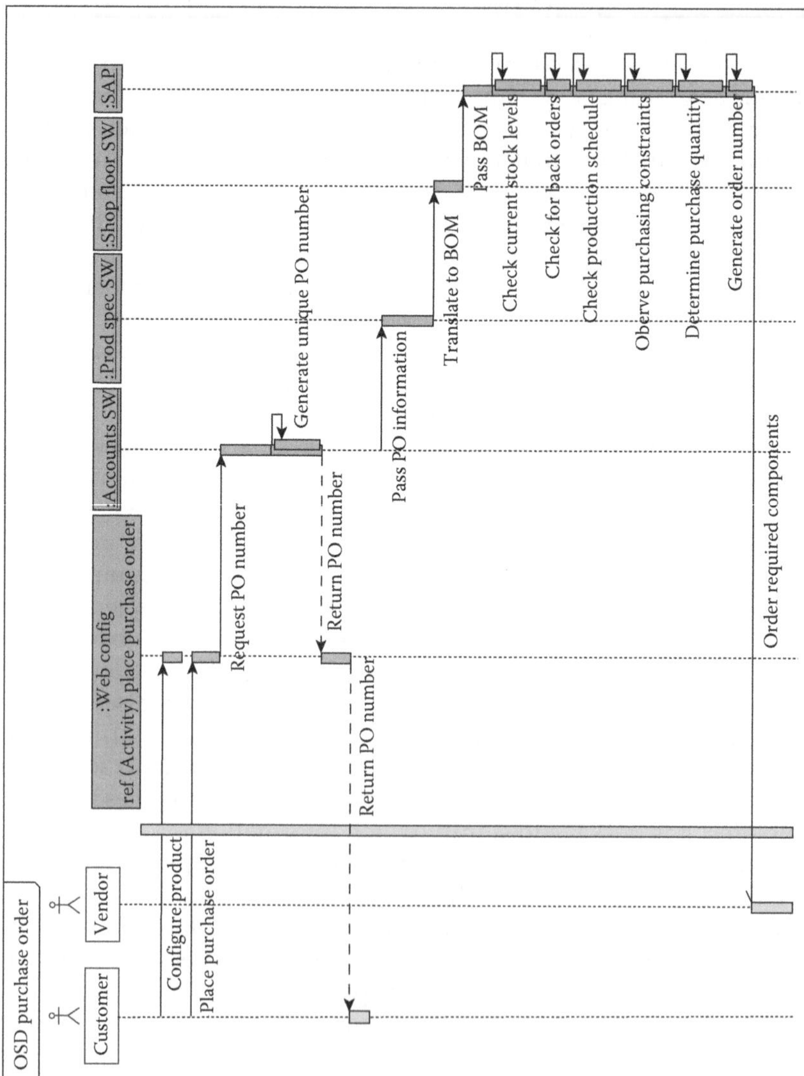

FIGURE 11.8
Sequence diagram of information flow in ordering process.

maintaining a SysML model more difficult. For example, the simple process shown in Figure 11.8 does not indicate what might happen if stock levels increased (perhaps due to components being reclaimed from scraped products) after a purchase order is generated. While the real system occasionally experiences this problem the model did not attempt to include this complexity. This and other issues are discussed in Section 11.4.

The original document that this SysML model was based on was in itself a useful resource for the company as it provided an end-to-end description of the process and it uncovered a number of interactions between processing areas that affected efficiency. One simple example of this was the realization that when certain data queries were ran in one section of the production facility it delayed the production order release process at the beginning of the line as it slowed down the process of identifying orders for which all components were available. This beneficial effect is widely reported in simulation studies, for instance Robinson 2004 suggests that possibly 50% of the benefit is obtained just from the development of the conceptual model; "The modeller needs to develop a thorough understanding of the operations system in order to design an appropriate model. In doing so, he/she asks questions and seeks for information that often have not previously been considered. In this case, the requirement to design a simulation model becomes a framework for system investigation that is extremely useful in its own right." Shannon (1975) even suggests that in some cases the development of a conceptual model may lead to the identification of a suitable solution and eliminate the need for further simulation analysis. Considering that some SysML tools allow for the diagrams to be "stepped through" (as discussed in Section 11.2.4), the use of SysML provides an even greater chance of resolving issues during the conceptual modeling phase as inconsistent information will be highlighted and cause and effect relationships can be explored. This SysML feature could have benefits for the validation of conceptual models and ensuring that a correct understanding of the system has been achieved.

To be successfully utilized in conceptual modeling, SysML needs to be compatible with the frameworks for conceptual modeling as discussed in section 11.1. Taking the most recent framework, Robinson 2008, the first step of *understanding the problem situation* can occur much more quickly if a SysML model of the system under investigation (SUI) already exists. Even if one does not exist, the process of developing a model would provide a structured means of gaining useful insight into the situation. The second step of *determining the modeling and general project objectives* is not directly solved by the use of SysML. This is an important step as it determines the direction of the simulation study and simply having a SysML model will not ensure that the correct model objectives have been identified. Techniques like the soft systems methodology (SSM) as discussed by Kotiadis 2006 and Kotiadis 2007 can be used to elicit the objectives from stakeholders. The purposeful activity models (PAMs) generated during SSM can be represented within a

SysML model using activity diagrams and the determined objectives can be recorded in SysML requirements diagrams. This would support central retention of information and would allow SysML relationships including "trace" and "satisfy" to be used to identify how model objectives are addressed in the simulation. As a simple example, an activity that records the number of products exiting a particular processing station could be traced to an objective to determine system throughput thus explaining why this activity is required in the simulation. It would be advantageous if SysML tools were able to highlight clearly any model elements that did not trace back to a requirement or similarly requirements that were not satisfied in the model. This did not appear to be possible in any of the SysML tools reviewed to date. The next steps of *identifying the model outputs* and *identify the model inputs* are readily supported by SysML object parameters and parametric diagrams. The final step of *determining the model content,* while identifying any assumptions and simplifications, can be successfully recorded in a structured manner in SysML as discussed earlier in regard to the kitting process example. By having a formal graphical model of the SUI, it is suggested that the difficult task of deciding on which assumptions and simplifications to make will be eased with natural selections and associated implications becoming more clearly recognisable.

On review of existing research in the area and the experiences gained while using the language, it is proposed that there is potential for using SysML as a common thread that could underlie all the activities undertaken in a simulation study from the initial requirements gathering phase through defining the conceptual model and on to the development of the simulation model (see Figure 11.9). Once information has been captured during one activity it would be available in a useable format for the next.

In this section it has been shown that there is merit in using SysML in the conceptual modeling process. It is capable of representing the type of information typically handled in this simulation phase such as information and material flows and moreover it brings structure and standardization that can greatly help knowledge transfer and reuse. There are nonetheless a number of challenges for the adoption of SysML as the standard conceptual modeling format. These are discussed in the next section.

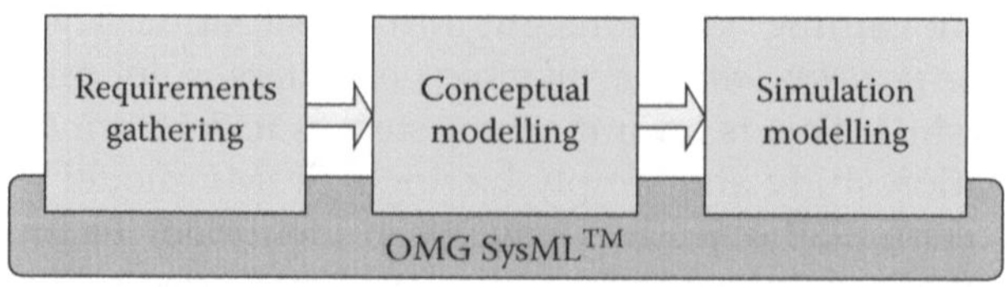

FIGURE 11.9
SysML: A common foundation.

11.4 Challenges for SysML-Based Conceptual Modeling

The key problems encountered when developing the SysML model discussed in the last section related to establishing: where to begin the model; which diagrams to use; and which way to present the data in these diagrams. Consulting available SysML literature on these fundamental questions typically returned answers that begin with the phrase, "It depends." While unsatisfactory at the time, a little SysML modeling experience proved this to be a justified prefix to the answers as the best diagram for representing a piece of information is dependent on many factors including the intended audience, the aspect of the information to be highlighted and if other additional information is to be represented. Consequently, it is more important to enter the information into the model than to deliberate on which diagram is best. As shown with the component category example in Figure 11.5, once the information has been captured it is possible to generate various diagrams and to show/hide different attributes of the model elements (e.g., Figures 11.6 and 11.7). In this way the most appropriate diagram/diagram layout can be selected for the circumstance at hand. A criticism of this capability would be that surplus information can be created in redundant diagrams, however, once the SysML modeling software has the appropriate consistency management features this should not pose a problem.

Another encountered problem was an inability to easily include a sketch of the production facility into the SysML model. The original document-centric model included such as sketch and it had proven effective for illustrating which areas were being referred to at any given point. Of course it was possible to describe the structure using block and internal block diagrams but a sketch that more closely resembles the physical structure would be beneficial when communicating with stakeholders. As noted in Section 11.2.4 certain SysML tools do have the capability to include alternative graphical representations but these can require considerable effort to develop and can be more complex than is generally required.

One consideration when using SysML for conceptual modeling is possible confusion around whether the model represents the SUI, the intended simulation model of this system, or indeed both of these. Two features of the SysML specification that can be used to address this issue are viewpoints and views. A viewpoint describes a perspective of the model that is of interest to a specific set of stakeholders. When defining a viewpoint, the reason for the perspective, the stakeholders of interest and the concerns to be addressed are specified. A view is then a type of package that conforms to a viewpoint and is intended to provide the model information that addresses the stakeholders concerns. By defining a real-world viewpoint of the system, a simulation model viewpoint of the system, and/or a simulation model viewpoint of how the system could/should be, this SysML functionality can be used to clarify the differences between various perspectives.

Specific guidelines for creating these viewpoints in a standardized format would be beneficial.

In terms of the diagram content, there are some aspects of existing modeling techniques that are not present in SysML. For instance in flow charts the cylindrical node is used to represent a direct data source, the SysML specification does not include specific nodes like this but the aforementioned freedom in the standard does allow custom node images to be added. The issues with this however are the loss of standardization and greater opportunity for misinterpretation.

SysML can however be tailored for use in specific domains. This is accomplished by developing specialized model libraries and/or profiles. Friedenthal et al. 2008b describe the difference between libraries and profiles as follows:

> Model libraries provide constructs that can be used to describe real-world instances represented by a model, be they blocks specifying reusable components or value types defining valid units and dimensions for block properties. Profiles, on the other hand, provide constructs that extend the modelling language itself; for example, stating that there is such a thing as a value type with units and dimensions in the first place.

Profiles are special types of packages that contain extension mechanisms called stereotypes (Note: SysML is a profile of UML). These stereotypes are based on one or more metaclasses in a reference metamodel, which in this case means that they based on existing SysML elements. For example, when modeling a production line it may be desirable to have a "Machine" object for representing the various machines in the system. This can be achieved by defining a stereotype called "machine" and applying it to the SysML blocks that represent machines. This stereotype can simply be used as a label for the purposes of clarity (i.e., <<machine>> in place of <<block>>) but further characteristics can also be manipulated. In the case of a "machine" stereotype properties such as capacity, feed rate or serial number could be defined in the stereotype. Any block to which this stereotype is applied will then inherit these properties and they can be populated with values specific to each instance. Constraints may also be added to a stereotype to specify rules about valid use of the newly defined properties. If a specific SysML Profile for the simulation modeling domain were to be specified then standard stereotypes for elements such as buffers and processing stations, and specific methods for identifying simplifications and assumptions, could be defined. The availability of a profile like this would address many of the standardization problems that the simulation modeling community face when creating/reading conceptual models.

Model libraries have obvious advantages in terms of cutting down the modeling workload. By saving model elements as reusable "building blocks" (e.g., a description of a prioritized queue or a detailed description of how the material reordering process is managed), they can be used to speed up the modeling process and reduce unnecessary repetition. If specific libraries for different domains such as supply chain analysis, electronics manufacturing or wafer fabrication were developed using a simulation-specific SysML profile and made publicly available, the inertia associated with getting simulation practitioners to use SysML could possibly be overcome. However, the economic benefit of developing libraries of reusable blocks may be questionable if additional effort is required of modelers to make the model elements reusable. This and the potential reuse rate would need to be investigated further before adopting this approach.

A high degree of collaboration is necessary when collecting information on a large or complex system. In the case example discussed in Section 11.3.1, it was necessary to hold meetings with six people with different responsibilities in the production facility, reflecting the localized knowledge that exists in such systems. The three days required to undertake and analyse these meetings reflects the difficulty associated with gathering knowledge when it is required. Attempts to maintain transferable information in process flow diagram format had only proven partially successful for the case company as they were not centrally stored and there were no process flow diagrams available for the assembly process steps even though they were the primary value adding activities in the factory. Again the difficulty with creating these process flow diagrams is associated with lack of collaborative supports for gathering the information and developing the diagrams. Many SysML tools already address these issues by offering multiuser features to help share the modeling effort. However, there are still many opportunities to take SysML software online and take advantage of the greater collaborative capability allowed by the Internet. This online collaboration would have significant benefits in the conceptual modeling space where often the simulation analyst is external to the company.

11.5 Conclusions

SysML is an accepted standard with a growing user base. The UML heritage and OMG adoption of the standard reflect the level of sophistication in the language while the development effort invested by practitioners and tool vendors alike shows the level of interest in the standard from both sides of the market.

SysML provides a standard that has potential to be used in DES. Such a standard would provide great benefit as it would provide a common language, which has been noted to be lacking in this domain.

All of the advantages that have been put forward supporting the use of UML for conceptual modeling still stand and indeed most are strengthened by the fact that SysML has a much broader scope than the software-specific

UML. SysML's ability to represent both physical and software aspects of a system allow the types of systems that are typically analyzed with simulation to be fully described in a formal model-centric format.

Compared to the document-centric approach, which is predominately used in conceptual modeling today, SysML models offer a much more useful format in terms of reusable blocks of information. Compartmentalizing information allows it to be offered to readers in more digestible quantities; different amounts and different sections of information can be offered to readers depending on the role they play in the study. Once built in appropriate software, a SysML model also allows for more intuitive navigation through the information, again aiding the communication process.

Although there may be a considerable learning period required to become proficient in SysML, this is not dissimilar to any programming language, natural language or software package. Indeed, SysML should be adopted in a similar fashion to any language; first understand some basic vocabulary and then expand upon this as and when required. The simulation packages already used by simulation analysts require even greater effort to learn so this in itself should not be a barrier to the uptake of SysML. It is therefore a matter of clearly conveying the benefits of using SysML for conceptual modeling so that simulation analysts can evaluate the return on their investment of time and effort.

It is acknowledged that SysML diagrams are easier to read than to write and collaboration will be essential if SysML models of large manufacturing (or other) systems are to be developed, maintained and made available for use in other activities. There is scope for further research around the use of online collaborative tools for the creation of SysML models.

More work is needed to fully evaluate the capabilities of SysML in the DES domain. There is scope to utilize the facilities of profiles and libraries to tailor SysML to the DES domain. Since it is of the utmost importance to maintain standardization, this work should be undertaken in a coordinated fashion.

References

Al-Ahmari, A.M.A., and K. Ridgway. 1999. An integrated modelling method to support manufacturing systems analysis and design. *Computers in industry*, 38(3), 225–238.

Alexander, D., S. Sadeghian, T. Saltysiak, and S. Sekhavat. 2007. *Quicklook Final Report*. Fairfax, Virginia: George Mason University.

Arthur, J. D., and R. E. Nance. 2007. Investigating the use of software requirements engineering techniques in simulation modelling. *Journal of simulation* 1:159–174.

Balduzzi, F., A. Giua, and C. Seatzu. 2001. Modelling and simulation of manufacturing systems with first order hybrid Petri Nets. *International journal of production research* 39(2):255–282.

Barjis, J., and B. Shishkov. 2001. UML based business systems modeling and simulation. In *Proceedings of 4th International Eurosim Congress*, Delft, The Netherlands, June 26–29.

Birta, L. G., and G. Arbez. 2007. *Modelling and Simulation: Exploring Dynamic System Behaviour*. London: Springer-Verlag.

Booch, G. 1999. UML in action: Introduction. *Communications of the ACM* 42(10):26–28.

Evans, J. B. 1988. *Structures of Discrete Event Simulation: An Introduction to the Engagement Strategy*. Chichester: Ellis Horwood.

Friedenthal, S., A. Moore, and R. Steiner. 2008a. OMG Systems Modeling Language Tutorial (revision b), INCOSE. http://www.omgsysml.org/ (accessed 4 Feb. 2009).

Friedenthal, S., A. Moore, and R. Steiner. 2008b. *A practical guide to SysML: The Systems Modeling Language*. Burlington, MA: Morgan Kaufmann Publishers.

Herzog, E., and A. Pandikow. 2005. SysML: An Assessment. In *Proceedings of 15th INCOSE International Symposium*, Rochester, NY, July 10–15.

Holt, J., and S. Perry. 2008. *SysML for Systems Engineering*. London: The IET.

Hu, Z., and S. M. Shatz. 2004. Mapping UML diagrams to a Petri Net notation for system simulation. In *Proceedings of International Conference on Software Engineering and Knowledge Engineering (SEKE)*, Banff, Alberta, Canada, ed. F. Maurer and G. Ruhe, June 20–24.

Huang, E., K. S. Kwon, and L. McGinnis. 2008. Toward on-demand wafer fab simulation using formal structure and behavior models. In *Proceedings of the 2008 Winter Simulation Conference*, Miami, FL, ed. S. J. Mason, R. R. Hill, L. Mönch, O. Rose, T. Jefferson, and J. W. Fowler, 2341–2349. Piscataway, NJ: IEEE.

Huang, E., R. Ramamurthy, and L. F. McGinnis. 2007. System and simulation modeling using SysML. In *Proceedings of the 2007 Winter Simulation Conference*, Washington DC, ed. S. G. Henderson, B. Biller, M.-H. Hsieh, J. Shortle, J. D. Tew, and R. R. Barton, 796–803. Piscataway, NJ: IEEE.

Jeong, K. Y. 2000. Conceptual frame for development of optimized simulation-based scheduling systems. *Expert systems with applications* 18:299–306.

Johnson, T. A., C. J. J. Paredis, and R. M. Burkhart. 2008. Integrating Models and Simulations of Continuous Dynamics into SysML. In *Proceedings of the 6th International Modelica Conference*, Bielefeld, Germany, 135–145. Modelica Association.

Kobryn, C. 1999. UML 2001: A standarization odyssey. *Communications of the ACM* 42(10):29–37.

Koriem, S. M. 2000. A fuzzy Petri Net tool for modeling and verification of knowledge-based systems. *The computer journal* 43(3):206–223.

Kotiadis, K. 2006. Extracting a conceptual model for a complex integrated system in health care. In *Proceedings of the Operational Research Society Simulation Workshop 2006 (SW06)*, Birmingham, UK, ed. S. B. J. Garnett, S. Robinson and S. Taylor, 235–245. Operational Research Society.

Kotiadis, K. 2007. Using soft systems methodology to determine the simulation study objectives. *Journal of simulation* 1(3):215–222.

Kwon, K., and L. F. McGinnis. 2007. SysML-based Simulation Framework for Semiconductor Manufacturing. In *Proceedings of the 3rd Annual IEEE Conference on Automation Science and Engineering*, Scottsdale, AZ,, Sept 22–25, 1075–1080.

Mattsson, S. E., H. Elmqvist, and M. Otter. 1998. Physical system modeling with modelica. *Control engineering practice* 6:501–510.

McGinnis, L. F., E. Huang, and K. Wu. 2006. Systems engineering and design of high-tech factories. In *Proceedings of 2006 Winter Simulation Conference*, Monterey, CA, ed. D. Nicol, R. Fujimoto, B. Lawson, J. Liu, F. Perrone, and F. Wieland. Piscataway, NJ: IEEE.

Nance, R. E. 1994. The conical methodology and the evolution of simulation model development. *Annals of operations research* 53:1–45.

Nance, R. E., and J. D. Arthur. 2006. Software requirements engineering: Exploring the role in simulation model development. In *Proceedings of the Third Operational Research Society Simulation Workshop (SW06)*, Birmingham, ed. J. Garnett, Brailsford, S., Robinson, S. and Taylor, S., 117–127. The Operational Research Society.

Nethe, A., and H. D. Stahlmann. 1999. Survey of a general theory of process modelling. In *Proceedings of International Conference on Process Modelling*, Cottbus, Germany, February 22–24, 2–16.

Niere, J., and A. Zundorf. 1999. Testing and simulating production control systems using the Fujaba Environment. In *Proceedings of International Workshop AGTIVE'99 (Applications of Graph Transformations with Industrial Relevance)*. Springer Publications, September 1–3

OMG. 2008. OMG SysML™ Specification Version1.1. http://www.omgsysml.org/ (accessed 4 Dec. 2008).

Onggo, B. S. S. 2009. Towards a unified conceptual model representation: A case study in healthcare. *Journal of simulation* 3:40–49.

Oren, T. I. 1981. Concepts and criteria to assess acceptability of simulation studies: A frame of reference. *Communications of the ACM* 24(4):180–189.

Ou-Yang, C., and C. M. Shieh. 1999. Developing a Petri-Net-based simulation model for a modified hierarchical shop floor control framework. *International journal of production research* 37(14):3139–3167.

Pace, D. K. 1999. Development and documentation of a simulation conceptual model. In *Proceedings of the 1999 Fall Simulation Interoperability Workshop (99F-SIW-017)*, Orlando, Florida: September 12–17.

Paredis, C. J. J., and T. Johnson. 2008. Using OMG's SysML to support simulation. In *Proceedings of 2008 Winter Simulation Conference*, Miami, FL, ed. S. J. Mason, R. R. Hill, L. Monch, O. Rose, T. Jefferson, and J. W. Fowler, 2350–2352. Piscataway, NJ: IEEE.

Perera, T., and K. Liyanage. 2000. Methodology for rapid identification and collection of input data in the simulation of manufacturing systems. *Simulation practice and theory* 7(7):645–656.

Perry, S. 2006. When is a process model not a process model: A comparison between UML and BPMN. In *Proceedings of The IEE Seminar on Process Modelling Using UML. (Ref. No. 2006/11432)*, The IEE, Savoy Place, London, 51–64. IEEE.

Richter, H., and L. Marz. 2000. Toward a standard process: The use of UML for designing simulation models. In *Proceedings of the 2000 Winter Simulation Conference*, Orlando, FL, ed. J. A. Joines, Barton, R.R., Kang, K. and Fishwick, P.A., 394–398. Piscataway, NJ: IEEE.

Robinson, S. 2004. *Simulation: The Practice of Model Development and Use*. Chichester: Wiley.

Robinson, S. 2008. Conceptual modelling for simulation part II: A framework for conceptual modelling. *Journal of the operational research society* 59:291–304.

Rosenblit, J., J. Hu, T. G. Kim, and B. P. Zeigler. 1990. Knowledge Based Design and Simulation Environment (KBDSE): Foundation concepts and implementation. *Journal of operational research society* 41(6):475–489.

Ryan, J., and C. Heavey. 2007. Development of a process modelling tool for simulation. *Journal of simulation* 1(3):203–213.

Schürr, A. 1994. Specification of graph translators with triple graph grammars. In *Proceedings of the 20th International Workshop on Graph-Theoretic Concepts in Computer Science*, 151–163. Springer-Verlag.

Shannon, R. E. 1975. *Systems Simulation: The Art and Science*. Englewood Cliffs, NJ: Prentice-Hall.

Shi, J. 1997. A conceptual activity cycle-based simulation modeling method. In *Proceedings of Winter Simulation Conference*, Atlanta, GA, ed. A. Andradottir, K. J. Healy, D. H. Withers, and B. L. Nelson, 1127–1133. Piscataway, NJ: IEEE.

Shih, M. H., and H. K. C. Leung. 1997. Management Petri Net: A modelling tool for management systems. *International journal of production research* 35(6):1665–1680.

Thomasma, T., and O. M. Ulgen. 1988. Hierarchical, modular simulation modelling in icon-based simulation program generators for manufacturing. In *Proceedings of Winter Simulation Conference*, San Diego, CA, Dec 12–14, ed. P. L. Haigh, J. C. Comfort, and M. A. Abrams, 254–262. Piscataway, NJ: IEEE.

van der Zee, D. J., and J. G. A. J. van der Vorst. 2005. A modeling framework for supply chain simulation: Opportunities for improved decision making. *Decision sciences* 36:65–95.

Van Rensburg, A., and N. Zwemstra. 1995. Implementing IDEF techniques as simulation modelling specifications. *Computer industrial engineering* 29:467–471.

Vojnar, T. 1997. Various kinds of Petri Nets in simulation and modelling. In *Proceedings of Conference on Modelling and Simulation of Systems MOSIS'97*, Ostrava, Czech Republic, 227–232.

Weilkiens, T. 2006. *Systems Engineering with SysML/UML*. Burlington, MA: The MK/OMG Press.

Willard, B. 2007. UML for systems engineering. *Computer standards & interfaces* 29:69–81.

12

Development of a Process Modeling Tool for Simulation

John Ryan and Cathal Heavey

CONTENTS

12.1 Introduction

Conceptual modeling or the precoding phases of any simulation project are crucial to the success of such a project (Wang and Brooks 2006). The problem definition, requirements gathering, and conceptual model formulation process is often a time-consuming one, as is the process of collecting detailed information on the operation of a system (Balci 1986). However, little substantive research on the subject has been reported in the literature (Brooks 2006, Robinson 2004).

Hollocks (2001) recognized that such premodeling and postexperimentation phases of a simulation project together represent as much or more effort than the modeling section of such projects and that software support for these phases of the wider simulation process would be valuable. Some of the particular areas of potential support highlighted by Hollocks included documentation, communication, and administration. Such areas are also discussed by Sargent (1999) in terms of model documentation and model validity. This lack of support for documentation in preference for rapid model production was further highlighted by Conwell and Enright (2000). This they ascribe points to the lack of development, documentation, maintenance and management practices for software development, which if in place can result in systems that can provide greater returns on investment and that can be used and evaluated for suitability without the need for costly rework. The difficulties of establishing model credibility due to the lack of good development practices and documentation are also discussed. Nethe and Stahlmann (1999) discuss the practice of developing high-level process models prior to the development of a simulation model. Such a method they feel would greatly aid in the collection of relevant information on system operations (i.e., data collection) and therefore reduce the effort and time consumed to develop a simulation model. Such a process modeling method for simulation could be used as a knowledge acquisition method for simulation studies. The above highlight both the importance of and lack of precoding support for simulation.

This chapter is structured as follows. The next section reviews current process modeling tools available to support simulation. This review concludes that current methods could be improved. Section 3 introduces a newly developed process modeling method that aims to overcome some of the weaknesses of current tools. In section 4 this new method is illustrated and compared with IDEF3 via a case study. Finally, the conclusions of the chapter are given.

12.2 Overview of Process Modeling Methods

During the initial stages of developing a simulation model, a means of presenting the current system and proposed simulation (or conceptual) model is typically required. This may be simply documentation of system description with diagrams or in some cases a process modeling tool may be used. A number of researchers have documented the benefits of using process modeling tools to support the initial stages of a simulation project (Van Rensburg and Zwemstra 1995, Nethe and Stahlmann 1999, Jeong 2000, Perera and Liyanage 2000). There are numerous process modeling tools available to aid in the modeling of a system. Kettinger et al. (1997) listed over a hundred in a survey that was not exhaustive. These tools are capable of modeling many different aspects of a system to varying levels of detail. Some of these tools allow simulation of process models developed within the tool (i.e., Mayer et al.

1995, Scheer 1998, and INCOME Process Designer 2003), and a number have been used to support simulation (i.e., Van Rensburg and Zwemstra 1995 and Al-Ahmari and Ridgway 1999). To ascertain the level of support given by current process modeling tools a selective review of a number of methods/tools was carried out (Ryan and Heavey 2006). The criteria used to conduct this review were as follows:

- Could the method/tool capture a detailed description of the various aspects of a discrete-event system for the purposes of a simulation project? Those being the following:
  - The flow of work, or change of state of a discrete-event system
  - The flow of information associated with the control of a discrete-event system
  - The activities that are associated with the execution of the flow of work and information within a discrete-event system
  - The resources necessary and their usage in the execution of the activities associated with both work and information within a discrete-event system
- Did the method/tool have a low modeling burden and was it therefore capable of being used and understood by nonspecialists during the conceptual modeling process? Aspects that were felt would facilitate this included these:
  - The modeling of a discrete-event system from the perspective of the user and their interactions with the system in the execution of activities within the system
  - The separation between the process modeling tool and the simulation engine to allow for the capture, representation, and communication of detailed interactions at a high level during the requirements gathering phase, as opposed to purely at the low-level code stage of a project
- Was the method/tool capable of modeling information in terms of concepts that were meaningful to system personnel such as resources and activities, as opposed to abstract terms? This was with a view to facilitating understanding and communication during conceptual modeling.
- The visualization capability of each method/tool was also examined with a view to ascertaining their abilities to facilitate communication between a model developer and system personnel. The following initiatives were examined in this regard:
  - The access to a means of elaborating graphical models to facilitate the communication of detailed information associated with such graphical representations

- The capabilities of hierarchically structuring a model to facilitate the decomposition of complex situations into related submodels;
- The graphical representation of the various tasks within a system and presentation of these tasks in a time phased sequence of execution within a system

The review focused on methods/tools that have been used to support simulation and/or exhibit characteristics desirable in a dedicated process modeling tool for simulation. The methods/tools were categorized into these methods:

Formal Methods: These are methods that have a formal basis and there are numerous software implementations of these methods. Methods reviewed under this category were: (i) Petri Nets (Ratzer et al. 2003); (ii) Discrete Event System Specification (DEVS) (Zeigler 1984); (iii) Activity Cycle Diagrams (ACD) (Tocher 1963), and (iv) Event Driven Process Chains (EDPC) (Tardieu et al. 1983).

Descriptive Methods: These methods have little formal basis and are primarily software implementations. Methods reviewed here were: (i) IDEF (NIST 1993); (ii) Integrated Enterprise Modeling (IEM) (Mertins et al. 1997); (iii) Role Activity Diagrams (RAD) (Ould 1995); (iv) GRAI Method (Doumeingts 1985), and (v) UML State Charts and Activity Diagrams (Muller 1997).

In summary, this review concluded that Petri Nets are to a certain extent capable of visually representing and communicating discrete-event-system logic, however such Petri Net models are not capable of visually accounting for complex branching logic or hierarchically decomposing complex models into submodels and as a result become very cumbersome as system complexity increases. The technique also does not account for a user's viewpoint, resources, information flows, or a means of elaborating the graphical model in a textual manner. However the technique is capable of accurately representing state transitions and the activities associated with the execution of such flows.

The DEVS formalism is capable of accurately representing the various changes in state of a discrete-event system along with being somewhat capable of representing resources, activities, and branching within its mathematical representation. However, the formalism is not visual in nature and does not account for the user's interactions with the system, information flows, or a user-friendly elaboration language.

ACDs are again somewhat capable of visually representing and communicating certain discrete-event–system logic. It achieves this by means of modeling state transitions and the activities that cause such state transitions to be executed. However, the technique fails to account for a user's perspective, resources, information modeling, branching logic or a means of textually elaborating graphical models.

EDPCs are a highly graphical process modeling technique that are capable of representing a discrete-event system as a series of activities. The technique is capable of representing branching logic and to a lesser extent information interactions within the system. Drawbacks of the system however include its lack of a representation of the user's perspective, state transitions, and resource interactions. The technique also does not have the capability to hierarchically decompose a model into submodels or have access to an associated elaboration language.

IDEF0 is a graphical modeling technique capable of representing a discrete-event system as a series of interrelated activities. The technique is capable of hierarchically decomposing a model into submodels and is also to a certain extent capable of accounting for both information and resource interactions. However the technique does not account for system branching, the elaboration of graphical models, state transitions, or the modeling of a user's perspective. The IDEF3 process modeling technique is capable of graphically representing the various states through which a discrete-event system can transition along with the various activities associated with each change of state. This technique also offers a means of representing complex system branching logic along with a means of hierarchically decomposing a model into related submodels. The technique is also capable of textually representing the graphical models; however, this representation language is abstract in nature. This representation language also offers a means of representing resources associated with the graphical models. However, the technique does not account for information flows or modeling from a user's perspective.

The IEM technique presents a highly visual and communicative model of a discrete-event system, which is capable of graphically representing state transitions, information, and resource elements. The technique is also capable of hierarchically decomposing a model into submodels along with having a detailed branching logic associated with it. However, the technique does not account for a user's viewpoint or have an associated elaboration language.

RADs are a highly visual modeling technique that accounts for the user's perspective in the development of a process model of a discrete-event system. The technique is to a certain extent also capable of representing the logical branching of such activities within a model. The technique, however, does not have the means of representing state transitions, information flows, resource interactions or a means of either hierarchically decomposing, or textually elaborating graphical models.

The GRAI model offers a means of modeling the detailed information and control interactions within a discrete-event system. This information model is also capable of representing discrete activities and model decomposition along with to a lesser extent both state transitions and resources. However, the model does not account explicitly for the user's perspective, branching logic, or an elaboration language.

UML statecharts are a highly visual and communicative modeling technique that are used represent a discrete-event system as a series of interrelated state transitions. This technique also has a means of graphically representing the logical flow of states and hierarchically decomposing a model into submodels. However the system does not account for information flows, resources, activities, and an inclusion of a user's interaction with the system or a means of textually elaborating the graphical model. UML activity diagrams are designed to represent a discrete-event system as a series of activities linked together to show the various phases of activity within a discrete-event system. The technique is highly visual and communicative and also has to a certain extent a means of visually representing the logical flow of activities. However the system does not account for a user's perspective, state transitions, information modeling, resource modeling, or a means of elaborating the graphical models.

Resources are a major issue in many simulation projects. Techniques such as IEM and EDPCs are capable of accurately representing such resources within a discrete-event system. To a lesser extent IDEF0, IDEF3, GRAI, RADs, and DEVS can represent aspects of resources within a discrete-event system. However techniques such as Petri Nets, ACDs, UML activity diagrams, and UML statecharts do not have such a means of representing such resources. Activities are also well represented within many techniques such as Petri Nets, ACDs, UML activity diagrams, RADs, GRAI, IEM, EDPCs, IDEF0, and IDEF3. While the DEVS technique is capable of representing activities to a lesser extent. Complex branching logic is well represented with techniques such as UML activity diagrams, UML statecharts, EDPCs, and IDEF3 by means of the branch types used in each. Techniques such as Petri Nets, DEVS, RADs, and IEM have the ability to represent such branching to a lesser extent. While techniques such as IDEF0, GRAI, and ACDs lack the capability to display such branching logic. Finally no technique examined apart from the IDEF3 technique was capable of presenting the user with an elaboration language to further explain the graphical model produced. Such elaboration languages are a textual means of describing complex information and interactions within discrete-event systems that cannot be readily or easily represented within a visual representation alone. While the IDEF3 technique did have this capability the elaboration language was abstract in nature and not easy to reason over.

From the analysis above it is concluded that while there are many process modeling techniques and software tools available that may be used to support the requirements gathering phases of a simulation project, none of the techniques reviewed fully support the conceptual modeling phase of a simulation project. As a result of this review research has been carried out into developing a process modeling specifically tailored to support the conceptual modeling phase of a simulation project. The following design objectives were used in developing the process modeling method:

- The technique has to be capable of capturing a detailed description of a discrete-event system.
- The technique should have a low modeling burden and therefore be capable of being used by nonspecialists.
- The technique should present modeling information at a high semantic level so that personnel can rationalize with it.
- The technique should have good visualization capabilities.
- The technique should support project teamwork.

The resulting process modeling tool is called Simulation Activity Diagrams (SAD) and is briefly described in the next section.

12.3 Simulation Activity Diagrams (SAD)

The SAD technique presented here endeavors to be a highly visual process modeling technique to aid in the process of communication between the model developer and system users involved in the process of developing a simulation model, while still aiding the model developer in the gathering of data for the creation of the model. As well as supporting the requirements gathering phase of a simulation project, another important function of the technique proposed here is to act as a knowledge repository. A brief overview of SAD is now presented.

12.3.1 SAD Action List

A discrete-event system consists of a series of discrete events, the outcomes of which when grouped together ultimately decide the progress of a particular system. In a simulation engine these events are stored in an event list and executed in order of their time of occurrence. To endeavor to graphically represent this scenario the SAD technique uses the concept of an activity, whereby an activity is any event that causes the change of state of a discrete-event system. Such aforementioned simulation events, however, can often be amalgamations of numerous real-world events in an attempt to lessen programming burden. This can lead to difficulties among personnel not intimately involved in the model coding understanding the simulation model. In an attempt to allow the model developer to account for such amalgamations of events graphically within the SAD technique each individual activity within a SAD diagram can be composed of a series of actions as shown in Figure 12.1.

The system is in state 1. Before it can transition to state 2, all actions, 1, 2 and 3 must be executed. In this way, an individual activity is considered a separate mini event list or action list within the SAD model. These actions

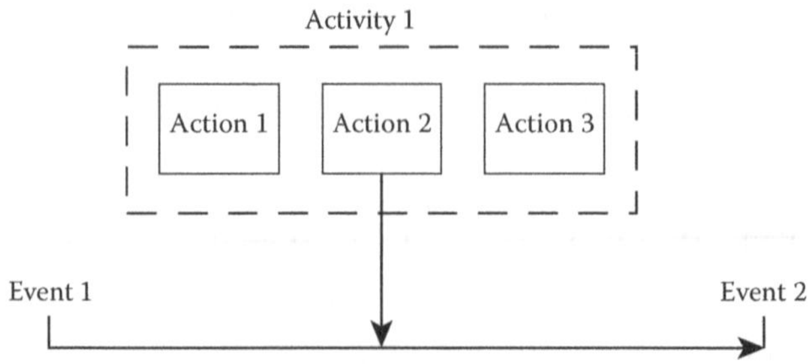

FIGURE 12.1
SAD actions.

are executed in a time ordered sequence from top to bottom and from left to right ensuring that each criterion is satisfied. Only when each action has been executed, can the full activity be executed and the system transition successfully to state 2. Taking this approach a SAD becomes a graphical representation of the various events in a simulation model. Each event is represented in a SAD by an activity. This activity is then further graphically represented by an action list. This will be further developed in the following section by the introduction of a series of modeling primitives that may be used in the detailing of such an activity.

12.3.2 SAD Modeling Primitives

Within most systems, actions such as those in Figure 12.1 are rarely executed without a number of other types of resources being used. These resources are briefly introduced below:

Primary resource element: A primary resource element represents any resource within a discrete-event system that facilitates the transformation of a product, physical or virtual, from one state of transition to another.

Queue resource element: A queue modeling element represents any phase of a discrete-event system where a product, virtual or physical, is not in an active state of transformation within the system.

Entity element: An entity element represents any product, physical or virtual, that is transformed as the result of transitioning through a discrete-event system.

Entity state element: An entity state represents any of the various states that a physical object or component explicitly represented within a system transitions through during physical transformation.

Informational element: An informational element represents any information that is used in the control or operation of the process of transition by a product through a discrete-event system.

Informational state element: An informational state represents any of the various states that information used in the operation or control of a discrete-event system transitions through during the support of the operation of the physical transformation.

Auxiliary resource element: An auxiliary resource represents any resource used in the support of a Primary Resource. For example, within a system being simulated a primary resource, such as a machine may be used in the transformation of an entity from state A to state B. However this primary resource may require an operator and a number of other tools that an operator may use to operate the machine, such auxiliary resources can be either of two varieties, namely an actor auxiliary resource or a supporter auxiliary resource.

Actor auxiliary resource: An actor auxiliary resource represents any auxiliary resource used in the direct support of the execution of an action or actions within the process of transitioning a system from one state to another.

Supporter auxiliary resource: A supporter auxiliary resource represents any auxiliary resource used in the direct support of an actor auxiliary resource in the execution of an action or actions within the process of transitioning a system from one state to another.

Branching Elements: Most discrete-event systems are complex in nature and are rarely, if ever, linear. To account for the representation of such situations the SAD technique uses the following branching elements. There are two general types of branching elements, fan in and fan out. Both of these branch types can be further subdivided into conjunctive and disjunctive branch elements. Where conjunctive branch elements represent the branching and joining of multiple parallel subsystems and disjunctive branch elements represent the branching and joining of multiple alternative subsystems. A logical, "AND," branch element is used to represent conjunctive branching. While there are two types of disjunctive branch elements, inclusive and exclusive, represented by an "OR" and an "XOR" respectively. Finally, the conjunctive branch and inclusive disjunctive branch elements may be either synchronous or asynchronous. Where a synchronous branch element signifies that all elements either preceding or proceeding the branch element depending on its type, fan in or fan out, must either begin or end simultaneously. Neither an exclusive disjunctive branch element or any asynchronous branch element require such simultaneous initiation or termination and are therefore the more commonly used.

Therefore a fan out, "AND" branch in a model means that when the execution of the model reaches that point in the process represented by such a

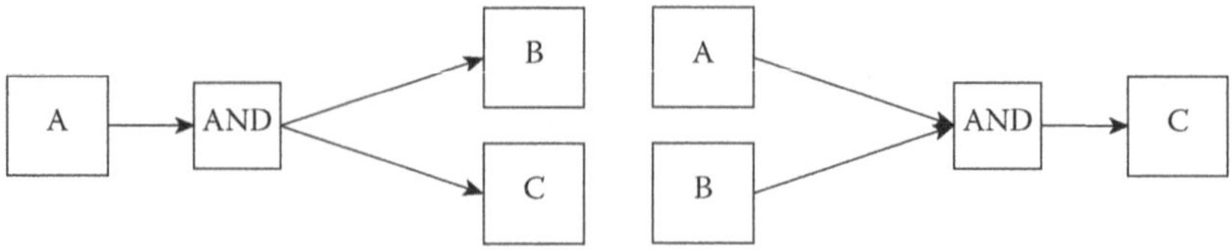

FIGURE 12.2
AND branches.

branch, all the elements that are immediate successors of the branch will be executed. If a synchronous, "AND(S)" branch is used then the execution of that branch will mean that all of the immediate successor elements must begin execution simultaneously.

Similarly in a model where a fan in, "AND," branch is executed all elements that immediately precede that branch will have been executed. If a synchronous, "AND(S)," branch is used, then, for that part of the model to execute all the elements preceding must all end simultaneously. Thus, an execution of the left-hand model in Figure 12.2 will consist of the execution of element, A, followed by elements B and C. Similarly the execution of the right-hand model in Figure 12.2 will result in the execution of element, C, preceded by the execution of elements A and B; if a synchronous, "AND(S)," branch is used, then for there to be an execution of the element, C, both elements, A and B must end simultaneously. For example the left-hand model of Figure 12.2 could represent a disassembly operation, element A could be broken down into two constituent parts, B and C. Similarly the right-hand model could represent an assembly where elements A and B are combined to create element C.

A fan out inclusive, "OR," branch in a model indicates that, in an execution of that branch there will be an execution of at least one of the elements connected to the branch to the right. Similarly, a fan out exclusive, "XOR" branch in a model indicates that, in an execution of that branch, there will be an instance of exactly one of the elements connected to the branch to the right, for example an element will either pass or fail inspection, it cannot do both. If a synchronous inclusive, "OR(S)" branch is used, then all elements that are executed must start simultaneously. This does not apply to exclusive, "XOR" branches, since there can only be one element executed in an "XOR" execution. Similarly with fan in inclusive "OR" branch, there will be at least one element executed to the left of the branch. If a synchronous inclusive "OR(S)" branch is used, then, those elements that are executed, if there is more than one, must all end simultaneously. Hence, an execution of the model to the left in Figure 12.3 consists of an instance of the element A proceeded by an instance of either B or C, or both. If the models in Figure 12.3 used "XOR" branches, then an execution of the first model could not include an instance in which the execution of both B and C occur while an execution

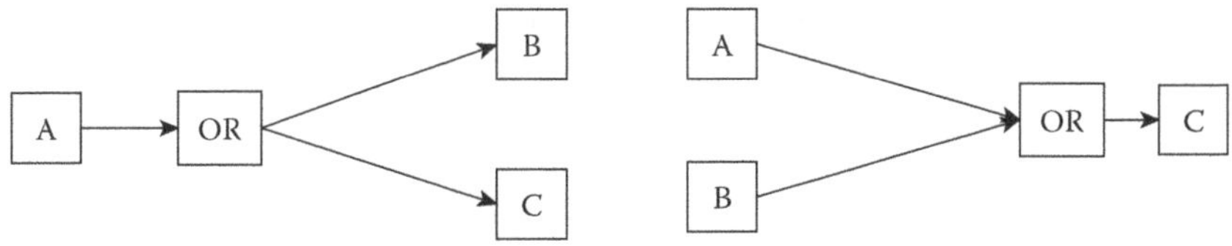

FIGURE 12.3
OR branches.

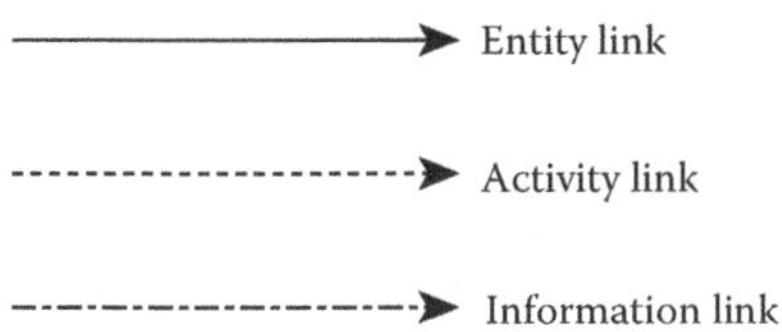

FIGURE 12.4
SAD link types.

of the second model could not include an instance where an execution of both A and B occur.

Link Types: Links are the glue that connects the various elements of a SAD model together to form complete processes. Within the SAD technique there are three link types introduced known as entity links, information links, and activity links. Arrows on each link denote the direction of the flow of each representative link. The symbols that represent each type are shown in Figure 12.4.

SAD Frame Element: The SAD frame element provides a mechanism for the hierarchical structuring of detailed interactions within a discrete-event system into their component elements, while also showing how such elements interact within the overall discrete-event system.

12.3.3 SAD Model Structure

A SAD model is executed in time sequenced ordering from left to right and from the center auxiliary resource area to the extremities of the model and is structured as shown in Figure 12.5. At the center of each model is the actor/supporter auxiliary resources. These are the auxiliary resources that are used in the supporting of any discrete processes being executed in either the physical or informational systems within any SAD model. The distinction between an actor and supporter type of auxiliary resource within this grouping of auxiliary resources allows for the separation

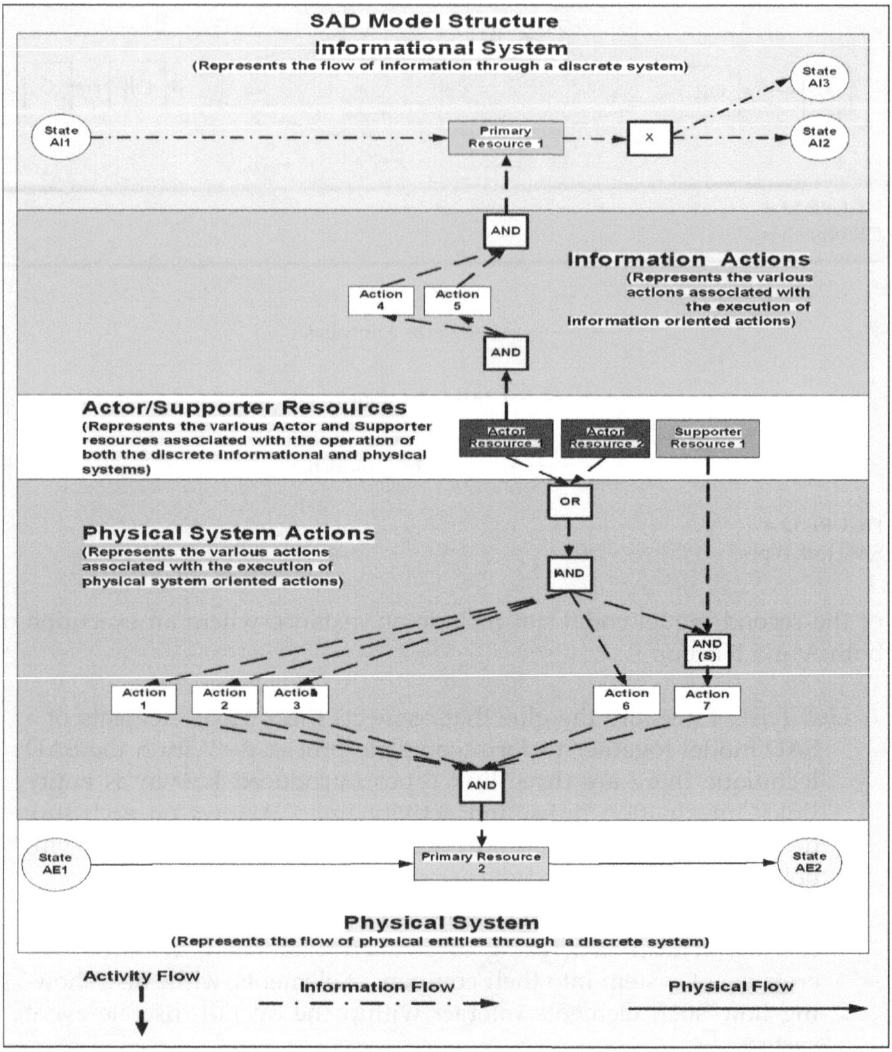

FIGURE 12.5
A simple SAD example.

between an operator resource and other auxiliary resources that may be modeled within a discrete-event system. This can be of advantage to communication during the requirements gathering phases of a simulation project as often the persons with whom the simulation model developer will be communicating will be actors within the process. In this way, a SAD model will be developed from the perspective of the persons interacting with the system. The interconnecting areas between both models contain the actions to be executed.

TABLE 12.1

Structured Language

Keyword	Description
USES	The supporter resource may at times make use of auxiliary resources to execute an action or actions, in other words a supporter USES auxiliary resources
TO	Details the action or actions that are executed by use of an auxiliary resource by a supporter resource
AT	Specifies the Locations where the action or actions are executed
TRANSITIONS TO	Specifies the change of state of entity or information from one state to another

A series of these actions and the associated interactions with other SAD modeling elements make up an action list. A series of these activities in turn make up a sequence of transition for physical or information entity.

12.3.4 Elaboration of SAD Models

Thus far, the modeling elements used to develop a SAD model have been introduced to provide a means of visually modeling discrete-event systems. However, such graphical models are capable of only representing a certain amount of detailed information and knowledge. Often, complex discrete-event systems contain detailed information and knowledge related to process interactions that cannot be captured well by such graphical representations. To provide a means of making such information available to a model user the SAD technique also makes use of an elaboration language with which each individual SAD diagram can be described in greater detail. This structured language makes use of a number of different reserved words to allow the description of SADs, see Table 12.1.

12.4 Evaluation of SAD: Case Study

A prototype software application called the Process Modeling Software (PMS) has been developed using Microsoft Visual C++ to implement the SAD methodology. The objective in developing PMS was to allow further evaluation, beyond evaluation using paper-based models. SAD was evaluated using the PMS software in a number of different production scenarios including a batch flow-shop type production system where the operators have a lot of decision making power in relation to the advancement of the system and the types of parts that are produced at a given time. This scenario is presented in the next section to further illustrate the SAD approach. In an effort to facilitate comparison the SAD description is first given and then is followed by an IDEF3 model. The case study ends with a discussion comparing SAD with IDEF3 and other modeling approaches currently available.

12.4.1 System Description

The manufacturing facility modeled produces mining rods and can be classified as a batch flow-shop, consisting of four major work regions. The first region consists of precarburising operations. The second work region relates to the carburising or induction-hardening phase of the production process. The third work region encompasses the postcarburising operations and finishing operations and the final work region represents the final inspection of the product before dispatch to the relevant customer. The second work area is quiet complex in terms of the decisions made by operators and the amount of control vested in them. It is on modeling this operator control and decision making process that this case study description concentrates on.

In work region 2 (carburising) parts arrive into the furnace area and wait until all operations such as roping, application of anticarburising paint and stamping of the batch number have been performed. Rods that require carburising are staged in the carburising area, where they are split up into separate holding areas based on their carburising setting. There are 12 carburising settings with the carburising times varying from 4.5 hours to 10.5 hours. Within each of these carburising setting holding areas the rods are separated according to length. Before the rods are carburised certain preparatory operations are performed, e.g., inserting a carburising rope. To enter the furnace the rods are manually loaded (using a crane if heavy) onto a carburising jig. The carburising jig consists of a column, made up of tiers of rods of which there are a maximum of four on each jig. Each tier on the jig consists of a six-sectioned "spider." Placed within each section of this spider is a honeycomb tray, which allows the rods to be hung vertically in each section. The spider, honeycomb trays, and rods contained therein are collectively known as a "tier," a schematic of such a carburising jig is shown in Figure 12.6.

The length of the rods being carburised determines the number of tiers on the jig. For very long rods only one tier is usable, for very short rods four tiers can be used. The diameter and shape of the rods determine the type of honeycomb tray that is used. Each tier has six trays containing honeycombs into which rods are slotted. There are four types of trays:

- Type A: Can hold a maximum of 16 rods
- Type B: Can hold a maximum of 12 rods
- Type C: Can hold a maximum of nine rods
- Type D: Can hold a maximum of three rods

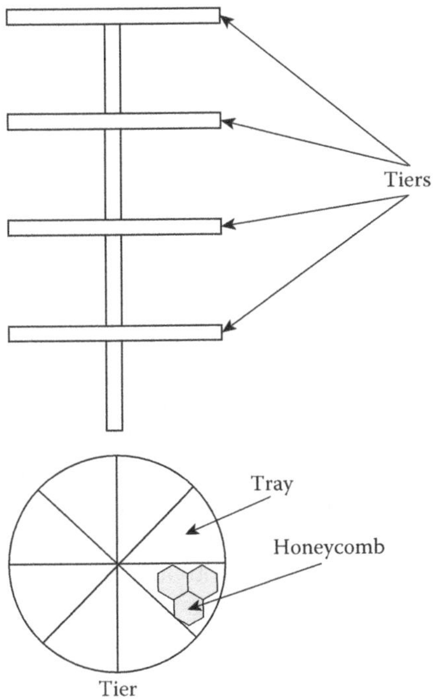

FIGURE 12.6
Schematic of a carburising jig.

There are two furnace operators who are required to carry out the following prioritized operations:

- Load/unload the furnace (Priority 1)
- Build/dismantle a jig (Priority 2)
- Load/unload the air cooling tower (Priority 3)
- Pre-jig building operations (Priority 4)

Pre-jig building operations consist of inserting rayon ropes, applying anti-carburising paint, and stamping the batch number on parts. Unloading the furnace occupies the operators for approximately 30 minutes. This task is assigned the highest priority in the model and therefore, whenever it occurs the operators stop working on all other tasks and are pulled to the furnace. Building or dismantling of jigs is given the next highest priority. All other tasks have very low priority and cannot be started unless the aforementioned operations are not possible. Operators will attempt to build a jig before dismantling one so as to ensure that a jig will be available when the furnace requires one. However, jigs are a limited resource in that there are only three jigs in the furnace area. Also, jig building may not be complete when the furnace next becomes empty. Finally, after the jig containing the rods is carburised, it must be transferred immediately to the cooling tower to be cooled under controlled conditions to ensure the required hardness is achieved by the carburising process. After the cooling tower the operators allow the jig to air cool until the rods are cool enough to be unloaded. The unloading operation is a manual operation, where the parts are unloaded and passed to the next work region.

12.4.2 SAD Model

The following section presents a SAD model developed to communicate the various interactions between the operators and the carburising part of the manufacturing system. Such interactions require the model developer to gather and communicate detailed information on a system. The SAD model developed using the PMS software for the carburising area is shown in Figure 12.7.

Each SAD diagram starts from the actor/supporter auxiliary resources section, in this way each SAD is developed and executed from the perspective of those using the system or interacting with it. In this case, either one or the other or both (the "OR" branch) of the operators can do either (the "XOR" branch) the "Rope & stamp parts" action or else they can either with or without a "Crane" supporter auxiliary resource (the next "OR" branch) can carry out any of each of the following individual actions, individually (the "XOR" branch), "Build a tray," "Build a tier." "Build a jig," "Move to jig waiting area," "Collect jig," "Load jig," "Carburise," "Unload jig," "Load jig," "Cool," "Temper," "Unload jig," "Move jig to holding area," "Dismantle jig," denoted by the yellow actions. Some of these actions can be done either separately or in conjunction with each other. For example either build a tray or a tier or an entire jig or possibly all of these or any combinations, denoted by the "OR" branch above the "Build a tray," "Build a tier," "Build a jig" actions. Also a number of tasks are always carried out in sequence with each other, for example "Build a jig" and "move jig to waiting area" are both carried out in sequence, denoted by the "AND" branch above these tasks. A number of "AND" branches are also located between the actions and the various queues and primary resources, these branches are used to indicate where each of the individual actions are executed, for example the actions "Rope & stamp parts," "Build a tray," "Build a tier," "Build a jig" are executed at the "jig holding area" queue element in this SAD model.

The physical system (located at the lower region of the SAD model) shows the change of state of the parts (entities) within the system having passed fully through the Furnace area, the parts denoted by the green entity state objects change from a "Pre anneal part" to an "Annealed part."

Coupled with this visual model the PMS software allows for the elaboration of the graphical models by means of the structured elaboration language, which is generated from the visual model. The elaboration facility of the PMS tool is shown in Figure 12.8. The full text of the elaboration is contained in Table 12.2. From this table it can be seen that information and

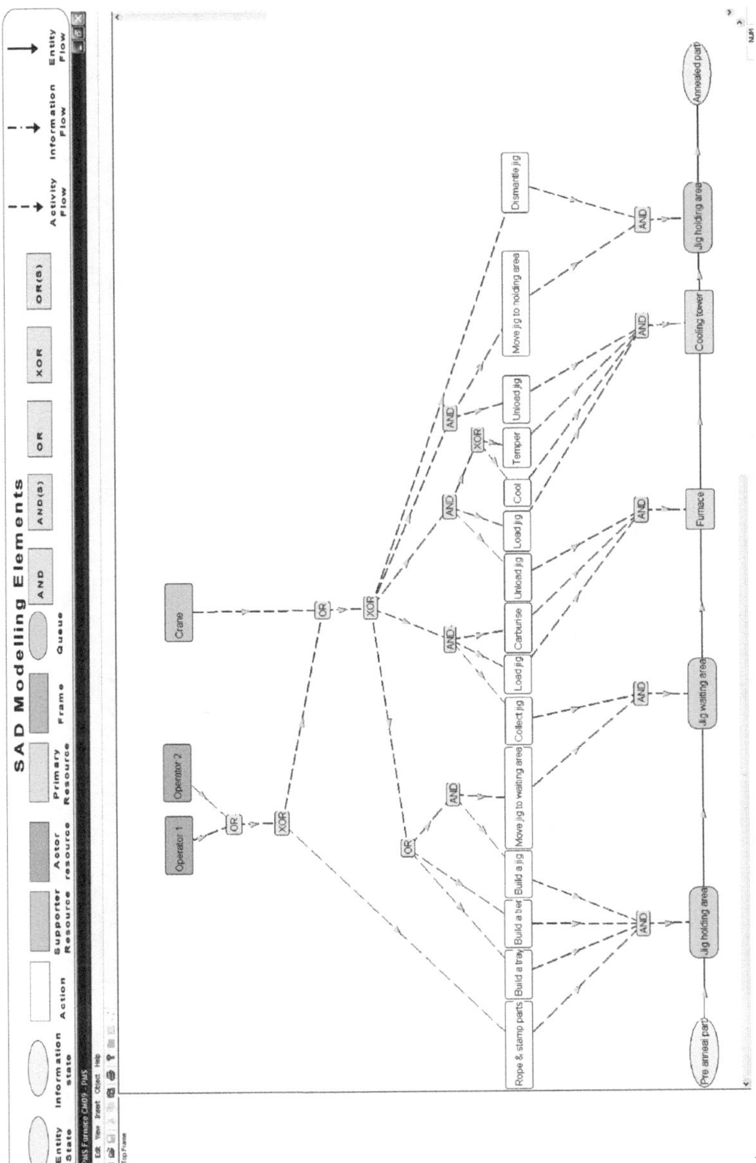

FIGURE 12.7
SAD model of Work Region 2.

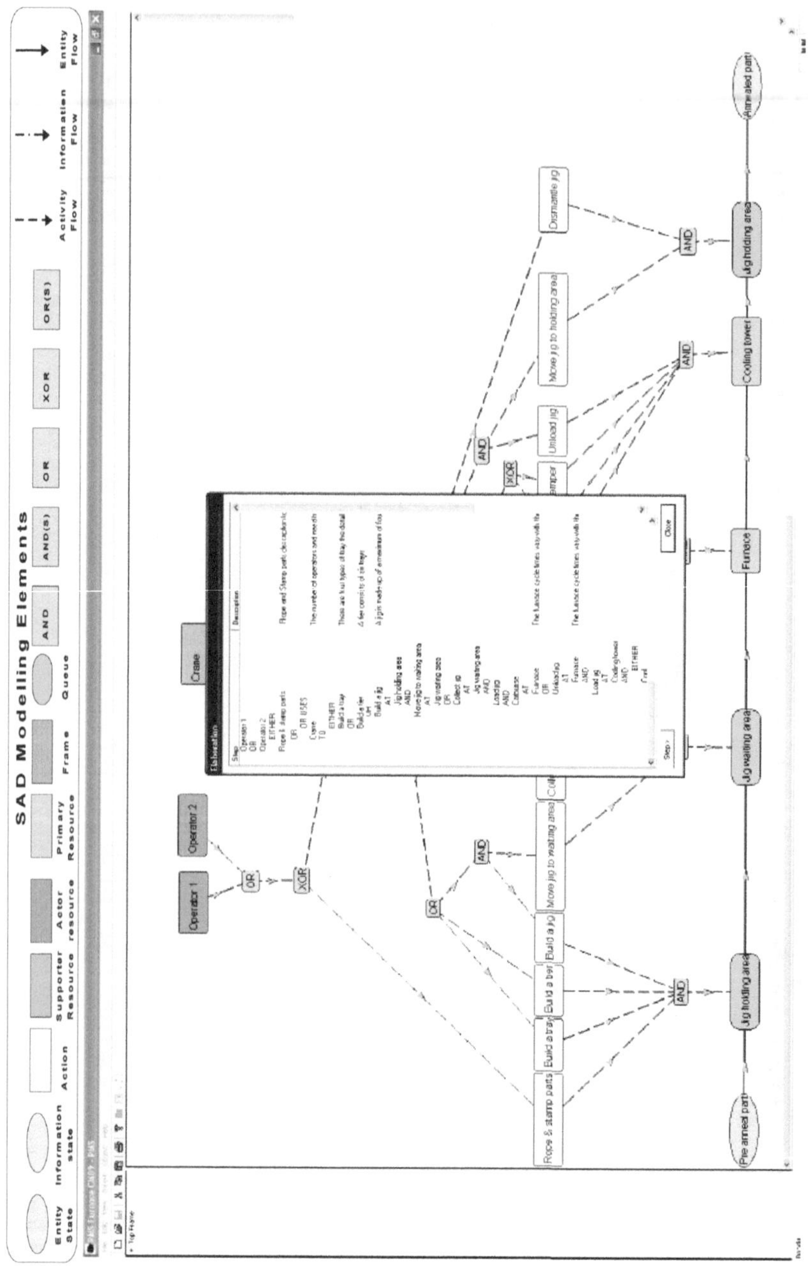

FIGURE 12.8
Elaboration of SAD model.

TABLE 12.2
Elaboration of the SAD Model

Elaboration of the Activity
Operator 1
OR
Operator 2
EITHER
The operations are outlined here in the sequence of execution to produce a part, however priority rules apply to the sequence of operations within the area and these priority rules are contained in an attached document (Furnace-operation- priorities.doc)
Rope & stamp parts
AT
Jig holding area
OR
OR
USES
Crane
The number of operators and need for a crane is dependant on the size of parts being placed on the tray/tier or jig. Details are contained in the following four attached documents.
(Load-requirements-hex-rods.xls)
(Load-requirements-round-rods.xls)
(Unload-requirements-hex-rods.xls)
(Unload-requirements-round-rods.xls)
TO
EITHER
Build a tray
There are four types of tray the details of which are contained in the attached document (tray-types.xls)
OR
Build a tier
A tier consists of 6 trays
OR
Build a jig
A jig is made up of a maximum of four tiers and each tier is made up of a number of trays. The number of tiers and trays used and the number of parts is dependant on the size and weight of parts with maximum limits on each. The details for this are contained within the following attached documents.
(Max-Furnace-utilisation.xls)
(Round-rod-weights.xls)
(Hex-Rod-weights.xls)
While fully built jigs are preferred, parts in the holding section for longer than 8 hours may be used on partially built jigs.

(Continued)

TABLE 12.2 (Continued)

Elaboration of the SAD Model

Elaboration of the Activity
AT
Jig holding area
AND
Move jig to waiting area
AT
Jig waiting area
Or
Collect jig
AT
Jig waiting area
AND
Load jig
AND
Carburise
AT
Furnace
The furnace cycle times vary with the details contained in the attached document (Furnace-cycle-times.xls)
OR
Unload jig
AT
Furnace
AND
Load jig
AND
EITHER
Cool
OR
Temper
AT
Cooling tower
OR
Unload jig
AT
Cooling tower
AND
Move jig to holding area
AT
Jig holding area

TABLE 12.2 (Continued)
Elaboration of the SAD Model

Elaboration of the Activity
OR
Dismantle jig
AT
Jig holding area
AND
THEN
Pre anneal part entity state
TRANSITIONS TO
Annealed part entity state

knowledge not suited to graphical representation such as priority rules or the usage rules for certain resources such as cranes, etc. can be easily linked to the visual model, explained, and accessed by means of the elaboration language.

12.4.3 IDEF3 Model

The IDEF3 Process Description Method provides a mechanism for collecting and documenting processes. IDEF3 captures precedence and causality relations between situations and events in a form natural to domain experts, by providing a structured method for expressing knowledge about how a system, process, or organization works. The resulting IDEF3 descriptions provide a structured knowledge base for constructing analytical and design models. These descriptions capture information about what a system actually does or will do, and also provide for the organization and expression of different user views of the system.

There are two IDEF3 description modes, process flow and object state transition network. A process flow description captures knowledge of how things work in an organization, e.g., the description of what happens to a part as it flows through a sequence of manufacturing processes. The object state transition network description summarizes the allowable transitions an object may undergo throughout a particular process. Both the process flow description and object state transition description contain units of information that make up the system description. In Figure 12.9 the IDEF3 model for the carburising area is shown. At the highest level in this model the carburise area is represented by a unit of behavior (UOB) named "CARB1 Carburise." UOBs can be used to represent a system, subsystem, or individual tasks within a model depending on the context and level at which they are used.

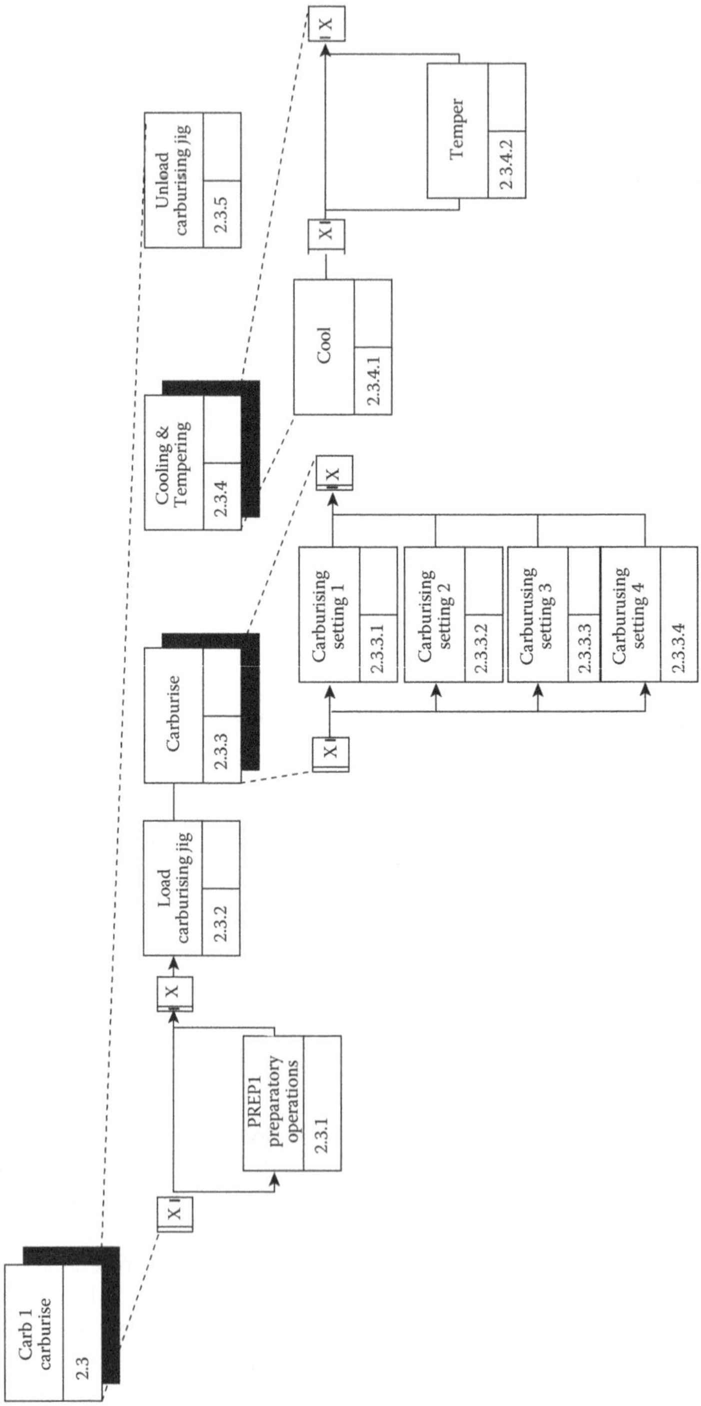

FIGURE12.9
IDEF3 model of Work Region 2.

The carburise area has a submodel or more-detailed description attached to further describe it. This is shown by means of the black background behind the aforementioned UOB. The submodel shows that within the carburise operation there are "preparatory operations" that are only conducted on some of the parts passing through this function, from here all parts both with and without preparatory operations proceed to the "Load carburising jig" operation and then in turn to the "Carburise," "Cooling & Tempering," and "Unload carburising jig" operations. Two of these final UOBs again have submodels associated with them. The "Carburise" UOB/operation has a submodel showing four carburising settings of which each part must pass though only one. The final submodel associated with the "Cooling & Tempering" UOB shows that all parts passing through this UOB must pass through a "Cool" operation and some parts will pass through a "Temper" operation. The IDEF3 process description developed allows for the capture, representation, and communication of the various states through which the system in question can transition. However, the information associated with the control of such systems and the use of resources are not graphically represented within the technique. This was especially pertinent in the case study presented as much of the model complexity emanated from the complex operator decision making processes embedded within the process as shown within the SAD model.

12.4.4 Differentiation of the SAD Technique from Currently Available Techniques

The SAD technique that has been presented in this chapter has been developed specifically to support the requirements gathering phases and conceptual model development within a simulation project. In facilitating this requirement the technique represents both what a discrete process is and likewise, how a simulation model represents such a process. As highlighted in this chapter there are various process modeling techniques available to a simulation model developer that can be used to aid in these precoding phases. The SAD technique has adopted certain aspects of a number of these techniques, which are outlined briefly below:

- The ACD and Petri Net approach of modeling a system as alternating phases of activity and waiting is represented within the SAD technique by means of the introduction of primary resource and queue elements.
- Two aspects of the IDEF3 technique are adopted within SADs those being the branching elements and hierarchical structuring of a process model.
- The EDPC style of modeling a discrete-event system as a series of events forms the basis of the concept of a SAD action list.

- The RAD viewpoint of placing a role or the person or persons charged with a task or series of tasks centrally within the model is implemented by means of the subdivision of auxiliary resources into actor and supporter resources with the actor resource capable of representing a person's role within a SAD.

While such similarities exist within the SAD technique, the overall modeling approach is radically different. The SAD technique endeavors to model complex interactions such as those that take place within an actual detailed simulation model of a real system. Again the SAD technique is designed to fulfill the design objectives outlined in Section 1 of this chapter. Each of these requirements are represented within the SAD technique. Both the physical and informational flows within a discrete-event system are modeled at either extremity of a SAD model as shown in Figure 12.5. Also modeled are the resources used in the execution of the various activities associated with the transitioning of both the physical and informational models through their various discrete states, again represented in Figure 12.5. In achieving these goals, the technique uses the various SAD modeling primitives to represent the various events that are listed in a simulation event list. To also represent more complex interactions, the SAD technique introduces the concept of an action list, which is used to represent detailed actions that collectively can make up any event within a simulation event list. Such a modeling approach allows for the modeling of a modern discrete-event system and in turn a simulation model of the same. Finally the use of a structured text-based elaboration within the SAD technique allows for the removal of any ambiguities that may arise within a complex model. Such an approach increases the user's access to the information and knowledge that would otherwise be lost in detailed simulation code. As a result of these modeling approaches the SAD technique uses a set of high-level modeling primitives that are capable of representing complex discrete-event systems. The modeling technique places a low modeling burden on the model developer while also promoting the capture, representation and communication of detailed information in a user-friendly manner for models users.

12.4.5 Discussion

The SAD technique while not yet supplying a full and definitive support tool for the requirements gathering phases of a simulation project does it is felt go some way toward acting as an initial solution space.

In its current guise the SAD technique endeavors to model complex interactions such as those that take place within an actual detailed simulation model of a real system. To achieve this the modeling method uses the various SAD modeling primitives to represent the events in a simulation model. To also represent more complex interactions the SAD method introduces the concept of an action list, which is used to represent detailed actions that collectively can make up any event within a simulation model. The SAD technique also allows for the modeling of both a physical and informational system that may make up a discrete-event system along with interactions between both (Ryan and Heavey 2006).

Each SAD diagram starts from the actor/supporter resources section, in this way each SAD is developed and executed from the perspective of those using the system or interacting with it. In this case either one or the other or both (the "OR" branch) of the operators can do either (the "XOR" branch) the "rope & stamp parts" action or else they can either with or without a "Crane" supporter auxiliary resource (the next "OR" branch) can carry out any of each of the following individual actions, individually (the "XOR" branch), "Build a tray," "Build a tier," "Build a jig," "Move to jig waiting area," "Collect jig," "Load jig," 'Carburise," "Unload jig," "Load jig," "Cool," "Temper," "Unload jig," "Move jig to holding area," "Dismantle jig," denoted by the yellow actions. Some of these actions can be done either separately or in conjunction with each other. For example either build a tray or a tier or an entire jig or possibly all of these or any combinations, denoted by the "OR" branch above the 'Build a tray," "Build a tier," "Build a jig" actions. Also a number of tasks are always carried out in sequence with each other, for example "Build a jig" and "Move jig to waiting area" are both carried out in sequence, denoted by the "AND" branch above these tasks. A number of "AND" branches are also located between the actions and the various queues and primary resources, these branches are used to indicate where each of the individual actions are executed, for example the actions "Rope & stamp parts," "Build a tray," "Build a tier," "Build a jig" are executed at the "Jig holding area" queue element in this SAD model. The physical system (located at the lower region of the SAD model) shows the change of state of the parts (entities) within the system having passed fully through the Furnace area, the parts denoted by the green entity state objects change from a "Pre anneal part" to an "Annealed part."

The SAD technique is not a definitive solution and currently needs further refinement, validation, and development. A number of issues are still in need of addressing. These include the incorporation of multiple modeling views, this would allow a model developer to initially model the system requirements "as is" model and from this develop a second system view or conceptual model. The facilitation of a process whereby both models could be developed in the same format and viewed simultaneously would it is felt further enhance communication and understanding. The implementation of a step through facility would also it is felt be advantageous. It is also felt that there is a need for the development of further techniques to support a simulation model developer in these precoding phases of a simulation project. It is hoped that further research will be carried out in this area with a view to the development of such techniques. The advantages that such techniques may offer while being difficult to accurately predict may include a number of the following. The development of detailed, valid and visual process models of complex discrete-event systems prior to the coding of simulation models may save time and ultimately money in the development of simulation models. The number of project failures could be reduced as a result of access to correct information and the development of valid and understandable models earlier in a simulation project. Such models should also facilitate better understanding of the process of simulation among non-simulation experts. This communication should allow for the reduction in the time taken to complete simulation projects, as model developers should be able to retrieve the necessary information for the project at an earlier stage in the project life cycle. The information gathered should also be more accurate and focused in relation to the problem areas being examined thus reducing project iterations at a later stage or in more extreme cases project failures. Graphical and accurate models of a problem area may even negate the necessity of simulation model development in certain cases as a solution may become apparent through the initial process modeling phase of a project.

12.5 Conclusions

The requirements gathering phase of a simulation project is important in relation to the overall success of a simulation project. This chapter highlights the fact that there is inadequate support currently available for this task. While numerous process modeling techniques are available and several have been used to support the requirements gathering of a simulation project, the chapter argues that the techniques available do not provide adequate support. The chapter presents an overview of a process modeling technique, SAD, developed to endeavor to overcome some of the current shortfalls highlighted. The SAD technique endeavors to model complex interactions such as those that take place within an actual detailed simulation model of a real system. To achieve this the modeling method uses the various SAD modeling primitives to represent the events in a simulation model. The SAD method has been evaluated on five case studies. The partial results of one case study (a batch flow line) was presented and using this case study a comparison with IDEF3 was made. It is important to note that SAD is not being presented as a "final" solution but results of work-in-progress research.

Acknowledgments

The authors wish to thank the following for permission to reproduce copyright material:

Ryan, J., and C. Heavey. 2007. Development of a process modeling tool for simulation. *Journal of Simulation* 1(3): 203–213. Reproduced with permission of Palgrave Macmillan.

References

Al-Ahmari, A. M. A. and K. Ridgway, 1999. An integrated modelling method to support manufacturing systems analysis and design, *Computers in industry,* 38: 225–238.

Balci, O. 1986. Credibility assessment of simulation results In J. Henriksen, S. Roberts, and J. Wilson (eds), *Proceedings of the 1986 Winter Simulation Conference,* Piscataway, NJ, New York: Association for computing machinery (ACM), 38–43.

Brooks, R. (2006), Some thoughts on conceptual modelling, performance, complexity and simplification. In J.Garnett, S. Brailsford, S. Robinson and S. Taylor (eds), *Proceedings of the 2006 OR Society Simulation Workshop (SW06),* Lemmington-Spa, UK, Birmingham :The Operational Research Society, 221–226.

Brooks, R., and Wang, W. 2006. Improving the understanding of conceptual modelling In J.Garnett, S. Brailsford, S. Robinson and S. Taylor (eds), *Proceedings of the 2006 OR Society Simulation Workshop (SW06),* Lemmington-Spa, UK, Birmingham: The Operational Research Society, 227–234.

Conwell, C. L., R. Enright, and Stutzman, M.A. 2000. Capability maturity models support of modelling and simulation verification, validation, and accreditation In R. R. Barton, P. A. Fishwick, J. A. Joines and K. Kang (eds), *Proceedings of the 2000 Winter Simulation Conference,* Orlando, Florida: IEEE Computer Society, 819–828.

Doumeingts, G. 1985. How to decentralize decisions through GRAI model in production management, *Computers in industry,* 6(6): 501–514.

Hollocks, B. W. 2001., Discrete event simulation: an inquiry into user practice, *Simulation practice and theory,* 8(67): 451–471.

INCOME Process Designer. 2003. URL: http://www.get-process.com/. Last accessed 6/12/2003.

Jeong, K.-Y. 2000., Conceptual frame for development of optimized simulation-based scheduling systems, *Expert systems with applications,* 18(4): 299–306.

Kettinger, W. J., J. T. C. Teng, and S. Guha, S. (1997), Business process change: A study of methodologies, techniques, and tools, *MIS quarterly,* 21(1): 55–80.

Mayer, R. J., C. P. Menzel, P. S. deWitte, T. Blinn, and B. Perakath, B. 1995. *Information Integration for Concurrent Engineering (IICE) IDEF3 Process Description Capture Method Report,* Technical report, Knowledge Based Systems Incorporated (KBSI).

Mertins, K., R. Jochem, and F. W. Jakel. 1997. A tool for object-oriented modelling and analysis of business processes, *Computers in industry,* 33: 345–356.

Muller, P. A. 1997. *Instant UML.* Birmingham, UK :Wrox Press.

Nethe, A., Stahlmann, H.D. 1999. Survey of a general theory of process modelling, In Scholz-Reiter, B., Stahlmann, H.D., Nethe, A. (eds), *Process Modelling,* Berlin: Springer-Verlag, 2–17.

NIST. 1993. *Integration Definition for Function Modelling (IDEF0).* Technical Report FIPS 183, National Institute of Standards and Technology.

Ould, M. A. 1995., *Business Processes: Modeling and Analysis for the Reengineering and Improvement.* Chichester, UK: Wiley.

Perera, T., and K. Liyanage, (2000). Methodology for rapid identification and collection of input data in the simulator of manufacturing systems, *Simulation practice and theory,* 7: 645–656.

Ratzer, A. V., L. Wells, H. M. Lassen, M. Laursen, J. F. Qvortrup, M. S. Stissing, M. Westergaard, S. Christensen, and K. Jensen, 2003. CPN Tools for editing, simulating, and analysing coloured Petri Nets. In W. van der Aalst and E. Best (eds) *Applications and Theory of Petri Nets 2003: 24th International Conference,* ICATPN 2003, Heidelberg, Eindhoven, The Netherlands: Springer-Verlag, 450–462.

Robinson, S. 2004. *Simulation: The Practice of Model Development and Use.* Chichester, UK: John Wiley and Sons.

Ryan, J., and Heavey, C. 2006. Process modelling for simulation, *Computers in industry,* 57: 437–450.

Sargent, R. G. 1999. Validation and verification of simulation models, In G. W. Evans, P. A. Farrington, H. B. Nembhard and D. T. Sturrock, (eds) *Proceedings of the 1999 Winter Simulation Conference,* Squaw Peak, Phoenix, Arizona: IEEE Computer Society, 39–48.

Scheer, A. W. 1998. ARIS In P. Bemus, K. Mertins, and G. Schmidt(eds), *Handbook on Architectures of Information Systems,* Berlin: Springer-Verlag, 541–565.

Tardieu, H., A. Rochfeld, and R. Colletti. 1983. *La methode MERISE, Principes et outils. / The MERISE method, Principles and tools.* Paris: Les editions d'organisation.

Tocher, K. D. 1963. *The Art of Simulation.* London: English Universities Press.

Van Rensburg, A., and N. Zwemstra, N. 1995. Implementing IDEF techniques as simulation modelling specifications, *Computers & industrial engineering,* 29(1–4): 467–471.

Zeigler, B. P. 1984. *Multifaceted Modelling and Discrete Event Simulation.* London: Academic Press.

13

Methods for Conceptual Model Representation

Stephan Onggo

CONTENTS

13.1 Introduction

Simulation conceptual model (or conceptual model, for brevity) representation is important in a simulation project because it is used as a tool for communication about conceptual models between stakeholders (simulation analysts, clients, and domain experts). There is a point in the simulation project when the conceptual modeling process happens inside the individual stakeholder's mind. This "thinking" process includes reflection on how to structure the problem and how the simulation model should be designed to help decision makers solve the problem at hand, subject to certain constraints. At some point in the simulation project, the conceptual model needs to be communicated to other stakeholders. Hence, the role of conceptual model representation is crucial. Moreover, different stakeholders may have different views on the system; their reasons may include different levels of understanding of the system, prior experience, and personal objectives. Nance (1994) refers to conceptual model representation for this purpose as the *communicative model*. When communication involves different types of stakeholders, a standard representation that can be understood by all stakeholders is essential. The fact that communication between stakeholders is important for the success of a simulation project (Robinson and Pidd 1998) makes the need for good conceptual model representation become even more essential.

The main challenge in designing conceptual model representation is to devise a representation that can be understood by all stakeholders and yet that is expressive enough to handle the varying levels of complexity in the conceptual model.* To complicate matters further, there is no single accepted definition of what a conceptual model is (see Chapter 1) as what is to be represented will surely affect its representation. Given the different definitions for a conceptual model, it is not surprising to see that a wide variety of conceptual model representations have been proposed.

One of the surveys conducted by Wang and Brooks (2007) listed the popularity of a number of methods for conceptual model representation. They are, in order of popularity, textual representations (e.g., list of assumptions and simplifications, component list and text description), process flow diagram, logic diagram (or flowchart), activity cycle diagram (ACD), and unified modeling language (UML). We can group these representation methods into three categories: textual representation, pictorial representation, and multifaceted representation. The objective of this chapter is to discuss the three methods for conceptual model representation and issues related to their use in practice. In the examples, we will demonstrate how the methods are applied to represent components of a conceptual model based on Robinson's definition. The same principle can be applied to other conceptual model definitions.

Robinson (2008) categorizes the components of a conceptual model into *problem-domain* components and *model-domain* components (see also Chapter 1). The problem-domain components are used as a means of communication mainly between clients/domain experts and simulation analysts, between clients, or between domain experts. These components include objectives, inputs, outputs, contents (scope/structure, level of detail, assumptions and simplifications), and data requirement. These components define parts of the system that are important for the objectives at hand. These components are independent of any modeling technique that is going to be used. At this stage, we need to decide whether simulation is the right tool to model the system.

Assuming that we have decided that simulation is the best option, we need to specify the model-domain components. At this stage, we need to decide the most suitable paradigms such as: discrete-event simulation, system

* Simulation analysts often deal with clients and domain experts who have little knowledge about simulation.

dynamics, and agent-based simulation. The choice between the different paradigms depends on the objective of the simulation project. Discrete-event simulation is suitable when it is necessary to track entities from their arrival in the system until they leave it (or until the simulation is completed). The results from individual entities are aggregated in the simulation outputs. System dynamics is suitable when the population of entities and the rates of entities moving from one place to another are more important than the individual entities. System dynamics also provides a way to explore complex feedback systems and it enables us to analyze the mutual interactions among entities over time. Agent-based simulation is particularly useful when the entities are adaptive, have the ability to learn, or can change their behaviors. Agent-based simulation is also useful when the behaviors of entities are affected by their spatial locations and the structure of their communication networks.

Each simulation-modeling paradigm views the system of interest differently. Discrete-event simulation sees a system as a collection of events, entities, resources, queues, activities, and processes. System dynamics views a system as a collection of stocks, flows, and delays. From an agent-based simulation perspective, a system is formed by a collection of agents and their environment. The communication at this stage, i.e., the development of a simulation model based on one of the paradigms, happens mainly between simulation analysts. The output of this stage is a simulation model that is independent of any software implementation.* The components of the simulation model are referred to as the model-domain components because they depend on the modeling paradigm used in the development process. Consistent with the theme of the book, this chapter focuses on the conceptual model representation in the discrete-event simulation. The examples given in this chapter are based on the District General Hospital Performance Simulation (DGHPSim) project to demonstrate how the methods discussed in this chapter could be applied in a real simulation project. DGHPSim is a collaborative research project that involves three British universities. The project aims to develop generic simulation models of entire acute hospitals so as to understand how hospital performance can be improved (Gunal and Pidd, 2009).

The remainder of this chapter is organized as follows. This chapter divides conceptual model representation methods into three categories: textual, pictorial, and multifaceted. Section 2 discusses the textual representation. Section 3 focuses on the most widely used pictorial representation in simulation, i.e., diagrams. Section 4 discusses the multifaceted representation. Finally, concluding remarks are made in Section 5.

* This may not be true in a simulation project where the requirement dictates the use of a specific implementation-dependent model representation (for reasons such as the familiarity to the simulation software). See Chapter 1 for the discussion on the importance of the software independency.

13.2 Textual Representation

As mentioned earlier, at some stage in the simulation project, a conceptual model needs to be communicated to other people. The communication can be done by passing the information verbally or via texts. In this chapter, we are more interested in written communication. A written document describing a conceptual model can become an important part of the simulation project. For example, the document can be used in any form of electronic communication and can even be used as part of the contract for the simulation project. The main objectives of the textual representation are to describe the content of each conceptual model component and to elicit visual imagery for the structure of the conceptual model components using narrative texts. The following excerpt shows how the conceptual model of a hospital simulation project is represented using narrative texts.

> The *objective* of this project is to improve overall hospital performance. The performance is measured based on the waiting times of patients at various departments at the hospital. The key departments included in the *model* are: Accident & Emergency (A&E), outpatients and in-patients. Patients arrive in the system through A&E and Outpatients. Depending on the condition, a patient can be admitted to hospital (in-patient) or discharged.

The excerpt describes a number of components in a conceptual model, such as: the objective, the output of the model, the scope of the model, and the flow of patients in the model. The main advantage of textual representation is its flexibility. Simulation analysts can write the description of a model in various ways and in different styles, for example, the previous excerpt could have been written in a bullet-point format or in tabular form. Textual representation can be done quickly, especially for some conceptual model components such as assumptions (and more naturally, perhaps). Most software that supports simulation modeling provides a facility for text annotations so that analysts can easily provide descriptions of the model and any part of it. This might explain why textual representation is very popular for documenting the assumptions in the survey carried out by Wang and Brook (2007). Robinson (2004, chap. 6, appendix 1, appendix 2) shows examples of how to specify conceptual model components using textual representation.

Textual representation is not without its disadvantages. First, the flexibility of textual representation may lead to an ambiguous description of the simulation model. As in any types of representation, the challenge here is to ensure that the mental model encoded in the text is decoded correctly by the target recipients. Effective textual representation should pay attention to the structure and content of the text and the assumptions about the target recipients (in this case, the stakeholders in a simulation project). Good organization of the text (sections, subsections, bullet-point lists, succinct description, etc.) may reduce ambiguity in the description. It may be necessary to develop a common understanding of a set of keywords (such as: objective, model, assumptions, etc.) among the stakeholders before the conceptual model is discussed. Another disadvantage of textual representation is that the correctness of the conceptual model cannot be verified elegantly using mathematical techniques. However, the conceptual model can still be validated using a more subjective validation technique such as the use of domain experts' opinions (see Chapter 15 for various validation methods in conceptual modeling). Finally, and rather obviously, communication can work only if all stakeholders understand the language used in the texts.

13.3 Pictorial Representation

The next type of conceptual model representation is pictorial representation where the conceptual model is communicated through pictures. Research in cognitive science has shown that a pictorial representation is very effective (for example, Larkin and Simon 1987). Unlike textual representation that presents information sequentially, pictorial representation can show the information in two dimensions, which allows nonsequential flows to be represented more effectively. In simulation, diagrams are the most widely used pictorial representation for conceptual models. A diagram is a special type of pictorial representation that represents information using shapes/symbols that are connected by links (such as arrows and lines). The use of diagrams in simulation modeling has increased, especially after graphical workstations became more affordable. Pooley (1991) conducted one of the earliest surveys on the use of diagrams in simulation modeling. Recently, Wang and Brooks (2007) conducted another survey that showed a number of popular diagrams used in simulation modeling. This section discusses two of the most popular diagrams in the survey, i.e., the ACD and the process flow diagram. We will also discuss another widely known diagram called the event relationship graph (ERG). The three diagrams are chosen because they focus on different aspects of a system that is to be modeled.

13.3.1 Activity Cycle Diagram (ACD)

ACD (Hills 1971) is an implementation-independent diagram that is used to model a system by focusing on the changes in the states of key entities in the system. When an entity arrives at the system, it must go through a set of activities that may change the state of the entity (for example, in service or waiting) until the entity leaves the system. In ACD, the change in the state of each entity is represented by a series of alternate dead and active states.

A dead state is represented as an oval and corresponds to a state where an entity must wait until the required resources are available. An active state is represented as a rectangle and corresponds to a state where an entity is in an activity with a specific duration (it may be sampled using a predefined distribution function).

Figure 13.1 shows how ACD can be used to represent an A&E simulation model. The diagram shows the cycle of entity patients. This simulation model assumes that the arrival of patients follows a certain distribution function (hence, an active state). Once a patient arrives at A&E, the patient waits until the clerk is ready for the registration process (a dead state). When the clerk is ready, the registration takes a certain amount of time, which may be sampled from a distribution function. The process continues until the patient leaves A&E.

In some cases, we may be interested in the state of a specific resource in the system over time. For this purpose, we can add the cycle of the resource to the diagram. For example, if we are interested in the utilization of each clerk, we can add a cycle for the clerk. The state of the clerk will constantly switch between being in a dead state of waiting for a patient to arrive, and being in an active state, registering a patient. The patient cycle meets the clerk cycle at the registration process. A complete ACD for the system under study should show the cycles of key entities and key resources. One of the modelers' main tasks is to decide which key entities and resources should be included in the

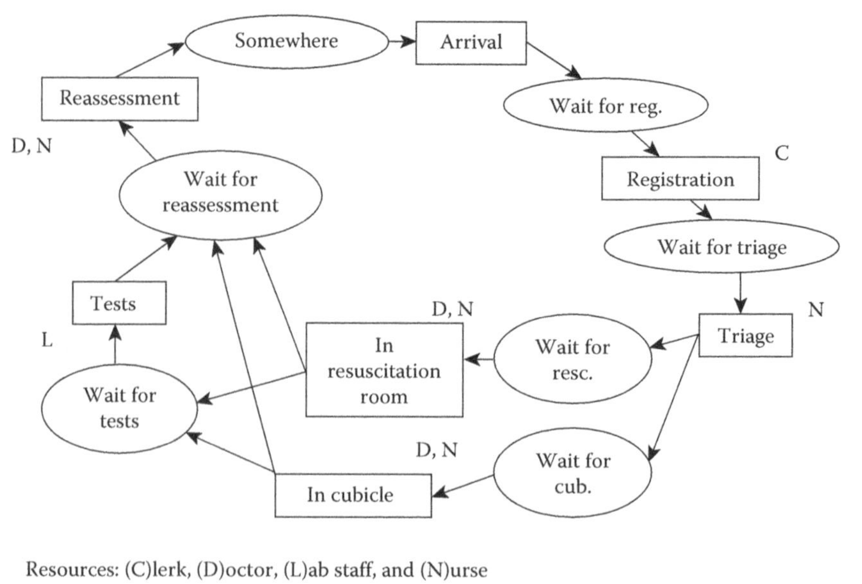

FIGURE 13.1
A&E: Activity cycle diagram. (From Onggo, B.S.S., *Journal of Simulation*, 3 (1), 46, 2009. With permission.)

model. ACD is commonly used to represent the model-domain component in a discrete-event simulation-modeling paradigm. One of the main reasons for this is that ACD could easily show key components such as: queues (all dead states), system state, resources, activities (most active states), and processes (all cycles).

13.3.2 Process Flow Diagram

A process flow diagram is commonly used to model the flow of processes in a system. A process in simulation is often defined as a sequence of activities (and events) in chronological order.

A process flow diagram focuses more on the sequence (or structure) of activities and the flow of entities from the point where they enter the system until they leave the system. This is different from ACD, which focuses more on the states that the entities and resources are in. Most commercial visual interactive modeling software (VIMS) that supports discrete-event simulation uses some sort of process flow diagram. The VIMS have their proprietary symbols to represent activities and their sequence in a process. In fact, some VIMS call the activities by other names such as tasks and machines.

In this section, we choose one of the widely known process flow diagrams called Business Process Diagram (BPD). BPD is the diagram that is specified by Business Process Modeling Notation (BPMN). BPMN is a standard that has been developed to provide a notation that is understandable by all business stakeholders (business people, business analysts, and technical developers) to model business processes. BPD is chosen because it is a widely known standard and is independent of any proprietary notations that may trap us into implementation-dependent components. Hence, it is suitable for our objective to provide a tool for communication about conceptual models between stakeholders that are independent of any software implementation. The four main BPD elements are activities (shown as rounded rectangles), events (circles), connectors (lines), and gateways (diamonds). As we know, a process is a sequence of activities and events. Hence, these four BPD elements can be used to model many different processes.

BPD activities are used to represent real-world activities. The activities can be further decomposed into subactivities. This facility is important to allow a hierarchical modeling process. In other words, the activity at one level in the hierarchy can be viewed as a process from a lower level in the hierarchy. The lowest-level activity, i.e., the activity that will not be decomposed further, is called a task. BPD events are used to represent events that happen in the real world. An event can start a process (i.e., start the first activity in the process), start an intermediary activity, or end a process (i.e., end the last activity in the process). BPD connectors are used to represent flows. BPD gateways are used to represent decisions in the process flow, i.e., joins, forks, and mergers. BPMN (http://www.bpmn.org/) provides a more detailed explanation

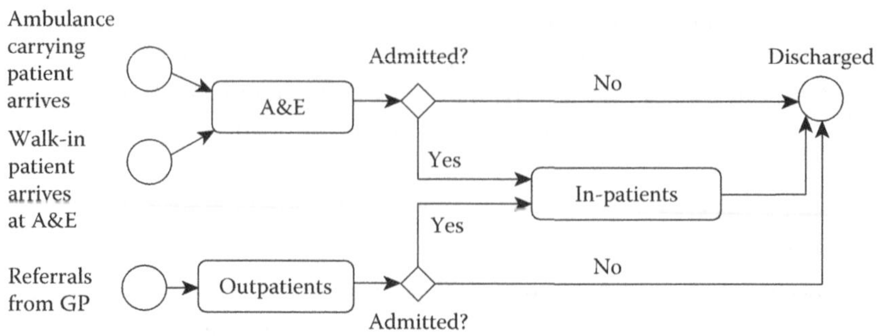

FIGURE 13.2
Hospital: Business process diagram. (From Onggo, B.S.S., *Journal of Simulation*, 3 (1), 45, 2009. With permission.)

of each element and other elements that are not mentioned here, such as pool and lane.

Figure 13.2 shows the BPD of a typical hospital operation, which includes three activities: Accident and Emergency (A&E), Outpatients, and In-patients. Patients arrive in the system through A&E and Outpatients. The arrivals of patients are events that start the processes. Depending on the condition, a patient can be admitted to hospital (In-patient) or discharged. Discharge is an event that terminates a process. If we want to add to the level of detail in the model, we can move to a lower layer in the system hierarchy and treat any of the activities as a process which can be decomposed further into a number of activities.

13.3.3 Event Relationship Graphs (ERG)

ERG (Schruben 1983) provides a concise representation of causality in a system. ERG is effective in representing model-domain components in discrete-event simulation-modeling paradigms. In an ERG, an oval represents state changes when an event occurs and an arrow shows that an event at the start of the arrow generates an event at the end of the arrow (hence, it shows the causality between the two events). The arrows also specify the conditions (/) and the times for events to be scheduled (arrows with time delays are drawn in bold).

Figure 13.3 shows the ERG of a typical Outpatient department. A "GP referral" event triggers the whole process. This event serves as a bootstrap event that will generate subsequent arrivals to the Outpatient department (with a specified time delay t_a). The "GP referral" event generates a "start first appointment" event when at least one consultant is available. The "GP referral" event also changes the system state, i.e., increases the number of patients waiting for their first appointments. The "start first

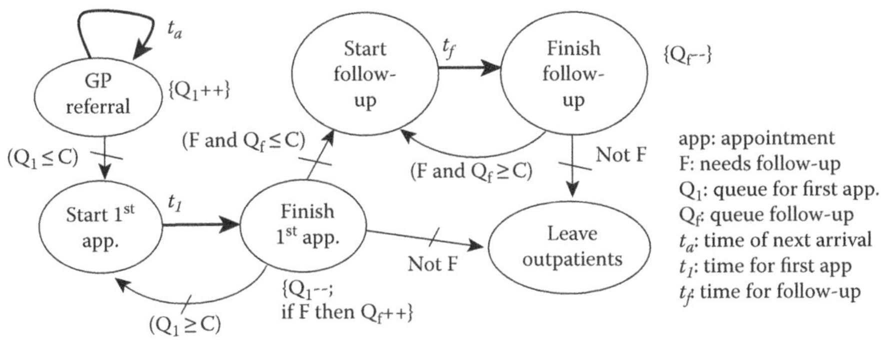

FIGURE 13.3
Outpatients: Event relationship graphs.

appointment" event leads to a "finish first appointment" event (with a specified time delay t_1). Subsequently, if the patient needs a follow-up appointment, a "finish first appointment" event may generate a "start follow-up appointment" event. Otherwise, treatment for the patient is complete. The "finish first appointment" event changes the system state, i.e., reduces the number of patients waiting for their first appointments and, in some cases, increases the number of patients waiting for their follow-up appointments.

13.4 Multifaceted Representation

Despite the differences in the definitions of a conceptual model, researchers agree that a conceptual model comprises a number of components. Hence, it is unlikely that a single diagram can be used to represent completely a conceptual model. For this reason, a multifaceted representation is more suitable for a more complete documentation of a conceptual model. In a multifaceted conceptual model representation, a set of diagrams and textual representation are used to represent different conceptual model components. Multifaceted representation has been used widely in software engineering. One of the most widely used multifaceted representations in software engineering is the UML. UML 2.0 defines 13 types of diagrams to represent three aspects of software system: static application structure, behavior, and interactions. A more detailed description can be found at http://www.uml.org. A multifaceted representation such as UML has the potential to provide a more comprehensive representation of a conceptual model.

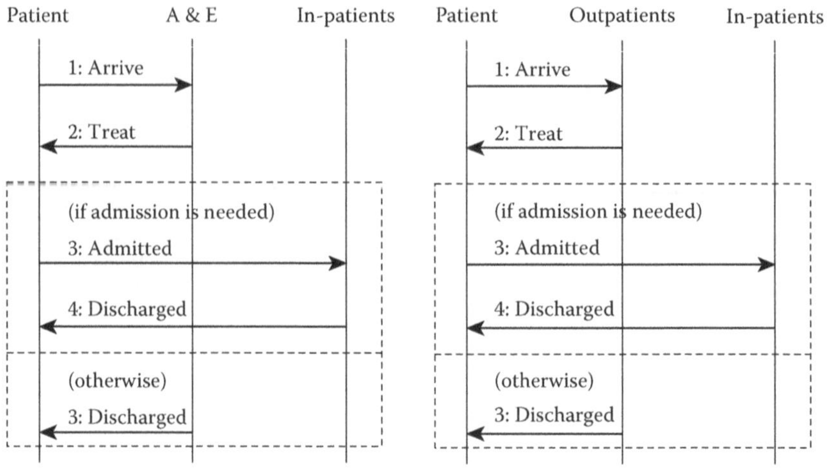

FIGURE 13.4
A&E: Sequence diagram.

13.4.1 UML and SysML

Richter and Marz (2000) proposed the use of four UML diagrams for documenting a simulation project. They use the "use case" diagram to document the interaction between users and the simulation model. The structure of a simulation model is represented using a class diagram. The dynamics of the model are represented using an interaction diagram and a state diagram. Vasilakis et al. (2009) used UML to specify the requirements for a patient flow simulation model. They use the activity diagram to specify the flow of patients, "use case" diagram to give the detail function of each activity in the activity diagram and state diagram to capture the state-transition of patients. These works show that we can use UML diagrams to provide a multifaceted representation of a conceptual model. We can extend their work to include the use of sequence diagrams and collaboration diagrams to show the dynamics of a model. Figure 13.4 shows a sequence diagram of the same hospital system that was shown earlier in Figure 13.2.

The Object Management Group (OMG) publishes the Systems Modeling Language (SysML) standard, which is an extension of UML and is designed to support system modeling. Huang et al. (2007) explored the use of SysML in representing conceptual models. One of the ultimate objectives is to provide a conceptual model representation (independent of any implementation software) that could be translated automatically to any simulation software (implementation dependent). SysML uses four UML diagrams (sequence diagram, state-transition, use case diagram, and package diagram), three modified UML diagrams (activity diagram, block definition diagram, and internal block diagram) and two new diagrams (requirement diagram and parameter diagram). These diagrams are used to specify a system's

TABLE 13.1

Diagrams Used in the Unified Conceptual Model

Domain	Component	Representation
Problem	Objectives	Objective Diagram, Purposeful Activity Model
	Inputs	Influence Diagram
	Outputs	
	Contents	Business Process Diagram with textual representation
	Data requirement	Textual representation, Data dictionary
Model	Discrete-Event	Activity Cycle Diagram, Event Relationship Graph
	System Dynamics	Stock and Flow Diagram, Causal Loop Diagram
	Agent-based	Flowchart, Business Process Diagram, UML Activity Diagram

Source: Adapted from Onggo, B.S.S., *Journal of Simulation*, 3 (1), 42, 2009. With permission.

structure, behavior, and requirements. SysML is discussed in greater detail in Chapter 11.

13.4.2 Unified Conceptual Model

Onggo (2009) proposed the use of another set of diagrams to represent the different conceptual model components. Table 13.1 shows Onggo's multifaceted conceptual model representation. In this chapter, we add a number of representation methods that were not part of Onggo's original methods. The first column gives the domains of a conceptual model's components. The second column lists the components of a conceptual model. The last column shows the diagrams selected to represent the conceptual model's components.

13.4.2.1 Objectives Component

The objective is the most important component in simulation modeling. Objectives are used to judge the success of a problem-solving exercise and to compare the quality of various decision alternatives. Onggo uses an objective diagram to represent the objectives component of a simulation conceptual model. Objective diagrams (Keeney 1992) are commonly used to structure objectives in decision science. They classify objectives into two categories: *fundamental objectives* and *means objectives*.

The fundamental objectives are the end result that we want to achieve and are organized into hierarchies. In an objective diagram, each fundamental objective is represented as a node in a tree. The higher-level fundamental objectives represent more general objectives and their measurement can be obtained from lower-level fundamental objectives. Thus, the lowest-level fundamental objectives provide the basis on which various design alternatives are measured.

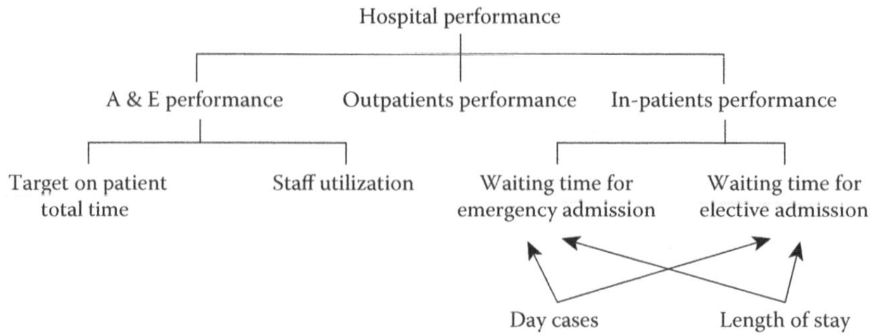

FIGURE 13.5
A&E: Objective diagram. (From Onggo, B.S.S., *Journal of Simulation*, 3 (1), 43, 2009. With permission.)

Consequently, the highest-level fundamental objective provides the ultimate measurable consequence that will be used to evaluate and compare various design alternatives. Figure 13.5 shows an example of fundamental objectives from a project that seeks to improve the performance of a hospital. The performance is linked to the waiting times of patients at the hospital, which are the averages of waiting times of patients in its various departments: A&E, Outpatients, and In-patients. These measurements will be used to compare alternatives. Second-level fundamental objectives can be further expanded if necessary. For example, A&E performance is obtained from two components: patient total time (98% of patients must spend less than 4 hours in A&E) and staff utilization.

Means objectives are important because they help us to achieve fundamental objectives and they are often used when the fundamental objectives are difficult to measure directly. In some cases, identifying means objectives can help us to characterize new alternatives. In the objective diagram, means objectives are organized into networks. Two examples of means objectives are shown in Figure 13.5. Maximizing the number of day cases and reducing patients' lengths of stay are important because they increase the number of available beds. Hence, this may reduce the waiting times for both emergency and elective admissions. In general, fundamental objectives can be differentiated from means objectives by continuously asking the question of why an objective is important. An objective is a means objective if it is important because it helps achieve another objective. The same question is repeated until we find an answer where an objective is important because it is important.

The objective diagram, however, only considers the structure of objectives. It may be useful to show the conditions under which the structure is built. Kotiadis (2007) presented interesting work on using soft systems methodology to determine simulation objectives (see Chapter 10 for more detail). In particular, she presented steps to extract simulation objectives from the

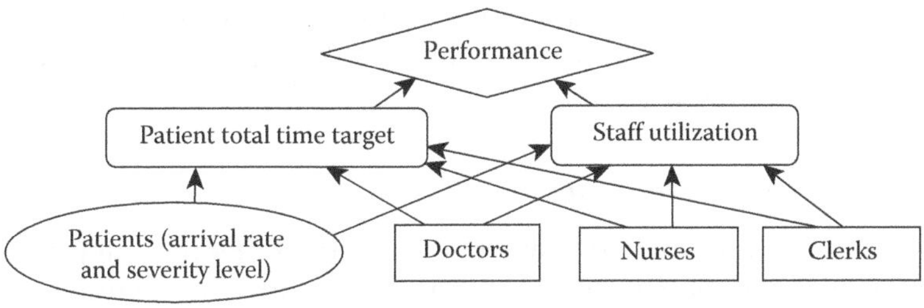

FIGURE 13.6
A&E: Influence diagram. (From Onggo, B.S.S., *Journal of Simulation*, 3 (1), 44, 2009. With permission.)

purposeful activity model (PAM). This work implies that PAM can be used to complement the objective diagram to show the conditions under which the objectives diagram is drawn.

13.4.2.2 Inputs and Outputs Component

Once the objectives have been defined, we need to translate them into output variables that can be quantified, and to identify the different alternatives (input variables) that will achieve the objectives. Outputs can be directly inferred from objectives. The controllable input variables are sometimes referred to as the decision variables. The inputs are sometimes specified explicitly in the objectives; otherwise they can be obtained from the clients. Onggo (2009) uses an influence diagram to represent the relationship between input variables and output variables.

The influence diagram (Howard and Matheson 1984) is commonly used to structure decisions by representing the relationship between key variables. An influence diagram consists of certain elements, as follows. Decision variables represent the decisions to be made (symbolized as rectangles in the diagram). Uncontrollable variables represent uncertainty or chance events (ovals). Outputs represent final consequences or payoffs (diamonds). Intermediary variables, including calculation nodes and constants, are used to compute the outputs (rounded rectangles). Relationships between nodes are represented using arcs. All arcs pointing to a rectangle (decision variable) show sequences. It means that the node at the beginning of the arc must be known before the decision can be made. All arcs pointing to ovals, diamonds, or rounded rectangles (non-decision variables) show the relevance relations. The node at the beginning of the arc is relevant for the node at the end of the arc.

Figure 13.6 shows the representation of inputs and outputs component from an A&E department using an influence diagram. The output of the A&E simulation model is the A&E performance. The A&E performance is calculated from two intermediary variables, namely the total number of patients who spend 4 hours or less in the A&E department and staff utilization.

The decision variables are the numbers of doctors, nurses and clerks. The uncontrollable variables (shown as ovals) are the arrival rate and severity of condition of the patients.

13.4.2.3 Contents Component

Once the inputs and outputs have been specified, the next step is to specify the transformation processes or the contents. The contents component of a conceptual model describes the scope of the model, the level of detail, assumptions and simplifications. The scope of the model specifies all relevant processes and their interactions within the boundary of the model. The level of detail specifies the required degree of detail for each process in the model and the required input data. Both scope and level of detail are determined based on the modeling objectives. Assumptions are necessary to address the uncertainty or unknown factors that may be important to the processes in the model. Simplifications are needed to handle the complexity of processes in the model. One of the possible diagrams that can be used to represent the contents component is the BPD that was discussed earlier.

The scope of a conceptual model can be represented easily by specifying the relevant activities, events that start these activities, and the process flows (including decisions or branching of flows). Figure 13.2 shows the scope of a hospital simulation model that excludes the general practitioner (GP). Figure 13.2 also shows the level of detail of the processes. It considers A&E, Outpatients, and In-patients as three black boxes. It is possible to show a more detailed model for each activity in Figure 13.2. For example, Figure 13.7 shows a detailed model of the A&E activity in Figure 13.2. The figure shows that the process in the A&E department starts with patient arrivals. There

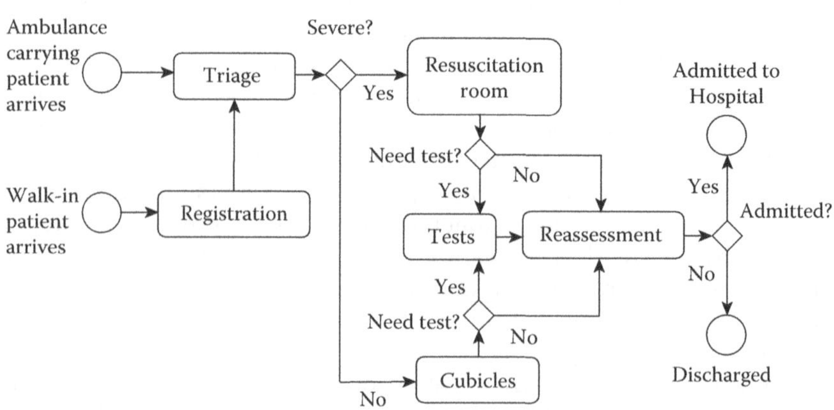

FIGURE 13.7
A&E: Business process diagram. (From Onggo, B.S.S., *Journal of Simulation,* 3 (1), 45, 2009. With permission.)

are two types of patient arrival: voluntary and by ambulance. A patient who arrives voluntarily at the A&E department will need to register before being evaluated by a nurse (triage) to determine the severity of the patient's condition. One who arrives by ambulance may, however, bypass registration (the triage is done on the way to the A&E department). Next, the patient will be seen and treated by a doctor and/or nurse (either in the resuscitation room or a cubicle). After treatment, patients will either be discharged or admitted to the hospital. Some patients may need tests and X-rays, and these patients then need a second session with a doctor and/or nurse before discharge or admission.

BPD provides three artifacts that can be used to provide additional information about an activity that is not directly related to the structure of the process flow. One of them is text annotation, which is suitable for representing the assumptions and simplifications used in the conceptual model. For example, in Figure 13.7, we can attach a text annotation to the activity "triage" that provides a list of assumptions, such as "the severity of condition of patients is modeled as a simple random sampling." Similarly, we can attach a text annotation to the activity "test" that provides a list of simplifications such as "the service time for tests does not differentiate between the types of test (X-ray, blood test, etc.)."

13.4.2.4 Data Requirement Component

Data are an important part of any modeling technique. Hence, it is important to recognize the required data early. At this stage, given the previous problem-domain components, we should be able to identify the data requirements. The required data should match the scope and level of detail of the conceptual model. Robinson (2004, chap. 7) discusses methods for dealing with unavailable data. The data requirement is often specified using textual representation. For example, Table 13.2 shows the data that need to be collected for entity patients in the A&E system in Figure 13.7.

TABLE 13.2

Data Requirement for Entity Patient

Field	Type	Note
Patient details	Name, address, patient identifiers, etc.	This can be useful to identify patients and, if the analysis requires it, profile patients.
Admission time	Date/Time	This is needed to determine the distribution of admissions.
Severity level	Minor or major	This is needed to find the proportion of patients needing minor treatment and major treatment.
Time in A&E	Minutes	This is needed to validate the output of the model.

13.4.2.5 Model-Dependent Component

As explained in the introduction, the method for representing model-dependent components depends on the modeling paradigm. In the discrete-event simulation-modeling paradigm, the method should be able to represent key components such as: entities, resources, system states, queues, activities, events, and processes. Onggo (2009) uses diagrams that are independent of any software implementation such as: ACD, BPD, and ERG. A discrete-event simulation model that is represented using these diagrams can be implemented in any simulation software. It is relatively straightforward to develop software that is able to read a model that is represented using any of the implementation-independent diagrams and either simulates the model (for example, Araujo and Hirata 2004, Pidd and Carvalho 2006) or converts the model to specific simulation software (for example, Huang et al. 2007).

The representation of components in system dynamics (such as stocks and flows) has been influenced by the notation given by Forrester (1961). Nowadays, the stock and flow diagram and the causal loop diagram (both are independent of software implementation) are widely accepted as standards in representing system dynamics models (Sterman 2004). This explains why many system dynamics VIMS use similar diagrams to represent system dynamics models. In the agent-based simulation-modeling paradigm, a simulation model is formed by a set of autonomous agents that interact with their environment (including other agents) through a set of internal rules to achieve their objectives. Much of the literature represents agent-based simulation models using flowcharts or pseudo codes. The flowcharts or pseudo codes are used to describe the internal rules of different agent types and the internal rules of a dynamic environment (i.e., its state is constantly changing, even if there is no action performed by any agent). Other than flowcharts, we can also use a BPD or UML activity diagram to represent an agent-based simulation model where each agent type is implemented in a swim lane.

13.5 Summary

We have discussed three categories of methods for conceptual model representation: textual representation, diagrams, and multifaceted representation. Textual representation can be used to give a brief description of a model. This is particularly useful when we have a repository of simulation models. The description allows others to decide quickly whether a model is suitable, or to search for the most suitable model to be used. The diagrams are effective during conceptual model development. A multifaceted representation is the best representation for the complete documentation of a conceptual model. Multifaceted representation has another advantage. It allows us to verify the consistency of conceptual model components (Onggo 2009). We have shown how to apply these methods to represent conceptual model components based on Robinson's conceptual model definition. The same principle can be applied to other conceptual model definitions since most of the definitions have overlapping components. Although the author believes that the representation methods discussed in this chapter could be applied to many applications in discrete-event simulation, it must be noted that the methods have been tested using a few business process models only. As noted in Robinson (2002), simulation applications are far from homogeneous; hence, it is possible that some of the methods may not be suitable for some applications.

Acknowledgments

Some sections of this chapter are based on: Onggo, B. S. S. 2009. Toward a Unified Conceptual Model Representation: A Case Study in Health Care. *Journal of Simulation* 3 (1): 40–49. © 2009 Operational Research Society Ltd. With permission of Palgrave Macmillan.

References

Araujo, W.L.F., and C.M. Hirata. 2004. Translating activity cycle diagrams to Java simulation programs. In *Proceedings of the 37th Annual Simulation Symposium*, 157–164. Piscataway, NJ: IEEE Computer Society Press.

Forrester, J. 1961. *Industrial Dynamics*. Cambridge, MA: MIT Press.

Gunal, M.M., and M. Pidd. (2009). Understanding target-driven action in A&E performance using simulation. *Emergency Medicine Journal* 26: 724–727.

Hills, P.R. 1971. *HOCUS*. Egham, Surrey, UK: P-E Group.

Howard, R.A., and J.E. Matheson. 1984. Influence diagram. In *The Principles and Applications of Decision Analysis*, vol. II, ed. R.A. Howard and J.E. Matheson, 719–762. Palo Alto, CA: Strategic Decisions Group.

Huang, E., R. Ramamurthy, and L.F. McGinnis. 2007. System and simulation modeling using SysML. In *Proceedings of the 2007 Winter Simulation Conference*, ed. S.G. Henderson, B. Biller, M.-H. Hsieh, et al., 796–803. Piscataway, NJ: IEEE Computer Society Press.

Keeney, R.L. 1992. *Value-Focused Thinking*. Cambridge, MA: Harvard University Press.

Kotiadis, K. 2007. Using soft systems methodology to determine the simulation study objectives. *Journal of simulation* 1 (3): 215–222.

Larkin, J.H., and H.A. Simon. 1987. Why a diagram is sometimes worth ten-thousand words. *Cognitive Science* 11: 65–99.

Nance, R.E. 1994. The conical methodology and the evolution of simulation model development. *Annals of operations research* 53: 1–45.
Onggo, B.S.S. 2009. Towards a unified conceptual model representation: A case study in health care. *Journal of simulation* 3 (1): 40–49.
Pidd, M., and A. Carvalho. 2006. Simulation software: Not the same yesterday, today or forever. *Journal of simulation* 1 (1):7–20.
Pooley, R.J. 1991. Towards a standard for hierarchical process oriented discrete event Simulation diagrams. *Transactions of the society for computer simulation* 8 (1):1–20.
Richter, H., and L. Marz. 2000. Towards a standard process: The use of UML for designing simulation models. In *Proceedings of the 2000 Winter Simulation Conference*, ed. J.A. Joines, R.R. Barton, K. Kang, et al., 394–398. Piscataway, NJ: IEEE Computer Society Press.
Robinson, S. 2002. Modes of simulation practice: Approaches to business and military simulation. *Simulation modelling practice and theory* 10 (8):513–123.
Robinson, S. 2004. *Simulation: The Practice of Model Development and Use*. Chichester, UK: Wiley.
Robinson, S. 2008. Conceptual modelling for simulation part I: Definition and requirements. *Journal of the operational research society* 59 (3): 278–290.
Robinson, S., and M. Pidd. 1998. Provider and customer expectations of successful simulation projects. *Journal of the operational research society* 49 (3): 200–209.
Schruben, L. 1983. Simulation modeling with event graphs. *Communications of the ACM* 26 (11): 957–963.
Sterman, J.D. 2004. *Business Dynamics: Systems Thinking and Modeling for a Complex World*. New York: McGraw-Hill.
Vasilakis, C., D. Lecznarowicz, and C. Lee. 2009. Developing model requirements for patient flow simulation studies using the Unified Modelling Language (UML). *Journal of simulation* 3 (3): 141–149.
Wang, W., and R. Brooks. 2007. Empirical investigations of conceptual modeling and the modeling process. In *Proceedings of the 2007 Winter Simulation Conference*, ed. S.G. Henderson, B. Biller, M.-H. Hsieh, et al., 762–770. Piscataway, NJ: IEEE Computer Society Press.

14

Conceptual Modeling for Composition of Model-Based Complex Systems

Andreas Tolk, Saikou Y. Diallo, Robert D. King, Charles D. Turnitsa, and Jose J. Padilla

CONTENTS

14.1 Introduction

Conceptual modeling is often understood as an effort that happens before systems are built or software code is written and conceptual models are no longer needed once the implementation has been accomplished. Conceptual models are primarily described as mental models that are used in an early stage in the abstraction or as a simplification process in the modeling phase. This early stage of abstraction makes conceptual models difficult to verbalize and formalize making them "more art than science," as Robinson (Section 1.1) mentions, given the challenging tasks to define applicable methods and procedures. This view is not sufficient for model-based applications. The goal of conceptual modeling in Modeling and Simulation (M&S) is not focusing on describing an abstract view of the implementation, but to capture a model of the referent, which is the thing that is modeled, representing a sufficient simplification for the purpose of a given study serving as a *common conceptualization of the referent and its context* within the study.

In this sense, conceptual modeling in M&S is slightly different from its traditional conception in which the focus is on capturing the requirements of a system in order to replicate its behavior. The M&S view has the additional requirement that the execution of the model will provide some additional insight into some problem while the traditional view mainly focuses on satisfying the identified requirements. In either view, the main challenge is to identify what should be captured in the conceptual model in order to enable users of the system to understand how the referent is captured.

In traditional conceptual modeling, this is less of a challenge because it is somewhat easier to look at the behavior of a system and identify its counterpart in the real world. The desired function can be captured in use cases and serve for validation and verification. For example, it is obvious that an Automatic Teller Machine (ATM) is representative of a real teller as it can perform many similar interactions including necessary inputs and outputs. In M&S systems, interactions in terms of inputs and outputs are not sufficient to identify a referent because many referents have similar inputs and outputs when abstracted, which makes it impossible to identify which one the conceptualization is referring to. A conceptual model of a teller designed to study the average processing time of a customer is different from an ATM. In this case, the customers and the teller may be abstracted into probability density functions and a queuing system, as this may be sufficient for the study. The validity of answers is therefore highly dependent on the context of the model. In other words, the conceptual model in M&S needs to capture data in the form of inputs and outputs, processes that consume the data and needs a way to distinguish conceptualizations of referents from one another by capturing the assumptions and constraints inherent to the model. Conceptual models must capture this information. The reason for this requirement is that this information is needed to be able to decide if a given model can be used to solve a problem, e.g., it is possible to reuse the model of the teller introduced above to calculate the average processing time in systems similar to banks (like fast food restaurants, or supermarkets). It could also be tailored to calculate the average waiting time or average time in system for a customer. In addition, given a modeling question, several models can be put together or composed to provide an answer. However, information is needed that was captured by the conceptual model in order to be able to decide if a model is applicable or not.

This observation becomes practically relevant when considering the use of models as services that answer specific questions similar to any real-world service (travel agency, bank) and orchestrate their execution to answer the modeling question. To be able to do this models need to be composed and orchestrated that can communicate with one another and provide the information needed to make the decision whether a model is applicable in the current application.

In general, orchestration, reuse, and composition are highly sought after capabilities in M&S, but they are currently perceived to be costly, error-prone, and difficult for many reasons, technical and nontechnical. The challenges increase in service-oriented environments, where applicable services need to be identified, where the best available solution in the context of the problem to be solved needs to be selected, all required services need to be composed, and their execution needs to be orchestrated. In traditional solutions, these tasks are conducted by system engineers. The ultimate goal in service-oriented environments is to have intelligent agents performing these tasks. In order to attain this goal, agents need to have access to the same information and knowledge used by system engineers. One of the first steps therefore should be to provide conceptual models that are machine understandable or computable. Yilmaz (2004) motivates this view in his research on defense simulation.

A formalization of conceptual modeling has direct implications within a system of systems perspective as well. Within a pluralist context (Jackson and Keys 1984), different viewpoints imply that questions about the referent made by different people carry their own perceptions of the system with a lack of a unifying consensus. These individual perceptions ultimately influence whether two systems will be composed or not. The resulting composition based on individual perceptions may be comprised of conceptually misaligned models and produce inappropriate results. Informal mental models allowing individual perceptions must therefore be replaced by formal representations of the conceptualization.

In order to support the composition of model-based solutions in service-oriented environments, assumptions, constraints, and simplifications need to be explicitly presented. This needs to be done using metadata formally representing the conceptualization. This metadata can be read by intelligent agents and used to identify, select, compose, and orchestrate model-based solutions as required before.

This chapter describes three engineering methods designed to capture data, processes, the assumptions and constraints of a conceptual model for model-based solutions, and shows how a computable conceptual model can be used particularly in support of composition, reuse, and orchestration. The focus of this chapter is on composability. The chapter is organized as follows:

1. The first part reviews composability and interoperability and shows how the addition of complex M&S systems affects both. This section also presents description frameworks that support conceptual modeling and what has to be taken into account in this process to support interoperability and composability.

2. The second part presents data engineering and model-based data engineering (MBDF) as complementary engineering methods that can be used to capture data and describe their meaning and relationships in support of interoperability. The section also discusses the difference between interoperation and interoperability. Process engineering is introduced as an engineering method that can be used to capture the description of processes and their relationships with other processes that are part of the conceptual model. Finally, constraint engineering is presented as an engineering method designed to document assumptions that the model is making about its data and processes and the constraints it puts on the use of processes. While data engineering and process engineering support interoperability, the addition of constraint engineering makes composability possible.
3. The next part discusses how the conceptual model is captured in terms of data, process, and assumptions and constraints that can be formally expressed in a machine-understandable language using ontology.
4. The final part concludes the chapter with a summary of the main points addressed and makes a case for the use of logic and mathematics to capture and expose semantic models so that they are discoverable and accessible to systems that can potentially use them.

14.2 Interoperability and Interoperation Challenges of Model-Based Complex Systems

As stated in the introduction, the engineering methods for conceptual modeling supporting the composition of model-based complex systems must ensure that the necessary information enabling the decision of whether two model-based solutions can work together is provided as a formal specification of the conceptualization. To this end, definitions are introduced for the two important concepts of interoperability and composability, which are followed by some supporting models that help to better understand what metadata needs to be provided and how.

14.2.1 Interoperability and Composability

In order for two systems to interoperate, they need to fit together. Traditionally, systems that are able to interoperate are referred to as interoperable systems. The Institute of Electrical and Electronics Engineers (IEEE 1990) defines *interoperability* as "the ability of two or more systems or components to exchange information and to use the information that has been exchanged."

Petty and Weisel (2003) discuss the differences and commonalities between interoperability and composability and show that the definitions are primarily driven by the challenges of technical integration and the interoperation of implemented solutions versus the ability to combine models in a way that is meaningful. Model-based complex systems further add a new category of challenges to the already difficult problem of composability and interoperability. A working definition for a *model-based complex system* can be derived from the definition of the combined terms: a system is made up of several components that interact with each other via interfaces; a complex system has many components that interact via many interfaces that represent typically nonlinear relations between the components; model-based systems use an explicit formal specification of a conceptualization of an observed or assumed reality. While complexity already plays a major role in the traditional view of interoperability, the model-based aspect is not often considered. The working definition of interoperation used in this chapter is simply: *two systems can interoperate if they are able to work together to support a common objective.*

To explicitly deal with challenges resulting from differences in conceptualization, the term composability is used. As with interoperability, the definitions used for the term composability are manifold. Petty et al. (2003) compiled various definitions and used them to recommend a common definition embedded in a formal approach. Fishwick (2007) proposed, in his recent analysis, the restriction of the scope of composability to the model level, and following the recommendations of Page et al. (2004) distinguishes between three interrelated but individual concepts contributing to interoperation:

- *Integratability* contends with the physical/technical realms of connections between systems, which include hardware and firmware, protocols, networks, etc.
- *Interoperability* contends with the software and implementation details of interoperations; this includes exchange of data elements via interfaces, the use of middleware, mapping to common information exchange models, etc.
- *Composability* contends with the alignment of issues on the modeling level. The underlying models are purposeful abstractions of reality used for the *conceptualization* being implemented by the resulting systems.

In other words, integratability ensures the existence of a stable infrastructure such as a reliable network, interoperability assures that simulation systems can be federated with each other, and composability assures that the underlying conceptualizations are aligned—or at least not contradictive.

The same recommendation is supported by Tolk (2006), where the importance of these categories for the domain of modeling and simulation is emphasized. An evaluation of current standardization efforts shows that the focus of these standards lies predominately on the implementation level of interoperability and doesn't consider conceptualization sufficiently. The modeling process purposefully simplifies and abstracts reality and constrains the applicability of the model in the form of assumptions and constraints. While interoperability deals with simulation systems, composability deals with models; hence, *interoperability of simulationsystems requires composability of conceptual models.*

14.2.2 Relevant Models Regarding Conceptual Modeling for Compositions

Conceptual models capturing the abstraction process are essential in order to evaluate whether systems can be composed. Ultimately, the goal is to make systems *semantically accessible* to another system so that they can make use of the conceptual model to select applicable solutions, choose the best available solution, compose the partial solutions to deliver the overall solution, and orchestrate their execution. This point is emphasized in (Benjamin, Akella, and Verna 2007, p. 1082):

> The semantic rules of the component simulation tools and the semantic intentions of the component designers are not advertised or in any way accessible to other components in the federation. This makes it difficult, even impossible, for a given simulation tool to determine the semantic content of the other tools and databases in the federation, termed the problem of semantic inaccessibility. This problem manifests itself superficially in the forms of unresolved ambiguity and unidentified redundancy. But, these are just symptoms; the real problem is how to determine the presence of ambiguity, redundancy, and their type in the first place. That is, more generally, how is it possible to access the semantics of simulation data across different contexts? How is it possible to fix their semantics objectively in a way that permits the accurate interpretation by agents outside the immediate context of this data? Without this ability—semantic information flow and interoperability—an integrated simulation is impossible.

In the remainder of this section, several relevant models that can be applied to support the fulfillment of related requirements enabling semantic transparency will be discussed.

14.2.2.1 The Semiotic Triangle

The view of many system developers is that systems supporting the same domain naturally are using very similar, if not the same, conceptualization. However, the principle documented by Odgen and Richards (1923) still holds. Odgen and Richards distinguish between referents, concepts, and symbols

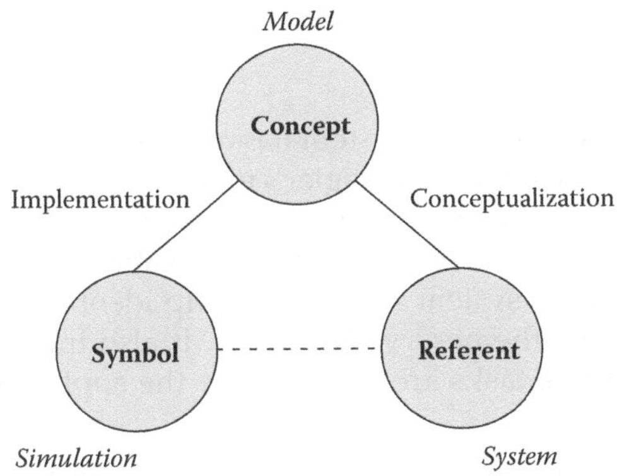

FIGURE 14.1
The semiotic triangle for M&S.

to explain why communication often fails. Referents are objects in the real (or an assumed or virtual) world. When communicating about the referents, perceptions or interpretations of these referents are used and captured in concepts that reflect the user's viewpoint of the world as object, etc., and then symbols are used to talk about the user's concepts.

Figure 14.1 shows the relation of this semiotic triangle to the M&S domain. This model is similar to the one presented by Sargent (2001), where real-world domain, conceptual model, and implemented model are distinguished, and to the framework for M&S as recommended in Zeigler et. al., (2000), where the experimental frame with the source model, the model, and the simulator are distinguished.

It should be pointed out that the implementation does not reveal why the conceptualization was chosen, only which one was chosen. A common conceptualization can result in different implementations. In order to ensure composability, conceptualization decisions need to be captured in addition to the implementation decision.

Furthermore, model-based solutions can only use their models and symbols, and no longer use the referent. The formal specification of the conceptualization is their view of the world. In order to decide if two model-based solutions are composable a decision needs to be made whether a lossless mediation between the conceptualization is possible (in the context of the task to be supported by the composition).

14.2.2.2 Machine-Based Understanding

As the process needs to be supported by intelligent agents, users need to understand how agents can gain machine-based understanding of tasks to be conducted and model-based solutions that may be composed to support this task. Zeigler (1986) identified three requirements that are applicable in the context of understanding the conceptualization and implementation:

- *Perception:* The observing system has a perception of the system that needs to be understood. In Zeigler's model, perception is not a cognitive process but is simply capturing sensor input. It implies that data and processes characterizing the observable system be captured by the observing system in order to help identify, select, compose, and orchestrate the observed systems. It also implies that all data needed for these tasks are provided by the applicable systems that are observed.
- *Metamodels:* The observing system has an appropriate metamodel of the observed system. The metamodels represent the categories of things the observing system knows about. As such, each metamodel is a description of data, processes, and constraints explaining the expected behavior of an observed system. Without such a metamodel of the observed system, understanding for the observing system is not possible. In Zeigler's model, the main part of machine-based understanding can be defined as identifying an applicable metamodel.
- *Mapping:* Mappings between observations resulting in the perception and metamodels explaining the observed data, processes, and constraints do exist, are identified, and are applied in the observing system

In the context of composition of model-based solutions this implies that the metadata describing the conceptual model must be perceivable by the composing agents and support the metamodel used to decide the composability. Data, processes, and constraints defining the models must be mapable to data, processes, and constraints defining the task to be supported.

14.2.2.3 Levels of Conceptual Interoperability Model

This leads to the question what information is needed to capture data, processes, and constraints in a formal specification of the conceptualization. The related work on the challenges of interoperability and composability enabling the attainment of these objectives led to the development of the Levels of Conceptual Interoperability Model (LCIM). As documented in Tolk (2006), the LCIM is the result of several composability and interoperability efforts. During a NATO Modeling & Simulation Conference, Dahmann (1999) introduced the idea of distinguishing between substantive and technical interoperability. In his research on composability, Petty (2002) enhanced this idea. In his work, he distinguished between the implemented model and the underlying layers for protocols, the communication layers, and hardware.

Realizing the need to explicitly address the conceptual layer, Muguira and Tolk (2003) published the first version of the LCIM, which was very datacentric. The discussions initiated by the LCIM work, in particular the work of Page et al. (2004) and Hofmann (2004), resulted in the currently used version, which was first published by Turnitsa (2005). Figure 14.2 shows the evolution of layered models of interoperation resulting in the LCIM.

The LCIM exposes six levels of interoperation, namely the following:

- The *technical* level deals with infrastructure and network challenges, enabling systems to exchange carriers of information.
- The *syntactic* level deals with challenges to interpret and structure the information to form symbols within protocols.
- The *semantic* level provides means to capture a common understanding of the information to be exchanged.
- The *pragmatic* level recognizes the patterns in which data are organized for the information exchange, which are in particular the inputs and outputs. These groups are often referred to as (business) objects.
- The *dynamic* level adds a new quality by taking the response of the system in form of context of the business objects into account. The same business object sent to different systems can trigger

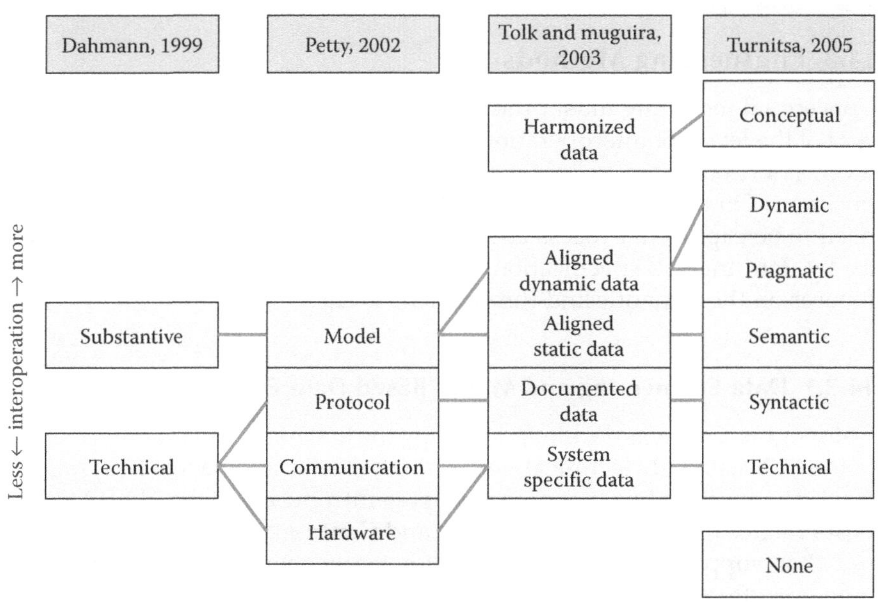

FIGURE 14.2
Evolution of levels of interoperation. (From Tolk, A., What comes after the semantic web: PADS implications for the dynamic web, *Proceedings of the 20th Workshop on Principles of Advanced Distributed Simulation,* 57, IEEE Computer Society, Washington, DC, 2006.)

very different responses. It is also possible that the same information sent to the same system at different times can trigger different responses.

- Finally, assumptions, constraints, and simplifications need to be captured. This happens on the *conceptual* level.

The LCIM has been applied in different communities. Wang et al. (2009) show the descriptive and prescriptive potential of the LCIM and evaluate a first set of specifications, in particular those defined by the simulation interoperability standards IEEE 1516, the High Level Architecture, and the Base Object Models (BOM) standard recently developed by the Simulation Interoperability Standards Organization (SISO). It is used in the following section to support the recommended engineering methods for conceptual modeling.

This section reviewed composability as it is currently understood and showed how it is related to yet different from integratability and interoperability. The next section discusses how to capture data, processes and constraints using engineering methods.

14.3 Engineering Methods

Conceptual modeling must produce a machine-readable description in support of the levels of interoperation identified in the LCIM. On the conceptual level, assertions need to be defined to avoid conceptually misaligned compositions. On the dynamic level, the system states governing the processes need to be captured. Process engineering lays the foundation for this activity by defining the specification for the pragmatic level. Data engineering focuses on the semantic and the syntactic levels.

14.3.1 Data Engineering and Model-Based Data Engineering

Data engineering was developed to study the first four levels of interoperation of the LCIM, namely technical, semantic, syntactic, and part of the pragmatic level. The notion of data engineering was introduced in the NATO Code of Best Practice (NCOBP) for Command and Control Assessment (NATO 2002, p. 232) in support of the integration of heterogeneous data sources for common operations and operational analysis. While the NCOBP was written by international NATO experts, its application is not limited to military systems. The NCOBP was created more as an application-oriented introduction on how to conduct operations research studies on complex, complicated, and wicked problems, such as the command and control challenge in a multinational organization with many independently developed information systems and not necessarily always aligned doctrinal viewpoints.

The NCOBP introduced data engineering as an engineering method to ensure that valuable resource data are best utilized. As defined in the NCOBP, data engineering consists of the following four main areas:

- *Data Administration*: Managing information exchange needs including source, format, context of validity, fidelity, and credibility. As a result of the processes in this area the data engineer is aware of the data sources and their constraints.
- *Data Management*: The processes for planning and organizing data including definition and standardization of the meaning of data as of their relations. Using the processes of this area, the data engineer unambiguously defines the meaning of the data.
- *Data Alignment*: Ensures that data to be exchanged exist in all participating systems. Using the results of data management, target data elements needs and source data abilities can be compared. The data engineer identifies particular gaps that need to be closed by the system engineers responsible for the participating systems.
- *Data Transformation*: Technical process of mapping data elements from the source to the target. If all data are captured in the first three processes, data transformation can be automated by configuring XML translators (Tolk and Diallo 2005).

Out of the four steps, data management is the most important step in data engineering and the most studied in the literature given that is the one that demands the most effort. To manage data, metadata registries have been defined to support the consistent use of data within organizations or even across multiple organizations. In addition, they need to be machine-understandable to maximize their use. Logically, the recommended structures for metadata registries are strong candidates for capturing the results of conceptual modeling for information exchange.

This work is supported by standards, such as Part III of the ISO/IEC 11179 "Metadata Registry (MDR)" standard that is used to conduct data management. Using this standard, four metadata domains are recommended to capture information on data representation and data implementation, which are summarized in Figure 14.3:

- The conceptual domain describes the concepts that were derived in the conceptualization phase of the modeling process. This domain comprises all the concepts that are needed to describe the referent or referents relevant to the information exchange.
- The property domain describes the properties that are used to describe the concept. Concepts are characterized by the defining

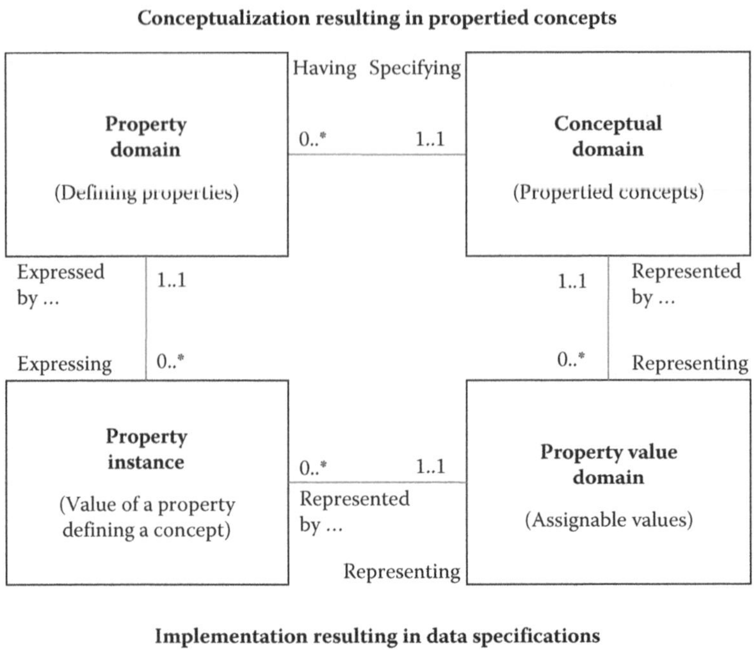

FIGURE 14.3
Domains of information exchange modeling.

properties. ISO/IEC 11179 refers to this domain as the data element concept.

- The property value domain comprises the value ranges, enumeration, or other appropriate definition of values that can be assigned to a property. ISO/IEC 11179 refers to this domain as the value domain.
- Property instances capture the pieces of information that can be exchanged. They minimally comprise the value of one property, which can be interpreted as updating just one value, or they can become an n-tuple of n properties describing a group of associated concepts, which represents complex messages or updates for several objects. ISO/IEC 11179 calls these property instances data elements.

Traditionally, only property instances are captured. From what has been specified in this chapter so far it becomes obvious that information needs to be specified in the context of the results of the underlying conceptualization to ensure the required semantic transparency. As the referent itself cannot be captured because it is replaced by its conceptualization, the information exchange model must at least capture the conceptualization of the model used and cannot be limited to the symbols used for its implementation. Tolk et al. (2008) introduce preliminary ideas on how to use these structures to enable self-organization of information exchange, if machines are not only able to understand how the implementations are related to the conceptualization, but how conceptualizations of different models are related to each other.

When talking about models as abstractions of a referent, it is unlikely that the same data are used in the same structure by the modelers involved. Model-Based Data Engineering (MBDE) introduces the notion of a Common Reference Model (CRM) to support data engineering for model-based solutions. In MBDE, the definition of a common namespace captured in the form of a logical model is the starting point. The CRM captures the objects, attributes, and relationships that are susceptible to being exchanged between two composed solutions. It is worth mentioning that, in theory, the four areas of data engineering are well-defined steps that can be conducted consecutively. In practical applications, however, model-based solutions are hardly ever documented according to data engineering, so they become an iterative process. Furthermore, it should be pointed out that the CRM is not a fixed model comparable to the predefined information exchange models. Rather, the CRM is gradually modified to reflect new concepts needed, resulting in *extensions*, or to reflect new properties or additional property values, resulting in *enhancement*.

MBDE can be explained using the same data engineering steps but must be supported by the use of domains of information exchange modeling.

14.3.3.1 Data Administration

If the data are not already structured and documented in form of an object or data model, this step is necessary. Conceptual modeling in support of data administration classifies each information exchange element either as a value (V) that can be grouped with other property values of its domain (D) that can be assigned to property (P), or as a property that can be grouped into a set of properties that identify a propertied concept (C), or as a concept that can be related to other concepts. At the end of this process, the domains of information exchange modeling for the information exchange capability (what can be produced as a data source) and the information exchange need (what can be consumed as a data target) for each system or service is documented. Figure 14.4 shows the result of the data administration process for two systems (or services) **A** and **B**.

14.3.3.2 Data Management

Using the information exchange modeling elements, the concepts (C) and the defining properties (P) of information exchange capability models and information exchange need models, common concepts and properties are identified. The result is a set of propertied concepts to which the elements

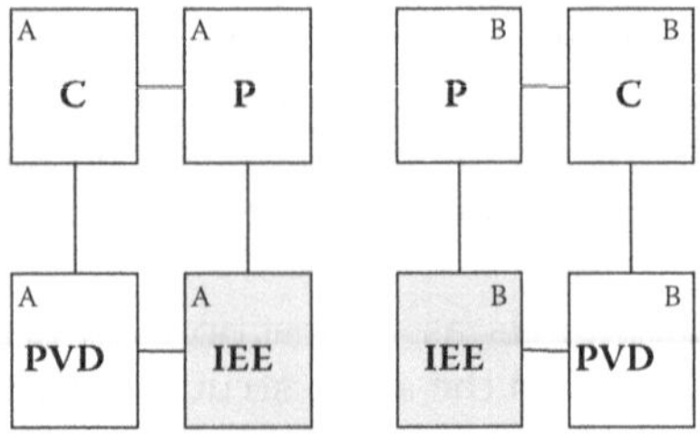

FIGURE 14.4
MBDE data administration.

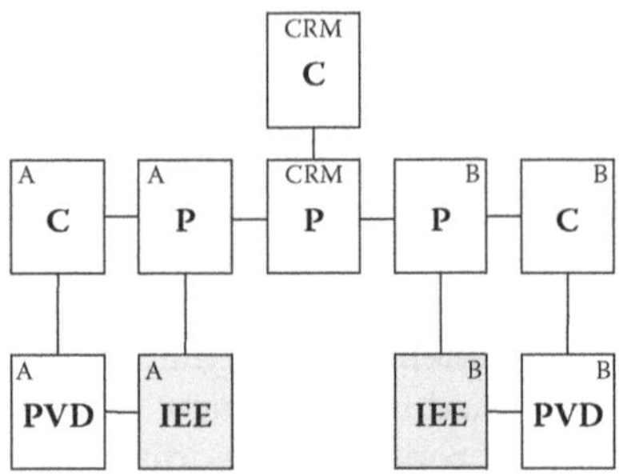

FIGURE 14.5
MBDE data management.

of the information exchange capability models can be mapped and that can be mapped to the elements of the information exchange need models. In the case of MBDE, these propertied concepts build the CRM, which is the logical model of the information exchange, or conceptualization, that can take place between the model-based solutions. It is worth stating that this shows that such a logical CRM always exists, whether it is made explicit or not, as whenever a property from system **A** is mapped to a property of system **B**. This constitutes a property that makes sense in the information exchange between the two systems, which is expressed by the CRM. The concepts of the CRM serve two purposes:

- They build the propertied concepts of the properties of the CRM and as such help the human to interpret the information exchange categories better. In particular when a CRM is derived from the information exchange requirements, this is very helpful.
- They conserve the context of information exchange for the receiving systems, which is their information exchange need.

Figure 14.5 shows the result of the data management process for two systems (or services) **A** and **B**.

14.3.3.3 Data Alignment

System **A** is said to be aligned with system **B** under the CRM if they are logically aligned and an injective function can be constructed from the property value domain of the source property in system **A** to the target property in system **B**. Figure 14.6 shows the result of the data alignment process for two systems (or services) **A** and **B**.

In this area of data engineering, it is necessary to distinguish between the logical and physical aspects of data alignment. Logically, information can be exchanged between two systems if the properties of system **A** have properties in system **B** that are logically equivalent, i.e., they have the same meaning and participate in the same relationships or simply put they map to the same subset of the CRM. Not every property in system **A** will have a counterpart in system **B**, and only for those properties that are connected to a property of the CRM that is part of the information exchange requirements model is the mapping operationally required. If a connection from system **A** to system **B** exists for every property of the CRM representing a piece of the information exchange requirement, system A and B are logically aligned under the CRM, i.e., **A** and **B** are equivalent under the CRM or, simply put, **A** and **B** mean the same thing. It is worth mentioning that logical alignment is mathematically symmetrical but remains dependent on the CRM.

The next step is the physical alignment. While two properties may be aligned under the CRM logically, their actual representations can be a challenge, as the property value domains must be equivalent classes, or at least the class of the information exchange need property value domain must be in the range of the information exchange capability property value domain. The modeling and simulation literature deals with problems derived from this step under the term multiresolution modeling (Davis and Huber 1992, Davis and Hillestad 1993, Reynolds et al. 1997, Davis and Bigelow 1998).

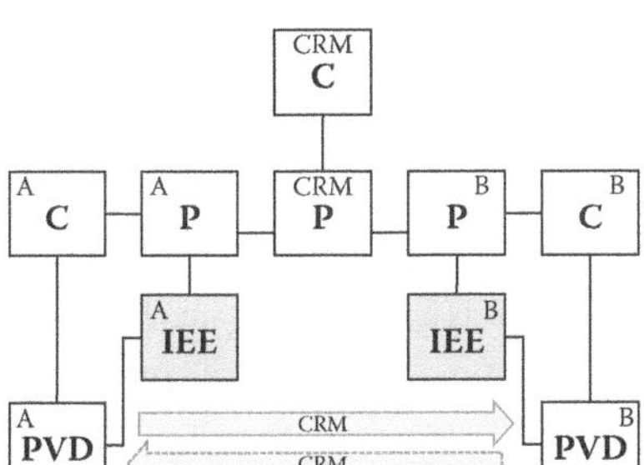

FIGURE 14.6
MBDE data alignment.

14.3.3.4 Data Transformation

Data transformation can map the information exchange elements to each other. This process of data mediation between the different viewpoints represented in the participating systems is already specified by the other three areas of data engineering, so that this step can be automated by using the results of the first three areas:

- Data administration provides the needed structures of the information exchange capabilities (system **A**) and information exchange needs (system **B**).
- Data management provides the identification of common properties on the logical level, resulting in the CRM. The CRM can be constrained to those concepts and properties that satisfy the information exchange requirements of the operation to be supported.
- Data alignment evaluates the logical alignment and the physical alignment. For each property in the CRM representing a part of the information exchange requirement, a logical counterpart is needed in the information exchange capability of system **A** (data can be produced) and in the information exchange need of system **B** (data can be consumed). In addition, the property value domains used to implement these properties must be mapped to each other.
- The result is a function that maps all relevant information exchange elements of system **A** under the CRM to the logically correct information exchange elements of system **B**. If the relations between **A** and **B** or not only injective, but also surjective, an inverse function exists as well.

Figure 14.7 illustrates the final result of data transformation:

Tolk and Aaron (2010) give application examples of MBDE using real-world CRM derived from projects conducted for the US Army and the US Joint

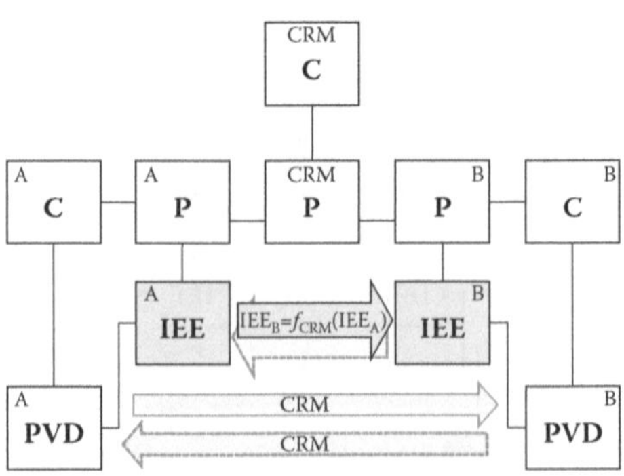

FIGURE 14.7
MBDE data transformation.

Forces Command and show how these applications can be generalized. In their contribution, they emphasize in particular the necessity of an efficient engineering management process in addition to the conceptual and technical constraints.

14.3.2 Process Engineering

While the emphasis so far has been on the information exchange model in the form of conceptually understanding what information exchange elements are exchanged between systems, the LCIM requires process engineering that captures the context and system changes that covers the pragmatic and dynamic layers of interoperability. To understand how the data are used within a system or what is needed to produce the data so they can be transmitted or how the system states are going to change once the data are produced or consumed requires a formal method for handling processes, similar to the one introduced for data in the previous section. While such a formal method is not yet standardized, such a method as described in this section will result in process engineering.

In order to apply process engineering, specifics concerning each process must be made available. A *process description language* and a *process-algebra* designed in support of process engineering are objects of ongoing research, resulting so far in the identification of the first process-defining attributes:

- *Initialization Requirements:* Unambiguous definition of process-specific object-attribute values that must exist, including object–object relationships that must be in place, for the process to be feasible. Additionally, system operational requirements for the initialization of the process are specified, if required.
- *Time:* Capturing the dynamic characteristics of the process, such as when does the process start and how long does it take to complete, and in terms of complex processes, the rate of progress. Such a method should be capable of identifying the relationship between a process, and the operational lifespan of the system it is supporting, and capturing the relationships previously identified (Allen 1983).
- *Effects:* A process affects some change in the attribution of an object in the system. This follows the definition of process, "a pattern of transformation that objects undergo" (Dori 2002). Complex processes may affect more than one attribute in a single object, or perhaps more than one object. In this category of defining attributes, the range of these effects that take place, including how these effects occur, are captured. The changed attributes can be identity attributes of the system or can possibly be coincidental attributes.
- *Halting Requirements:* Some processes will terminate given a certain passage of operational time, others require specification of what

conditions cause the process to halt. This specification is more likely to be required for a complex process (where more than one attribute or object change is part of the process specification) than for a simple process.

- *Postconditions:* The state of the system once a process has halted including specifying the nature of the process or related processes once it has expired, such as termination, waiting in idle mode, etc.

Once a process description language that captures these attributes is formally presented and used to define processes within a model of a system, then the application of process-algebra will enable the four steps of process engineering that can be understood and supported by machines.

Tolk et al. (2009) describe the four areas of process engineering, similar to data engineering, as an approach to align the process of separate systems. As with data engineering, process engineering follows:

- *Process Administration*: The processes included in the systems to be made interoperable must be identified, including information concerning the original system that each process belongs to, and the operational context in which it acts.
- *Process Management*: This involves capturing and organizing processes. Each process must be specified (capturing the pertinent attributes of the process, as described earlier) in order to enable the following steps.
- *Process Alignment*: The defining attributes of the processes, previously mentioned, as organized within process management, determine where, in the life of the system, the process will occur, and what affect it has internally on the system, as well as what the process means externally for interoperability. These attributes must be aligned so that the resulting effects of interacting processes can be determined.
- *Process Transformation*: Finally, if the resulting effects of interacting processes do not produce a desired outcome, it may be necessary to perform some transformation to one or more attributes of some of the processes in question. The information concerning the processes that results from the first three steps will enable this to happen.

Current research focuses on using the web-ontology description for services, such as OWL-S (Martin et al. 2004). Related ideas have been published by Rubino et al. (2006) and others.

14.3.3 Constraint Engineering

The assumptions, constraints, and simplifications of a model are part of conceptual modeling, and are important elements that need to be evaluated if two model-based solutions are to be composable. These assumptions are often not reflected in the implementation and unfortunately are rarely documented. Spiegel et al. (2005) present a simple physics-based model of the falling body problem to show how many assumptions are implicitly accepted within models, and how hard it is for experts to reproduce them "after the fact." While already challenging for human system engineers, reproduction of such knowledge is currently perceived to be impossible for machines. Capturing constraints and assumptions in machine-understandable form result in constraint engineering enabling machines to address challenges on the level of conceptual interoperability.

King (2009) describes the foundation for constraint engineering, which covers the conceptual layer of the LCIM. In his work, he outlines a process for capturing and aligning assertions as well as a formalism to enable the envisioned machine support. Using the falling body problem described by Spiegel et al. (2005), he captures all applicable forces in an ontology using the Protégé Tool. He next develops a formalism to represent assertions, which are critical modeling decisions the model-based systems rely on. In his work, King (2009) distinguishes between four assertion types, namely assumptions, constraints, considerations, and competencies. Assertions must be encoded in a knowledge representation language to make them machine-understandable. Each proposition consists of its axioms and logical assertions that relate it to other concepts and propositions. Having the constraints and the assertion formalized, it then becomes possible to compare the simplifications, assumptions, and constraints using logical reasoning to identify incompatibilities on the conceptual level, as envisioned in King (2007). The idea is captured in Figure 14.8.

As stated above, building and manipulating these lists requires an ontology to express the assertions, and a formalism for representing them. The formal model of assertion has four components that are described in the following list:

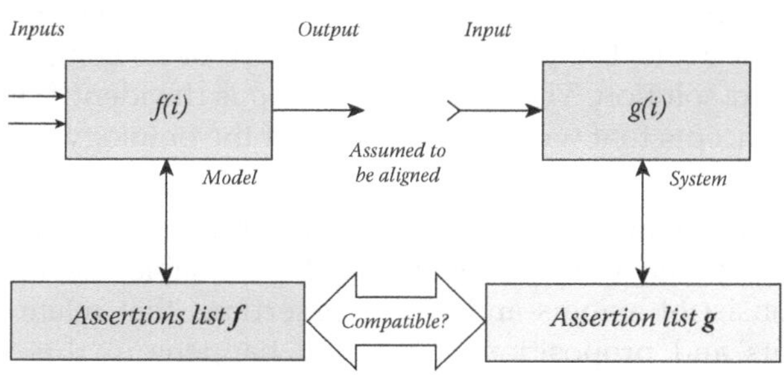

FIGURE 14.8
Evaluating compatibility of assertion lists. (Adapted from King, R. D., On the Role of Assertions for Conceptual Modeling as Enablers of Composable Simulation Solutions, PhD thesis, Old Dominion University, Norfolk, VA, 2009.)

- *Use function:* Describes what assertions are used for.
- *Referent:* The referent of an assertion is the entity to which it refers. A referent can be an object, a model, a process, a data entity, a system, or a property of one of these. When an assertion acts as a constraint, the referent is what is being limited by the proposition.
- *Proposition:* The proposition of an assertion is the statement that it is making. Propositions are not restricted to simple concepts—they may encompass the content expressed by theories, books, and even whole libraries.
- *Scope:* This is an optional description that extends the portions of the overall system to which the assertion applies. Scope can limit consideration to a system component, the system as a whole, the environment of the system, or to combinations of these (e.g., *component-environment* scope means that the scope of assertion is the relationship between the component and its environment.) If scope is not specified, then the assertion has *component* scope. Scope can be stated explicitly or implicitly.

The formal description of an assertion as described above is therefore a tuple with three mandatory components and one optional component structured as:

```
Assertion<=>(referent useFN Proposition<scope>)
```

In order to apply these ideas, three steps need to be conducted, that are captured in the following list. The viewpoint is slightly different from the data and process engineering areas, but the results are comparable.

- *Capture assertions:* The first step is capturing the assertion propositions for the model, system and environment that are known within the scope or that are otherwise important. Each proposition represents a concept that is initially expressed as a natural language statement about the problem, one or more of its components, or a particular solution. The result of this step is the identification of the main concepts that will form the basis of the ontology.
- *Encode propositions:* The list of propositions expressed in natural language statements must be encoded in a knowledge representation language to make them machine-understandable. Each proposition will consist of axioms and logical assertions that relate it to other concepts and propositions. Encoding happens in this step. King (2009) used the Protégé Tool, but every other tool supporting the encoding in logical form is applicable. The second step finally consists of assigning the use function, referent and scope to each proposition in both the model and the system lists. Experience shows that

the analyst should be prepared to make several iterations through this process step as the assertion lists are refined. The output of this step is list of statements encoded in a knowledge representation language.

- *Compare assertion lists:* The task of comparing assertion lists requires a multilevel strategy that is described in detail in King (2009). The full details go beyond the scope of this chapter except to note that the comparison is steered by the use function assigned to each proposition in the previous step. The method is used to compare the list of assertions about the model to be composed with the statements about the system. Each proposition represents a concept and there are different ways that concepts can match. The topic of semantic similarity—deciding if, and how closely concepts match—is the subject of much current study, particularly with respect to research into the Semantic Web. The issue is a complex process influenced by many different factors or characteristics, however analysis such as performed by Kokla (2006) reveals four possible comparison cases between concepts: equivalence, when the concepts are identical in meaning; subsumption (partial equivalence), when one concept has broader meaning than the other; overlap (inexact equivalence), when concepts have similar, but not precisely identical meanings; and difference (nonequivalence), when the concepts have different meanings. The first three are potentially useful for comparing assertions lists. Testing for equivalence is straightforward and involves searching for text or label matches. Subsumption is handled by first order logic or a subset of FOL such as a Description Logics reasoner operating on an OWL-DL ontology. Determining overlap requires greater sophistication in reasoning, such as analogical reasoning described by Sowa and Majumdar (2003) or the agent-based metalevel framework for interoperation presented by Yilmaz and Paspuletti (2005).

King (2009) shows with these steps that it was possible to capture conceptual misalignments of services that did not show up on the implementation level. In other words: without capturing the results of conceptual modeling appropriately, these services would not have been identified as not composable, and a composition of them would have been technically valid but conceptually flawed, leading to potentially wrong results. Depending on the application domain the simulation composition uses, such conceptual misalignment can lead to significant harm and even the loss of human lives or economic disasters, in particular when decision makers base their decisions on flawed analysis.

It is worth mentioning that the rigorous application of mathematics ensures the consistency of conceptualizations and their implementations and not that the conceptualizations are correct. In other words, it is possible to evaluate if two different conceptualizations can be aligned to each other and if transfer functions exists between resulting implementations. The three engineering methods described here support together this evaluation and enable the support thereof by machine, if all needed artifacts are available and can be observed as discussed earlier in this chapter.

14.4 Technical and Management Aspects of Ontological Means

As stated in the introduction, in the context of service-oriented solutions and reuse of system functionality in alternative contexts users need machine support to identify applicable services, select the best available set of solutions, compose all selected services to provide the solution, and orchestrate their execution. If machines are to help a user with these tasks it is essential to express all the findings captured so far in machine-understandable form. The means provided by current simulation standards, such as IEEE1278 and IEEE1516, are not sufficient. The results of data, process, and constraint engineering need to be captured in machine-readable form. The conceptual model of a model-based solution must capture the concepts and their relations (data), transformations (processes), and rules and axioms (constraints). Without data engineering, the mediation between alternative viewpoints captures in model-based solution is not possible, which is needed to identify applicable solutions. Without process engineering, composition and orchestration cannot be achieved. Without constraint engineering, conceptual misalignment may occur.

Ontological means have been applied for similar tasks in the context of the Semantic Web. Obrst (2003) introduced a spectrum of ontological means that is captured in Figure 14.9 (which merges the two figures originally used by Obrst into one).

The ontological spectrum emphasizes the viewpoint that ontologies are not a radically new concept, but that they are a logical step in the process of increasingly organizing data: starting with pure enumeration, thesauri relate similar terms and taxonomies relate them in an order. Capturing assumptions and constraints results in CM. Formulating the CM in the form of logical expressions and axioms leads to models and makes the represented specifications understandable for software systems, such as intelligent agents. It is worth mentioning that the "conceptual" model in this spectrum represents the implementation-independent abstraction of entities, as they can be captured in UML and OWL, which is less powerful than the viewpoint of conceptualization presented in this chapter.

A common definition for an ontology is *a formal specification of a conceptualization* (Gruber, 1993). It can be argued that the use of ontological means to represent conceptual models in machine-understandable form is quasi motivated *per definitionem*: ontologies are the required specification of the

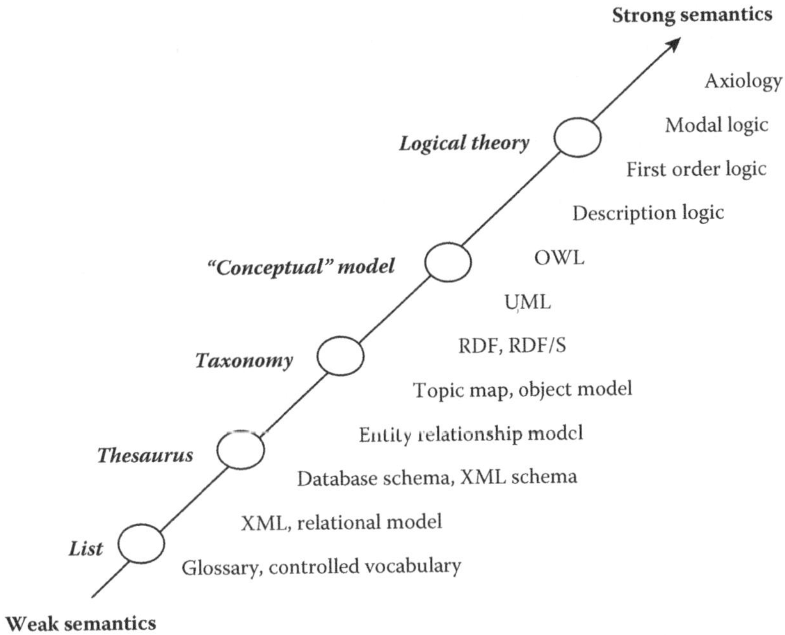

FIGURE 14.9
Ontological spectrum and methods.

conceptualization, and as they are formal, machines can read them and reason about them. West (2009) gives an example for practical applications of ontologies in connection with modeling of data for a significant business. A more theoretic introduction to the use of ontological means has recently been published by Guizzardi and Halpin (2008) in their special issue on conceptual modeling in the *Journal of Applied Ontologies*. The discussion on ontological means in support of systems engineering are ongoing.

Recker and Niehaves (2008) broadened the scope, limits and boundaries regarding ontology-based theories for conceptual modeling beyond the pure technical discussions by focusing on three additional questions:

- What does it mean to engage in conceptual modeling?
- What does it mean to evaluate the outcome of conceptual modeling?
- What does it mean to achieve quality in conceptual modeling?

Following the work of Recker and Niehaves (2008), it is possible to distinguish between conceptual models that are defined to *collectively construct an artifact that reflects subjectivity and purpose* on one hand, and conceptual models that are defined to *produce a direct representation of an external reality* on the other hand. This chapter focuses on the second viewpoint, as a machine-understandable description of the system or the service in order to identify applicable solutions is requested, the best one of these is selected, the solutions are composed and orchestrated in their execution. Furthermore, Recker and Niehaves (2008) distinguish between testing of conceptual models by comparing them to a reference model of reality versus consensus building of experts. Finally, the quality needs to be measured by the degree to which properties of the modeled portion of reality are represented in the conceptual model. The alternative presented by Recker and Niehaves (2008)—its perception as a good model by a social community—is not objective-driven. Nonetheless, remembering the lessons recorded in Tolk and Aaron (2010), a consensus between engineers and managers is mandatory for successful procedures for data engineering, including the conceptual modeling aspects. As such, the observations of Recker and Niehaves (2008) deserve special attention for setting up the necessary management structures.

14.5 Conclusion

To be able to decide if two model-based solutions are composable to support a common objective, implementation details of those solutions alone are not sufficient. In order to identify applicable solutions, select the best ones in the context of the task to be supported, compose the identified set of solutions, and to orchestrate their execution, conceptual models in the form of formal specifications of conceptualizations of data, processes, and constraints are mandatory. This formal specification must be captured as metadata. The methods in the ontological spectrum can be used to capture the metadata in machine-readable form. Without such annotations for model-based solutions that capture the results of the conceptual modeling phase as machine-understandable metadata, the concepts of system of systems and service-oriented architectures will remain incomplete.

Conceptual modeling for composition of model-based complex systems supported by the methods of data engineering, process engineering, and constraint engineering produces the metadata necessary to enable the lossless mediation between viewpoints as captured in the conceptualization of different models. We documented the necessary step of data engineering in detail and motivated the feasibility of similarly detailed methods for processes and constraints.

The rigorous use of mathematics to produce metadata for the annotation of model-based solutions is a necessary requirement to enable consistent system of systems solutions or service-oriented architectures. The solutions provided by current standards as discussed in this chapter are not sufficient.

References

Allen, J. 1983. Maintaining knowledge about temporal intervals. *CACM* 26 (11): 832–843.

Benjamin, P., K. Akella, and A. Verma. 2007. Using ontologies for simulation integration. In *Proceedings of the Winter Simulation Conference*, 1081–1089. Washington, DC: IEEE Computer Society.

Burstein, M. H., and D. V. McDermott. 2005. Ontology translation for interoperability among Semantic Web services. *AI Magazine* 26(1) : 71–82.

Dahmann, J. S. 1999. High Level Architecture Interoperability Challenges. Presentation at the *NATO Modeling & Simulation Conference*, Norfolk, VA, October 1999, NATO RTA Publications.

Davis, P. K., and R. K.Huber. 1992. Variable-resolution combat modeling: Motivations, issues, and principles. *RAND Notes*, Santa Barbara, CA.

Davis, P. K., and J. H. Bigelow. 1998. *Experiments in MRM*. RAND Report MR-100-DARPA, Santa Barbara, CA.

Davis, P. K., and R. Hillestad. 1993. Families of models that cross levels of resolution: Issues for design, calibration and management. In *Proceedings of the Winter Simulation Conference*, 1003–1012. Washington, DC: IEEE Computer Society.

Dori, D. 2002. *Object Process Methodology: A Holistic Systems Paradigm*. Berlin, Heidelberg, New York: Springer Verlag.

Fishwick, P. A. 2007. *Handbook of Dynamic System Modeling*. Chapman & Hall/CRC Press LLC.

Gruber, T. R. 1993. A translation approach to portable ontology specification. *Journal of knowledge acquisition* 5:199–220.

Guizzardi, G., and T. Halpin. 2008. Ontological foundations for conceptual modeling. *Journal of applied ontology* 3(1–2):1–12.

Hofmann, M. 2004. Challenges of model interoperation in military simulations. *Simulation* 80:659–667.

Institute of Electrical and Electronics Engineers. 1990. *A Compilation of IEEE Standard Computer Glossaries*. New York: IEEE Press.

Institute of Electrical and Electronics Engineers IEEE 1278 Standard for Distributed Interactive Simulation

Institute of Electrical and Electronics Engineers IEEE 1516 Standard for Modeling and Simulation High Level Architecture

International Organization for Standardization (ISO)/International Electrotechnical Commission (IEC). 2003. Information technology metadata registries part 3: registry metamodel and basic attributes. ISO/IEC 11179–3:2003.

Jackson, M. C., and P. Keys 1984. Towards a system of systems methodology. *Journal of the operations research society* 35(6): 473–486.

King, R. D. 2007. Towards conceptual linkage of models and simulations. In *Proceedings of the Spring Simulation Interoperability Workshop*. Washington, DC: IEEE Computer Society.

King, R. D. 2009. On the role of assertions for conceptual modeling as enablers of composable simulation solutions. PhD thesis, Old Dominion University, Norfolk, VA.

Kokla M. 2006. *Guidelines on Geographic Ontology Integration.* ISPRS Technical Commission II.

Martin, D., M. Burstein, J. Hobbs, O. Lassila, D. McDermott, S. McIlraith, S. Narayanan, M. Paolucci, B. Parsia, T. R. Payne, E. Sirin, N. Srinivasan, and K. Sycara. 2004. OWL-S: semantic markup for Web Services. *Technical Report UNSPECIFIED,* Member Submission, World Wide Web Consortium (W3C).

Muguira, J. A., and A. Tolk. 2003. The levels of conceptual interoperability model (LCIM). In *Proceedings Fall Simulation Interoperability Workshop.* Washington, DC: IEEE Computer Society.

NATO. 2002. *NATO Code of Best Practice for Command and Control Assessment.* RTO-TR-081 AC/323(SAS-026)TP/40. Washington, DC: CCRP Press.

Obrst, L. 2003. Ontologies for semantically interoperable systems. In *Proceedings of the Twelfth International Conference on Information and Knowledge Management,* November 2003, 366–369.

Ogden, C. K., and I. A. Richards. 1923. *The Meaning of Meaning,* 8th ed. New York: Harcourt, Brace & World, Inc.

Page, E. H., R. Briggs, and J. A. Tufarolo. 2004. Toward a family of maturity models for the simulation interconnection problem. In *Proceedings of the Spring Simulation Interoperability Workshop.* Washington, DC: IEEE Computer Society.

Petty, M. D. 2002. Interoperability and composability. *Modeling & Simulation Curriculum* of Old Dominion University.

Petty, M. D., and E. W. Weisel. 2003. A composability lexicon. In *Proceedings of the Spring Simulation Interoperability Workshop.* Orlando, FL: IEEE Computer Society, 181–187.

Petty, M. D., E. W. Weisel, and R. R. Mielke. 2003. A formal approach to composability. In *Proceedings of the Interservice/Industry Training, Simulation and Education Conference,* 1763–1772, Orlando, FL National Training Simulation Association (NTSA)/National Defense Industrial Association (NDIA).

Recker, J., and B. Niehaves. 2008. Epistemological perspectives on ontology-based theories for conceptual modeling. *Journal of applied ontology* 3(1–2):111–130.

Reynolds Jr., P. F., A. Natrajan, and S. Srinivasan. 1997. Consistency maintenance in multi-resolution simulations. *ACM transactions on modeling and computer simulation* 7(3):368–392.

Robinson, S. 2007. Conceptual modelling for simulation part I: Definition and requirements. *Journal of the operational research society* 59 (3):278–290.

Rubino, R., A. Molesini, and E. Denti. 2006. OWL-S for describing artifacts. In *Proceedings of the 4th European Workshop on Multi-Agent Systems EUMAS'06,* Lisbon, Portugal.

Sargent, R. G. 2001. Verification and validation: Some approaches and paradigms for verifying and validating simulation models. In *Proceedings of the Winter Simulation Conference,* 106–114. Washington, DC: IEEE Computer Society.

Spiegel, M., P. F. Reynolds Jr., and D. C. Brogan. 2005. A case study of model context for simulation composability and reusability. In *Proceedings of the Winter Simulation Conference,* 437–444. Washington, DC: IEEE Computer Society.

Sowa, J., and A. Majumdar. 2003. Analogical reasoning. In *Conceptual Structures for Knowledge Creation and Communication,* eds. A. de Moor, W. Lex, and B. Ganter, 16–36. Berlin: Springer.

Tolk, A. 2006. What comes after the Semantic Web: PADS implications for the dynamic Web. In *Proceedings of the 20th Workshop on Principles of Advanced and Distributed Simulation*, 55–62. Washington, DC: IEEE Computer Society.

Tolk, A., and R. D. Aaron. 2010. Model-based data engineering for data-rich integration projects: Case studies addressing the challenges of knowledge transfer. *Engineering management journal* 22(2) July (in production).

Tolk, A., and S. Y. Diallo. 2005. Model-based data engineering for Web services. *IEEE internet computing* 9(4) July/August: 65–70.

Tolk, A., S. Y. Diallo, and C. D. Turnitsa. 2008. Mathematical models towards self-organizing formal federation languages based on conceptual models of information exchange capabilities. In *Proceedings of the Winter Simulation Conference*, 966–974. Washington, DC: IEEE Computer Society.

Tolk, A., S. Y. Diallo, and C. D. Turnitsa. 2009. Data engineering and process engineering for management of M&S interoperation. In *Proceedings of the Third International Conference on Modeling, Simulation and Applied Optimization*, Sharjah, U.A.E, January 20–22, 2009.

Turnitsa, C. D. 2005. Extending the levels of conceptual interoperability model. In *Proceedings Summer Computer Simulation Conference.* Washington, DC: IEEE Computer Society.

Wang, W., A. Tolk, and W. Wang. 2009. The levels of conceptual interoperability model: Applying systems engineering principles to M&S. In *Proceedings of the Spring Simulation Multiconference.* Washington, DC: IEEE Computer Society.

West, M. 2009. Ontology meets business. In *Complex Systems in Knowledge-Based Environments: Theory, Models and Applications*, SCI 168, ed. A. Tolk and L. C. Jain, 229–260. Heidelberg: Springer.

Yilmaz, L. 2004. On the need for contextualized introspective simulation models to improve reuse and composability of defense simulations. *Journal of defense modeling and simulation* 1(3):135–145.

Yilmaz, L., and S. Paspuleti. 2005. Toward a meta-level framework for agent-supported interoperation of defense simulations. *Journal of defense modeling and simulation* 2(3): 61–175.

Zeigler, B. P. 1986. Toward a simulation methodology for variable structure modeling. In *Modeling and Simulation Methodology in the Artificial Intelligence Era*, ed. M. S. Elzas, T. I. Oren, and B. P. Zeigler, 195–210. Amsterdam: Elsevier Scientific Pub.

Zeigler, B. P., H. Praehofer, and T. G. Kim. 2000. *Theory of Modeling and Simulation: Integrating Discrete Event and Continuous Complex Dynamic Systems.* Amsterdam: Academic Press, Elsevier.

15

UML-Based Conceptual Models and V&V

Ö. Özgür Tanrıöver and Semih Bilgen

CONTENTS

15.1 Introduction

Although there is no consensus on a precise definition of a conceptual model (CM), it is generally accepted that a CM is an abstract representation of a real-world problem situation independent of the solution. This representation may include entities, their actions and interactions, algorithms, assumptions, and constraints.

Recently, there has been a growing tendency to adapt UML for different modeling needs and domains. Having various representation capabilities, being a multipurpose modeling language and allowing extension mechanisms, UML seems to be promising for conceptual modeling as well. However, as there is a lack of an agreed definition of a conceptual model, it is difficult to define a best set of UML views for representing it.

Nevertheless, in the military simulation domain, which constitutes one of the major areas of use of conceptual modeling, three approaches that support simulation conceptual modeling based on UML have emerged: The first one, Syntactic Environment and Development and Exploitation Process (SEDEP 2007) is HLA (High Level Architecture) oriented. Two UML profiles have already been developed toward tool support (Lemmers and Jokipii 2003). The second one, BOM (Base Object Model) (BOM 2006) has been developed by SISO (Simulation Interoperability Standards Organization). BOMs are defined to "provide an end-state of a simulation conceptual model and can be used as a foundation for the design of executable software code and integration of interoperable simulations" (BOM 2006). Hence, BOMs are closer to the solution domain and the developer rather than the problem domain and the domain expert. The third one is the KAMA notation (Karagöz and Demirörs 2007), which is more CMMS (Conceptual Model of the Mission Space) oriented and platform independent. CMMS is defined as "simulation-implementation-independent functional descriptions of the real world processes, entities, and environment associated with a particular set of missions" by DMSO (U.S. Defense Modeling and Simulation Office) (2000b, Karagöz and Demirörs 2008). KAMA has been revised through experimental processes and empirically shown to be fit for CMMS purposes (Karagöz and Demirörs 2008). A CMMS serves as a bridge between subject matter experts (SME) and developers. In CMMS development, SMEs act as authoritative knowledge sources when validating mission space models.

The roles of various parties involved in the scope of CM verification as discussed in this chapter will be defined as follows:

Modeler: Whoever prepares the UML-based CM from the formal, informal, written, or verbal description of the problem domain.

Inspector: Whoever carries the responsibility of verifying the UML-based CM, possibly applying the techniques and performing the processes described in this chapter.

SME: (Also referred in different texts as "domain expert," "knowledge engineer," or " domain engineer.") The party or parties who describe the problem domain formally, informally, in writing, or verbally based on their domain expertise and who may be consulted, if and when necessary, to enhance the preparation of UML-based CM or to validate an existing UML-based CM.

In this chapter, we will mostly deal with the verification of CMMS models developed with the UML-based KAMA notation. The general problem with utilizing a UML-based notation is that, in addition to defects and omissions that may be introduced during translation of problem domain to a conceptual model, semiformality of UML (Kim and Carrington 2000, Ober 2004), its support of multiple views, and its extension mechanism increases the risk of inconsistency, incorrectness, and redundancy. Furthermore, the specification of UML (UML Superstructure 2005) does not provide a systematic treatment of correctness, consistency, completeness and redundancy issues in models.

The rest of the chapter is organized as follows: First, we describe verification and validation (V&V) in the context of CM. Then, desirable properties to be used in verification of UML models are presented along with a summary of research in two main streams: formal and informal approaches. Then, as a candidate for informal V&V, an inspection process to address semantic properties is presented. Next are two case studies, illustrating how that inspection approach helps to identify logical defects as well as some important semantic issues to be consulted a SME. The last section concludes the chapter with an evaluation of the described techniques.

15.2 Verification and Validation of Conceptual Models for Simulations

15.2.1 V&V

Boehm (1984) describes software validation as a set of activities designed to guarantee that the right product from a user's perspective is being built and verification as activities that guarantee the product is being built correctly according to requirements specifications, design documentation, and process standards. In particular, verification consists of mostly the static examination of the intermediate artifacts such as requirements, design, code, and test plans. Boehm's well-known maxim puts this as "validation is building the right system whereas verification is building the system right."

On the other hand, DOD (2001) defines simulation conceptual model validation as determining that the theories and assumptions underlying the conceptual model are correct and the representation of the validated

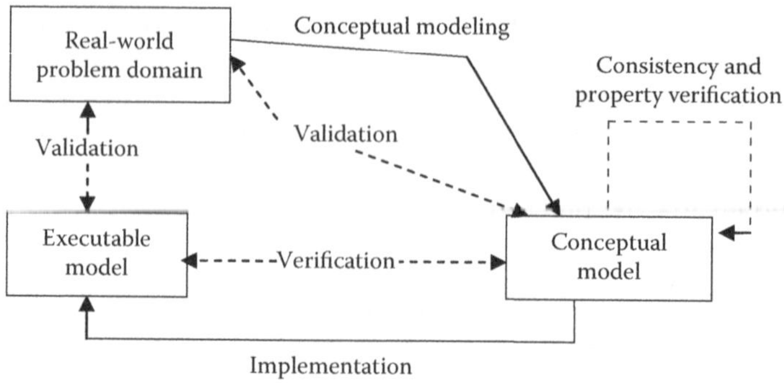

FIGURE 15.1
Conceptual model validation and verification.

requirements is reasonably accurate for the intended purpose of the model. In particular, the conceptual model's structure, logic, mathematical and causal relations, and the processes need to be reasonably valid with respect to the real system.

Therefore, CMs should be internally complete, consistent, coherent, and correct. That is, the CM should not include conflicting elements, entities, and processes. Redundant elements need to be avoided to establish a coherent concept of the simulation in which all components have certain functions and that all model components are reachable (Pace 2000). In addition to these criteria, Lindland et al. (1994) suggest a set of conceptual model quality evaluation criteria such as modularity, implementation independence, maintainability, and generality that may be considered in V&V activities.

In Figure 15.1, an overview of a generally accepted simulation conceptual modeling, verification and validation framework is shown. It is seen that conceptual model development can contribute to overall validity of the system in two ways. First, before the simulation is developed, the conceptual model may be checked against problem definition or domain of interest for validation such that defects can be eliminated earlier in the lifecycle. Second, a valid conceptual model enables computerized verification of the model.

15.2.2 V&V Process

Although the conceptual modeling activity has been mentioned within generally accepted simulation development methodologies, the details of a standard validation and verification process for CMs have not been explicitly stated (Hue et al. 2001) even though CM V&V is expected to adhere to a set of established principles. For this purpose, a set of international standards can be considered as resources.

Among these, first, IEEE 1014 (Software Verification and Validation std.) can be used as a general guideline for V&V activities for SQA (Software Quality Assurance) of computerized simulation development projects. On the other hand, NATO (2007), for instance, focuses specifically on V&V for federations that are developed according to the FEDEP (2000) (Federation Development and Execution Process). It considers verification, validation, and accreditation (VV&A) activities as an "overlay" process to FEDEP, whereas the ITOP (2004) approach aims at supporting the capture, documentation, reuse, and exchange of V&V information. Finally, the REVVA 2 (2005) methodology is intended to provide a generic VV&A framework. In spite of having different focuses, these resources share some common concepts.

They agree that, in the first stages of M&S development, a vague intended purpose must be formulated, which is refined into a set of subpurposes, which again must be decomposed to a set of Acceptability Criteria (AC). Hence, passing the AC implies fitness for the intended purpose (for UML-based models these may be stated as properties). From AC, V&V objectives can be derived. V&V objectives are usually decomposed into more concrete V&V tasks. The execution of these tasks produces items of evidences to substantiate the AC.

When a modeling notation such as UML or KAMA is used, AC and associated V&V objective formulation for the conceptual models should also take into account the set of representational and abstraction capabilities of the modeling notation. For example, if the purpose of conceptual modeling is just to provide a generic repository for reuse then the set of criteria will not focus on implementation requirements but rather on understandability, easy adaptation for reuse etc. On the other hand, if the conceptual model is to be used straight in FEDEP, run-time criteria should be defined also.

The following set of general principles for simulation V&V, originally suggested by Balci (1998), can be followed during CM V&V:

1. V&V must be conducted at each phase of modeling.
2. The outcome of V&V should not be considered correct or incorrect.
3. V&V must be conducted by personnel other than the developer.
4. Exhaustive model testing is not possible.
5. V&V must be planned and documented.
6. Errors should be detected as early as possible.
7. Multiple views and interpretations of models must be identified and resolved properly.
8. Testing of each submodel does not guarantee integral model quality.

One striking observation one can make is that the accepted methodology or guidelines by the simulation V&V community (DMSO 2000a, FEDEP 2000, ITOP 2004) is that validation and accreditation of simulations is addressed in detail and extensively. However, internal verification of conceptual models is not explicitly addressed. This may be partly justified by the fact that during simulation software development projects, SQA is already performed for requirements, design and code verification. However, due to experimental intentions of simulations in general, conceptual models are also used to represent a domain or problem entity to be simulated. Hence, in addition to a model of the software system, a model of the simulation domain must be represented. Therefore, there exists a need for verification to assure that the conceptual models for simulations are represented as to respond to the intended purpose of the simulation, in addition to verification of the model of software running the simulation. This distinction clarifies why only SQA is not enough.

On the other hand, as simulation projects may deal with a wide range of experimental domains, SMEs with specific domain expertise are consulted for validation of conceptual models. However, verification is required even if a validation activity is planned. Because, first, it would be unwise to wait until model validation to find out whether specified requirements related to the domain of interest have defects such as inconsistency, redundancy, and incompleteness. Furthermore, during model development, defects and inconsistencies may be introduced by the modeler. So it is necessary to assure that the model is at least internally correct and consistent before the validation process. Second, verification techniques identify illicit or inconsistent representations and undesired behavior. For example, a deadlock in a Petri Net model representing a production process can be identified by verification techniques. Hence, verification identifies the issues validation may not. In this way during the validation, for example, the SME will not be distracted by possible concurrency issues that may be introduced when utilizing Petri Nets as a representation technique and can concentrate more effectively on how well the model represents the real world and its fitness for the intended use.

15.2.3 V&V Techniques

Verification and validation of conceptual models is part of the overall verification and validation of simulations. A list of techniques for verification and validation that can be used in different phases of modeling and simulation (requirements, conceptual modeling, design, development, use, assessment) is listed by DoD VV&A RPG (2000, 2001).

Within the modeling and simulation literature, a variety of specific techniques for V&V have been suggested by authors such as Law and Kelton (1999). Balci (1998) offers a collection of 77 verification, validation, and testing techniques. Furthermore, more specific V&V techniques for object-oriented simulation models are presented (Balci 1997). Conventional methods are classified into four main streams; informal, static, dynamic, and formal in order of the level of formality required. Techniques, however, vary extensively, from alpha testing, induction, cause and effect graphing, inference, predicate calculus, and proof of correctness to user interface testing.

Informal techniques are easy to use and understand with checklists, manuals, and guidelines. They may be effective if applied consistently and are relatively less costly. Furthermore, informal V&V techniques may be used at any phase of the simulation development process including conceptual modeling. Static techniques can reveal a variety of information about the structural inconsistencies, data and control flow defects and syntactical errors. On the other hand, dynamic techniques find defects in behavior and results of the model execution. Finally, formal techniques rely on a formal process of symbol manipulation and inference according to well-defined proof rules of the utilized formal language. They are very effective, but costly due to their complexity and sometimes due to the size of the model under consideration. Many formal techniques are either unusable except in trivial examples or require an understanding of complex mathematics (Garth et al. 2002).

Validation of conceptual models is usually informal and consists of SME reviews or audits, self inspection, face validation, etc. In addition, Sargent (2001) mentions the use of traces, again an informal technique. The use of traces is the tracking of entities through each submodel and the overall model to determine if the logic is correct and if the necessary accuracy is maintained. If errors are found in the conceptual model, it must be revised and conceptual model validation should be performed again. Furthermore, CM validation can be also performed using Simulation Graph Models (Topçu 2004). These are mathematical representations of the state variables, events that change the state, and the logical and temporal relationships between events.

On the other hand, UML-based CMs are also prone to errors; however, well-known resources (DMSO 2000a, NATO 2007, ITOP 2004, REVVA 2005) do not provide any guidance specific to UML-based CM verification. In the following section we review a wider category of research related to UML model verification.

15.3 Verification of UML-Based Conceptual Models

In general, formal and informal approaches complement one another, in addressing V&V challenges. However, in order to talk about verification, first, the desired properties should be defined. In the following section, we present the types of desirable properties for UML models.

15.3.1 Desirable Properties for UML-Based Models

Like any other language, UML has its syntax and semantics (UML Superstructure 2005). Syntactic correctness or well-formedness rules of a UML model are specified in the abstract syntax through metamodels or by object constraints language (OCL) constraints. For example, the properties such as (1) "every class should have a unique name" or (2) "an initial node in an activity diagram has at most one outgoing flow" are desired syntactic properties for a UML model. On the other hand, some of the semantics for UML elements are described informally in natural language in the specification. A simple example of semantic property is, "All generalization hierarchies must be acyclic." However, the specification is quite voluminous and there is not a systematic treatment of semantic properties.

According to Sourrouille and Caplat (2002), there are five levels of properties a UML model must possess: The first level defines the semantics of the modeling primitives of the UML. Well-formedness rules that express invariant property of metaclass instances are of this level. These constraints link attributes and associations of the metamodel. At the second level are constraints added by the extension mechanism of UML when UML is extended for a specific domain. These constraints must not conflict with other level constraints. Third level properties ensure conformance to modeling norms and standards. They are used to make models more precise, developed with preferred modeling practices and style. For example, Berenbach (2004) describes heuristics and processes for creating semantically correct models that are presented for analysis and design phases. Examples are (1) there will be at least one message on a defining sequence diagram with the same name as each included use case since a set of sequences diagrams are represented by a use case, and (2) Use an activity diagram to show all possible scenarios associated with a use case. The fourth level contains properties specified by the modeler for the specific model in development. The last two are implementation and coding constraints. These are to be considered as a part of the simulation framework rather than conceptual models.

One can also distinguish between static and dynamic properties of a UML model (Sourrouille and Caplat 2003). The most interesting properties of the dynamic type in the literature (Berardi et al. 2005) are based on the semantics of class diagrams. Similar properties are used by Queralt and Teniente (2006) however they add OCL constraints in class diagrams.

The following are some examples of properties of this type:

1. *Consistency of the class diagram:* A class diagram is *consistent* if its classes can be populated without violating any of the constraints in the diagram.
2. *Class and relation equivalence:* Two classes are *equivalent or redundant* if they denote the same set of instances whenever the constraints imposed by the class diagram are satisfied.

Some of semantic properties are applicable only for specific kinds of diagrams. Csertan et al. (2002), for example, verify general properties defined for state diagrams. In their study, properties defined by Levenson (1995) are used. Among the defined properties are (1) "All variables must be initialized," and (2) "All states must be reachable."

Engels et al. (2001) mention horizontal and vertical UML consistency properties. They acknowledge that horizontal consistency properties are desired and may be a means to reduce contradictions that might exist due to overlapping information residing in different views of the same model. An example of a property related to horizontal consistency is: (1) "Each class with object states must be represented with a state-chart diagram." They also mention about vertical consistency properties used to reduce inconsistencies that may exist among different abstraction levels. An example for this type of property is: (2) "The set of states of an object defined by a parent class must be a subset of the set of states of an object of the child class." Some research studies (Kurzniarz et al. 2002, Kurzniarz et al. 2003, Van der Straten et al. 2003) havc formally defined these kinds of properties.

Ambler (2005) lists a collection of conventions and guidelines for creating effective UML diagrams and defines a set of rules for developing high-quality UML diagrams. In total, 308 guidelines are given with descriptions and reasoning behind each of them. It is argued that, applying these guidelines will result in an increased model quality. However, inter-view properties are not considered at all. Some examples of properties are (1) "Model a dependency when the relationship is transitory in a structural diagram," (2) "Role names should be indicated on recursive relationships," and (3) "Each edge leaving a decision node should have a guard." A similar approach is used by SD Metrics tool (2007), which checks adherence to some UML design rules. Rules extend from well-formedness rules of UML to object-oriented heuristics collected from the literature. Most of the rules are simple syntactic rules. Some examples of errors detected are (1) the class is not used anywhere, (2) use of multiple inheritance class has more than one parent, and (3) the control flow has no source or target node, or both.

On the other hand, a perspective-based reading method for UML design inspection, so-called object-oriented reading techniques, have been presented by Travassos et al. (2002). Examples of properties provided are (1) there must be an association on the class diagram between the two classes between which the message is sent, and (2) for the classes used in the sequence diagram, the behaviors and attributes specified for them in the class diagram should make sense (*sic* Travassos et al. 2002). Another informal approach is suggested by Unhelkar (2005). Quality properties within and among each diagram type have been described along with checklists for UML quality assurance. Although conceptual modeling (CIM, Computation Independent Model) is considered separately and verification and validation checklists in different categories such as aesthetics, syntax and semantics are provided, most of the checklist items are related to validation and completeness. Examples of the properties are

(1) the notation for fork and join nodes should be correctly used to represent multithreads, and (2) the aggregations should represent a genuine "has a" relationship.

This section has briefly described different type of properties by giving examples from various studies in the literature. It is clear that each of these studies consider only certain type of properties and there is a lack of agreement on a set of desirable properties for quality UML CMs.

15.3.2 Formal Techniques for UML CM Verification

There are many studies that rely on the transformation of UML models to a formal language (Amalio and Polack 2003) for checking desirable properties in the target formalism. This type of research work on verification of UML models either emphasizes the structural perspective or behavioral perspective. The following section summarizes the research based on these two perspectives.

15.3.2.1 Approaches with Structural Emphasis

According to Mota et al. (2004), First Order Logic (FOL) is quite suitable for representing UML class diagrams for consistency verification purposes because, lately XMI (Meta Data Interchange XML [eXtensable Markup Language]) is being used for model exchange. And any valid XML description may be associated to a DOM (Document Object Model). As DOM descriptions are easily mapped into FOL expressions, all modern UML-based case tools that export XMI can be used for this purpose.

The example class diagram in Figure 15.2 is represented in FOL assertions in Figure 15.3. However, this is a syntactically correct (well-formed) but inconsistent class diagram. Intuitively, "mobile launcher" and "fixed

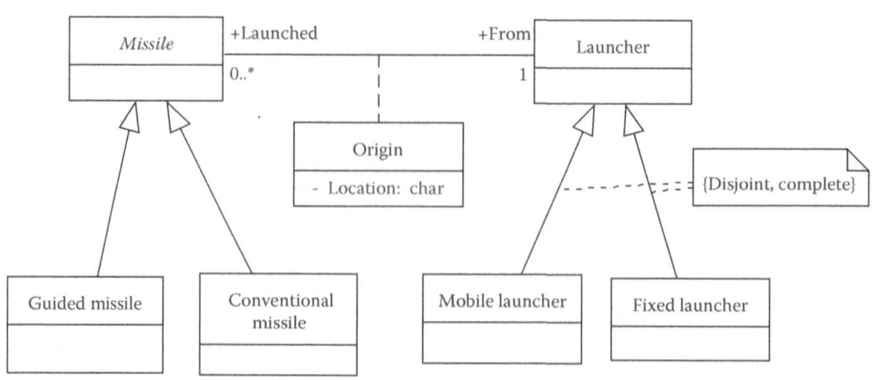

FIGURE 15.2
Semantically incorrect UML class diagram example.

1) $\forall$ x. Mobile launcher(x) $\rightarrow$ launcher(x)
2) $\forall$ x. Mobile launcher(x) $\rightarrow \neg$ fixed launcher(x)
3) $\forall$ x. Fixed launcher(x) $\rightarrow$ launcher(x)
4) $\forall$ x. Mobile launcher(x) $\rightarrow$ fixed launcher(x)
5) $\forall$ x. Launcher(x) $\rightarrow$ mobile launcher(x) $\vee$ fixed launcher(x)

FIGURE 15.3
A part of FOL representation of UML class diagram shown in Figure 15.2.

launcher" classes are disjoint, i.e., they can not have common instances as imposed by the generalization relation. But formally, because of assertions 2 and 4, the knowledge base in Figure 15.3 becomes inconsistent.

It is in general possible to translate FOL statements to an input language of an inference engine such as Prolog to check incrementally the consistency conditions. However, the general decision problem of validity in FOL is undecidable. In order to overcome this, a fragment of FOL, called Description Logics can be used for representing CMs. As an example, Berardi, Calvanese, and De Giacomo (2005) and Van der Straten et al. (2003) rely on the transformation of UML models into description logics. As opposed to FOL, subsets of description logics, which can be used for semantic consistency of only a restricted subset of class diagrams, have decidable inference mechanisms. By exploiting the services of description logic inference engines for example (ICOM 2000), various kinds of checks for properties can be performed. Among these are the properties such as *consistency of a class diagram*.

Inference engine Sherlock, for example (Caplat 2006), linked to a UML case tool is used. In this work, models are built using a UML-modeling tool with tags and constraints. As a lighter alternative, instead of first describing the MOF (Meta Object Facility), they have chosen to describe the UML metamodel directly in the inference engines language. Next, models are expressed in terms of this metamodel. UML metamodel and generic rules are added. Finally, the UML model is loaded and checked.

Dupey (2000) has proposed to generate formal specifications in the Z language with proof obligations from UML diagrams. This is done automatically with the RoZ tool. UML notations and formal annotations reside together: the class diagram provides the structure of Z formal skeleton while details are expressed in forms attached to the diagram. Either OCL or Z-Eves constraints are used. Then the Z-Eves theorem prover is used to validate a given diagram.

Similarly, Marcano and Levy (2002) describe an approach for analysis and verification of UML/OCL models using B formal specifications. In this work, a systematic translation of UML class diagrams and OCL constraints of a system into a B formal specification is given. They propose to manipulate a UML/OCL model and its associated B formal specification in parallel. At first a B specification is derived from UML class diagrams. Then, OCL constraints of the model are automatically translated into B expressions. Two types of constraints are taken into account: invariants specifying the static properties, and pre-/post-conditions of operations specifying the dynamic properties. The objective is to enable the use of automated proof tools available for B specifications in order to analyze and verify the UML/OCL model of a system.

Andre, Romanczuk, and Vasconcelos (2000) have presented a translation of UML class diagram into algebraic specification in order to check consistency. They aim to discover inconsistent multiplicities in a class diagram and deal with important concepts of UML class diagrams: class, attribute, association, generalization, association constraints and inheritance. The theorem prover used discovers some of the inconsistencies automatically; others require the intervention of the user.

15.3.2.2 Approaches with Behavioral Emphasis

Works under this category focus on property checking in mostly behavioral diagrams such as activity, state-chart, and sequence diagrams. For verification of behavioral properties, first a suitable formal verification formalism (e.g., a Petri Net) has to be chosen capable of verifying the aspects associated to the property. For example, for the property of deadlock freedom, the formalism has to support the aspects of concurrency, communication, and interaction of processes. A UML model must first be translated into such a specification language.

There are various types of formalisms used in different studies that deal with verification of behavioral properties in activity and state-chart diagrams. For example, Eshuis and Wieringua (2004) describe a tool for verification of workflow models specified in UML activity diagrams. The tool, based on a formal semantics, translates an activity diagram into an input format for a model checker. Also, techniques are used to reduce an infinite state space to a finite one. With the model checker, any propositional property can be checked against the input model. If a requirement fails to hold, an error trace is returned by the model checker. They illustrate the whole approach with a few example verifications.

In Chang et al. (2005) and in Zhao et al. (2004), for example, UML models are translated to Petri Nets for analyzing the behavioral aspects. In the former study, the goal is to use the graphic interface and the mathematical analysis methods of the Petri Net to verify the logic correctness of the flow control mechanism. In the latter study, deadlock, liveness, boundedness, and reachability properties are verified using Petri Net properties defined by Murata (1989).

Apart from approaches using formal environments, algorithmic approaches have also been proposed. (Litvak 2003) describes an algorithm to a check consistency between UML Sequence and State diagrams. The algorithm also handles complex state diagrams such as those that include forks, joins, and concurrent composite states. They have implemented BVUML, a tool that assists in automating the validation process and consistency check algorithm.

Recently, Gagnon et al. (2008) presented a framework supporting formal verification of concurrent UML models using the Maude language. In spite of its relatively limited scope of applicability, the interesting aspect of this research is that both static and dynamic features of concurrent object-oriented systems depicted in UML class, state and communication diagrams are considered, unlike the majority of similar studies that adopt a single perspective.

15.3.3 Tool Support for UML-Based Conceptual Model Verification

UML tools that can be used for property checking of UML-based CMs are available. However, many of them are based on syntax (Rational 2004) and some of the well-formedness rules of static semantics (Argo 2002, OCLE 2005, Poseidon 2006). Basic consistency checks for UML can be done with CASE tools, which are becoming more and more sophisticated (Egyed 2006). In these conventional case tools, however, properties of a behavioral nature such as the absence of deadlocks and livelocks can not be checked.

For example, Argo UML tool has many well-formedness rules implemented. Furthermore, the tool performs these checks on the fly and categorizes them under three priorities. However, many of the checked properties are based on well-formedness rules of static semantics and heuristics of object-oriented development specifically for producing JAVA code.

MDSD (Model Driven Software Development) tools have been developed to support metamodeling. These include graph-transformation-based editors like DiaGen (Minas et al. 2003), ATOM (De Lara et al. 2002), and Meta Edit (2007), which generate domain-specific editors from language specifications based on graph transformation. However, graph-transformation-based editors are usually purely syntax directed, i.e., each editing operation yields a syntactically correct diagram (Taentzer 2003). Meta Edit for example is a commercial case tool for domain-specific software development. It allows users to define both basic rules and checking rules depending on the graph type. By the help of this tool, the developer can define the modeling elements, the types of graphs and the relations between modeling elements, so the metamodel rules are enforced while developing the model. These rules depend on the graph type and can vary between graphs. The model checker is a powerful tool for enforcing metamodel rules. So by using Meta Edit (2007) for modeling, various verification activities can easily be performed and also the injection of various kinds of defects can be prevented. Hence, when the domain rules are mostly static the tool may be helpful for verification. Other environments such as Open Architecture-ware (2007) and GME (2006) can be used to check properties related to syntax and simple consistency rules of the domain-specific notation.

Lilius and Paltor (1999) for example developed vUML, a tool for automatically verifying UML models. UML models are translated into the Process Meta Language (PROMELA) and model-checked with the SPIN model checker. The behavior of the objects is described using UML statechart diagrams. The user of the vUML tool neither needs the know how to use SPIN nor PROMELA. If the verification of the model fails, a counterexample described in UML sequence diagrams is generated. The vUML tool can check that a UML model is free of deadlocks and livelocks as well as that all the invariants are preserved. In general, the translation employed is not trivial.

Other tools for verification exist and each one here implements a particular kind of semantic property checking (Statemate-Magnum 2007, Tabu 2004, Eishuis and Weringua 2004, Schinz et al. 2004), adopting a particular formalism. Hence, complexity and semantic correspondence problems remain to be tackled.

15.3.4 Inspections and Reviews for UML Model Verification

Inspections and reviews are considered a fundamental way to achieve software quality. Fagan (1976) is one of the pioneers in this field. He defines an *inspection* as "formal, efficient, and economical method of finding errors in design and code." A *defect* is defined as "any condition that causes a malfunction or that precludes the attainment of expected or previously specified results." It is argued that inspections have evolved into one of the most cost-effective methods for early defect detection and removal (Laitenberger and DeBaud 2000). It has been claimed that inspections can lead to the detection and correction of between 50 and 90% of software defects (Gilb and Graham 1993).

During defect detection phase of inspections, inspectors read the software document to determine whether quality requirements, such as correctness, consistency, testability, or maintainability, have been fulfilled. The defect detection and defect collection activities can be performed either by inspectors individually or in a group meeting. Since findings reveal that the effect of inspection meetings is low (Johnson and Tjahjono 1998), defect detection is considered as an individual activity (Basili 1996).

The defect detection activity can be conducted using three types of methods. The most widely used defect detection method is ad hoc review, which provides no explicit support to the inspectors. The inspectors have to decide on how to proceed, or what specifically to look for during the activity. Hence, the results of the review activity in terms of potential defects or issues are fully dependent on inspectors experience and expertise. Checklist-based reading on the other hand (Gilb and Graham 1993) provides some guidance about what to look for in a review, but it does not describe how to perform the required checks. Third, due to effectiveness problems in ad hoc and checklist methods, Porter et al. (1995) developed a scenario-based method to offer more procedural support in which scenarios are derived from defect types. A scenario describes how to find the required information to reach a possible defect.

Few studies have been published in the area of inspection of UML models. Travassos et al. (2002) describe a family of software reading techniques for the purpose of defect detection of high-level object-oriented designs represented using UML diagrams. This method is a type of perspective-based reading for UML design inspection and can be considered as following the line of techniques discussed by Basili et al. (1996). Object-Oriented Reading Techniques consist of seven different techniques that support the reading of different design diagrams. This method is composed of two basic phases. In the horizontal reading phase, UML design artifacts such as class, sequence and state chart diagrams are verified for mainly interdiagram consistency. In the vertical reading, design artifacts are compared with requirements artifacts such as use case description for design validation. Hence most of the properties checked in these studies are related to validation and the main artifact considered is software design rather than a conceptual model.

An important book on UML quality assurance (Unhelkar 2005) describes quality properties within and among each diagram type along with checklists for UML quality assurance. The foundation for quality properties are set by the discussion on the nature and creation of UML models. This is followed by a demonstration of how to apply verification and validation checks to these models with three perspectives: syntactical correctness, semantic meaningfulness, and aesthetic symmetry. The quality assurance is carried out within three distinct but related modeling spaces: (1) model of problem space (CIM in MDA terms), (2) model of solution space (Platform independent model), and (3) Model of background space (Platform-specific model). This makes it easier for the inspectors to focus on the appropriate diagrams and quality checks corresponding to their modeling space. Although CIM (Computation Independent Model) is considered separately and verification and validation checklists in different categories such as aesthetics, syntax, and semantic are provided, most of the checklist items are related to completeness. Items related to verification are mostly syntax, static semantic, or simple cross-diagram dependency checks.

15.4 An Inspection Approach for Conceptual Models in a Domain-Specific Notation

15.4.1 Need for a Systematic Inspection Method

Generally, formal techniques for verification, while being very effective are often very costly due to their complexity and sometimes due to the size of the model under consideration (Garth et al. 2002). And many of the studies based on transformation to formal languages are restricted to one or two types of diagrams. Hence, only certain dynamic aspects are analyzed with Petri Nets for example. Moreover, the formalism also restricts the type of properties to be checked.

On the other hand, when a UML-based notation is used for conceptual modeling, mapping of UML diagrams into a formal notation yields the semantic correspondence issue. Besides, most of the formal techniques assume at least a predefined completeness in models. However conceptual models, unlike design models, are developed in a sketchy manner at the initial phase of the requirements elicitations and may be incomplete in various ways that are difficult to determine in advance.

Furthermore, as conceptual models are in general not executable, it is not easy to use dynamic techniques either. Conceptual models are used primarily as a means of communication, and the term *conceptual* inherently implies tractable abstraction levels and size. Consequently, techniques such as walkthroughs and inspections can be used rigorously for assuring conceptual model quality. It may also be cost effective to integrate the verification tasks with the validation tasks that require human interpretation.

Figure 15.4 summarizes the advantages and disadvantages of both the formal approaches and informal approaches for CM verification. An inspection approach may be preferred to a formal approach due to various advantages: First, informal techniques are easy to use and understand.

Their application is straightforward. As checklists and guidelines are the main sources, they can be performed without any training in mathematical software engineering. Inspections may be very effective if applied

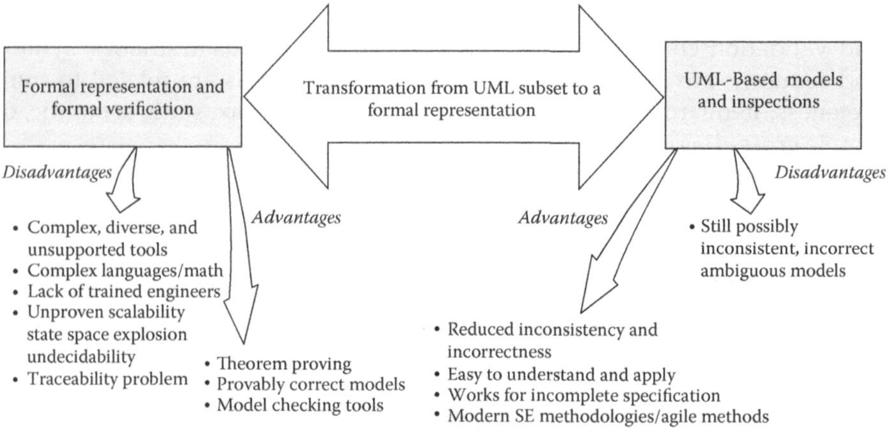

FIGURE 15.4
Comparison of inspections to formal verification for UML-based CM.

rigorously and with structure and they are relatively less costly and they can be used at any phase of the development process. Hence, a systematic and holistic approach, rather than using formalisms, may provide significant practical results. In the following section, an appropriate inspection process is described to assure the quality of conceptual models in a notation derived from UML.

15.4.2 Desirable Properties for UML-Based KAMA Notation

In this section, we describe the properties identified for a domain-specific notation for conceptual modeling, namely KAMA (Karagöz and Demirörs 2007) (see Chapter 7 of this book for details about the notation). KAMA models are independent of the simulation environment, infrastructure and implementation. Since most syntactic properties of KAMA are reused from UML and can be eliminated through UML CASE tools, only desirable semantic properties are identified (Tanrıöver 2008) and presented in this chapter.

Briefly, class consistency, multiplicity consistency, relation and class liveness, consistency of inherited constraints, lack of transitive cycles, and lack of redundant relations are identified as desirable properties of the structural views. For example, "A class is *consistent*, if it can be populated without violating any of the constraints in the diagram" (Berardi, Calvanese, and De Giacomo 2005). Liveness of tasks, deadlock freedom in task flows, lack of dangling tasks in task flows, and completeness and consistency of guard conditions are identified as desirable properties of the behavioral views.

Various interdiagram consistency issues have been considered in the literature (e.g., Ambler 2005, Briand, Labische, and O'Sullivan 2003, SD Metrics 2007, Killand and Borretzen 2001, Ohnishi, 2002). Interdiagram properties identified in the context of KAMA were mission versus task flow dependency, ontology versus subontology dependency, task flow versus sub-task flow dependency, mission and task flow refinement consistency, refinement consistency of entities in task flow and entity ontology views, consistency of actor in mission space and organization views, and consistency of attributes in entity state and entity ontology views. Note that these properties are related to the most commonly used modeling elements and views.

15.4.3 An Inspection Process

Inspection tasks for checking desirable properties are presented in this section in the order of execution. It should be noted that the inspection tasks, in general, may not identify all possible violations of an identified property. Properties of the structural perspective for example, can be only partly checked by the help of deficiency patterns presented in the following subsection. However, we think that by the help of these inspection tasks, it will be possible to identify a set of semantic issues that would otherwise be left undetected. Many issues identified in the inspection process have to be checked for validity in the context of the specific conceptual model being developed.

15.4.3.1 Intradiagram Inspection

Structural diagrams inspection phase: In the structural diagram inspection phase, contradictions and redundancies in diagrams derived from UML class diagrams are checked. For structural diagram checks, a set of deficiency patterns have been identified. All patterns have been validated with two modeling experts. Our aim was to provide examples and simply guide the inspector toward the types of structural deficiencies that we would like to identify. We have formulated patterns including two fundamental relations namely *association* and *generalization*. However, similar patterns may also be used for more specific, derived relations used in domain-specific notations. For example, "request" or "transacts with" relations used in business domains are subclasses of the association relation.

Our observation is that inconsistencies may occur because a view of the model may be represented in multiple diagrams connected with extension points or simply each diagram may represent the viewpoint of a stakeholder participating in the conceptual modeling process. Most contemporary CASE tools allow a given model element to appear in multiple diagrams. When the same model element is used in more than one diagram of the same view at the same abstraction level, contradictions and redundancies may be introduced and remain undetected. Specifically, we observed that transitivity and asymmetry of derived relations in the domain-specific notation may cause redundancy or contradiction, because in domain-specific modeling the same type of relations may be used many times. For example, if we model symmetry by using an asymmetric relation (e.g., $A \rightarrow B$ and $B \rightarrow A$) this may be the indication of a contradiction, and if we explicitly assert a relation that is already implied (e.g., $A \rightarrow B \rightarrow C$ together with $A \rightarrow C$), this will result in a redundancy.

The second observation is that conceptual models are developed in a sketchy manner, at a high level of abstraction early in the development life cycle. Hence, only basic modeling constructs such as classes and various relationships are used in the models at this phase. Furthermore, usually domain-specific notations allow only a limited number of types of relations and model elements in each diagram type. For this reason, a limited number of deficiency patterns can be helpful.

Figure 15.5 presents patterns mostly based on generalization type of relations and Figure 15.6 depicts patterns mostly based on association type of relations. Structural diagrams are entity ontology, command hierarchy,

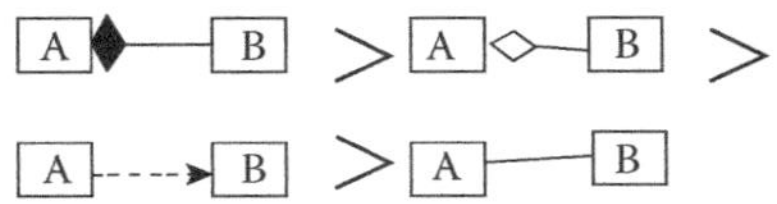	*1.1 Strength of association type of relations:* It is observed that there is a partial order between strength of association type of relations. The re-occurrence of a weaker kind of association between two classes can be considered as a redundancy to be reviewed. For example; given that A is composed of B, A is associated to B may be signaled as a redundancy warning to be reviewed.
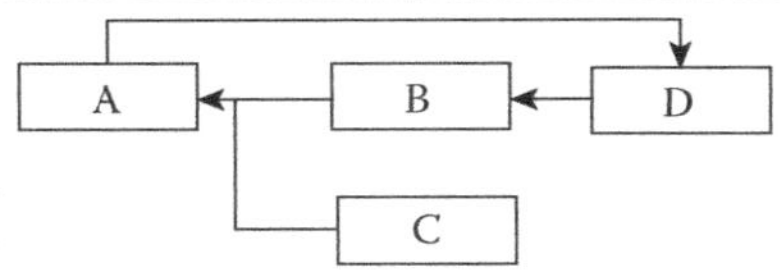	*1.2 Circular transitive relation:* This deficiency is the well known UML wff-rule which corresponds to the circular inheritance problem. This occurs if a class transitively inherits from itself, in this pattern, with the generalization relation between A and D and D and A. However, in addition to circular inheritance, we identify any circularity formed by any transitive relations (such as aggregation, composition, and dependency) as a semantic issue to be reviewed.
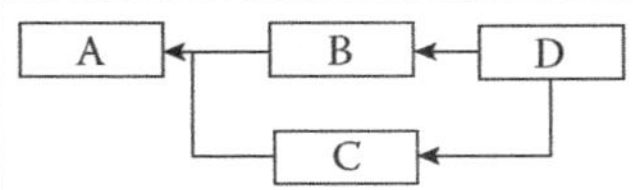	*1.3 Lattices-multiple inheritances:* Although permissible in UML and allowed in C++ like object-oriented interpretations, handling of this pattern depends on how the model is interpreted in the target domain. Note that, any relation which is a subtype of generalization relation (such as <extend> and <include> relations) can form this pattern. This pattern may be formed by aggregation, composition, or dependency relations. Hence, lattices and multiple inheritances should be reviewed.
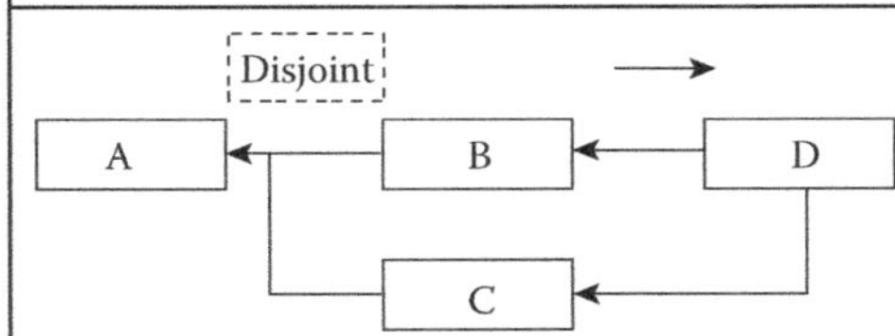	*1.4 Disjoint inheritance:* This pattern is based on the possible (overlapping, disjoint, complete, incomplete) constraints that can be applied to a set of generalization relations. This example shows two cases of contradictions. The first one is formed since A and C are disjoint, they can not have same instances but on the other hand, since D is inherited from both B and C, there should be at least some common instances. So this forms a contradiction. The second one is formed by inheriting B from D rather than D from B. In this case, since D is a type of C and C is a type of A, B becomes a type of A by transitivity. However, by disjointness constraint B cannot have instances that C has, hence this again forms a contradiction. Such situations should be reviewed.

FIGURE 15.5
Patterns developed based on strength of relations, generalization, and transitivity.

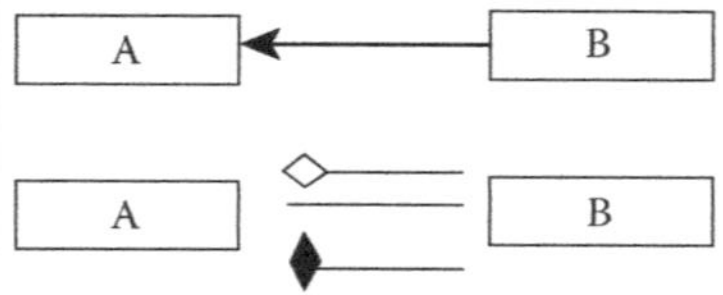	*1.5 Generalization with aggregation:* This pattern is based on the observation that utilization of different types of relations for same concepts may be a source of redundancy or contradiction. In this case, if there is a generalization relation between A and B and if an aggregation or a composition from A to B is defined, this forms a very rare pattern. E.g., a chicken-and-egg kind of ontology. So, this should be identified as a redundancy warning to be reviewed.
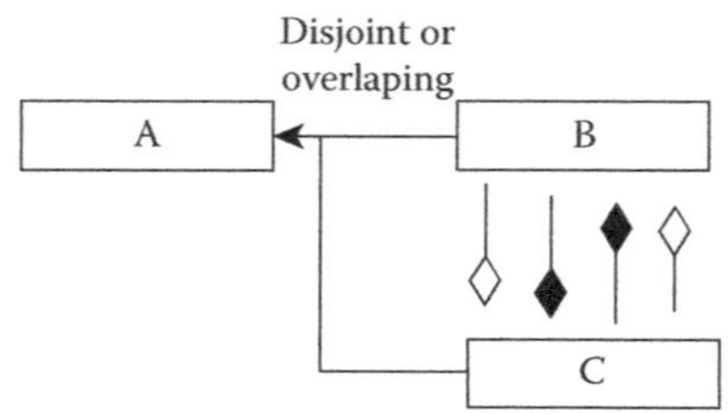	*1.6 Disjoint or overlapping with aggregation:* This pattern is based on disjointness and overlapping constraints which can be defined on a set of generalization relations. There are two main cases. The first case occurs when classes B and C are disjoint, but there is a composition or aggregation relation between them. This forms a possible contradiction because this is equivalent to saying that B and C have no common instances but B is composed of C. Remark that if B is composed of C and only C, this pattern will result in a contradiction. If there were other classes that B is composed of and which are not inherited from A, this pattern would not cause a contradiction. The second case occurs when class B and C are overlapping, that is they have common instances and there is a composition relation between B and C. Hence, overlapping constraint becomes a redundancy. The first case should be reviewed.

FIGURE 15.5 (CONTINUED)
Patterns developed based on strength of relations, generalization, and transitivity.

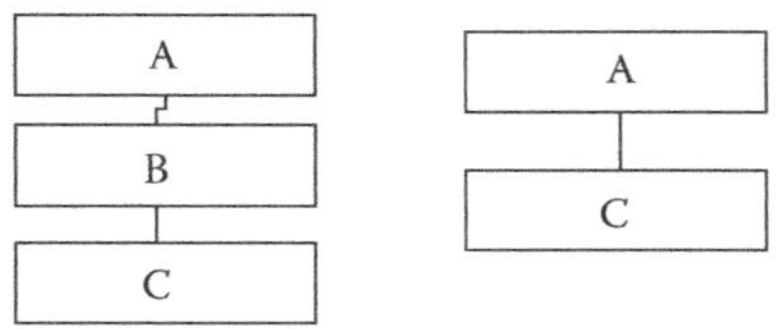	*2.1 Redundancy by transitivity:* This pattern is based on the transitivity property of dependency, composition, aggregation and generalization relations. In fact, many of the relations in UML can be considered in this category since they are a sub-type of generalization relation. In this case since a class is transitively related to another one, there is no need to specify a direct relation between two. Remark that the strength of relation concept can be used to identify transitive redundancies by considering all weaker forms of relations.
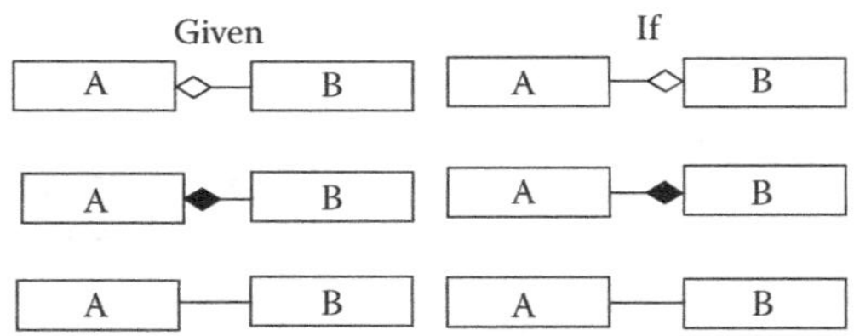	*2.2 Contradiction by asymmetry:* This pattern is based on the asymmetry property of directed relations, such as composition, aggregation and generalization. For classes A and B, all the classes of the inheritance hierarchy of A and B should be checked for identification of this pattern. In this case, the utilization of a relation with direction between classes in opposite directions may be signaled as a possible contradiction and should be reviewed.
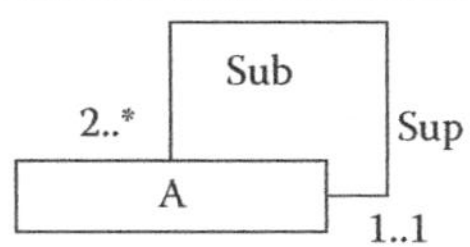	*2.3 Recursive association multiplicity:* This pattern is based on possible contradictions when recursive relations are used. Since the lowest multiplicity of one end is greater than the others' highest multiplicity, this forms a contradiction.
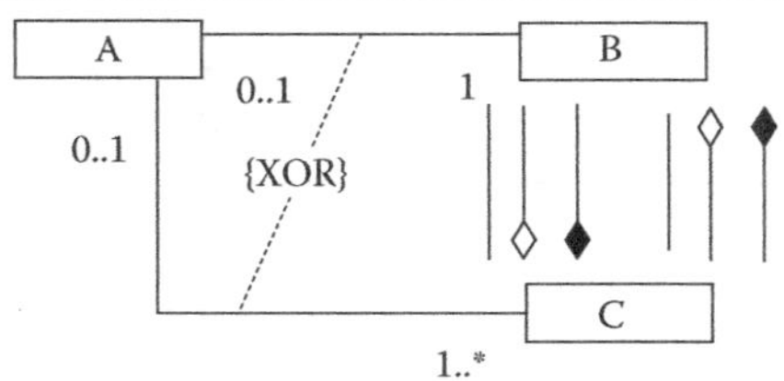	*2.4 Association constraint (XOR) with association:* This pattern is developed considering the possible constraints that can be used on a set of association relations. These are XOR, NAND, or similar constraints. If any association relation between B and C is defined, then an immediate contradiction arises because B and C can never exist at the same time because of the XOR constraint. Hence, these situations should be identified as contradictions.

FIGURE 15.6
Patterns developed based on asymmetry and deep inheritance.

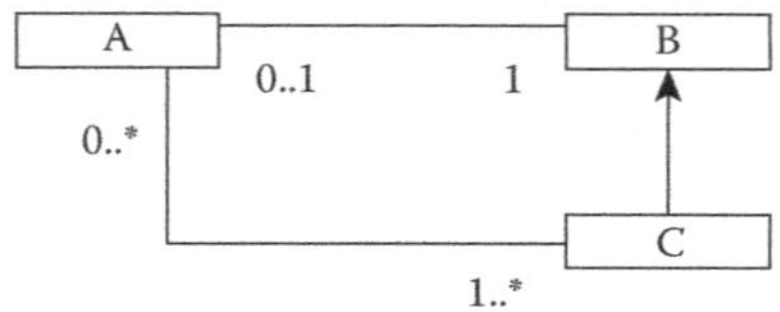	2.5 *Inherited constraint:* This pattern is developed based on the concept of deep inheritance. In this case, it is assumed that the multiplicities should be inherited from the base class. Hence, the multiplicity range of the base class should be wider than that of the inherited class. If this kind of structure is encountered, the lowest(l) and highest(h) multiplicities of B and C must be checked. If "$l(B_c) \leq l(C_b)$ and $h(B_c) \geq h(C_b)$ and $l(A_b) \leq l(A_c)$ and $h(A_b) \geq h(A_c)$" ≠ TRUE than multiplicities of B and C form a contradiction.
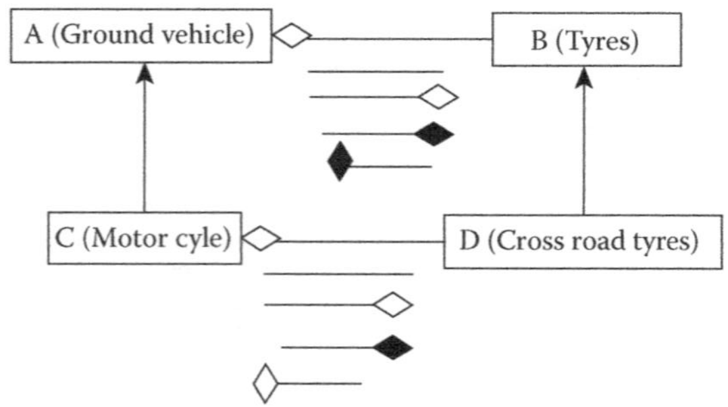	2.6 *Inherited constraints II:* This pattern is similar to "inherited constraint I." If this kind of structure is encountered, the lowest and highest multiplicities of A and C and B and D must be checked. If $l(B_a)$ = < $l(D_c)$ and $h(B_a)$ => $h(D_c)$ and $l(A_b)$ =< $l(C_d)$ and $h(A_b)$ => $h(A_c)$ <> TRUE, multiplicities of B and D or A and C form a contradiction. On the other hand, remark that given A is composed of B, if D is composed of C than this is identified as a contradiction because of contradiction by asymmetry pattern.

FIGURE 15.6 (CONTINUED)
Patterns developed based on asymmetry and deep inheritance.

TABLE 15.1
Structural Diagram Inspection Phase

1. Check syntactical errors such as omissions, missing attributes, and name clashes, based on the syntactic rules.
2. Look for deficiency patterns in the class model.
2.1 Look for a match with each pattern for a contradiction or a redundancy. Consider the transitive closure of the relations for pattern matching.
2.2 Depending on the matched pattern validate the issue with the SME.
3. Identify complex structures (structures with central classes participating in more than one relation and/or relationship type) not considered in task 2 by using the semantics of the modeling elements forming the structure.

TABLE 15.2
Mission Space Diagram Inspection

1. Check syntactic errors such as duplicate names, dangling missions without actors.
2. Check for patterns 1.2 and 2.2 to identify contradictions.
3. Check the < inclusion > and < extends > relations for semantically correct usage.
3.1 Trace and check the relation to the refining task flow diagram of the use case to make sure they are properly used.

entity-relation and mission space diagrams in KAMA notation. For checking these diagrams, the inspector is presented with the deficiency patterns and their descriptions to familiarize him with the kind of defects he will be looking for. Table 15.1 summarizes the inspection tasks for structural diagrams.

Mission space diagram inspection phase: In this phase, diagrams derived from use case diagrams, i.e., KAMA mission space diagrams, are inspected. The tasks in Table 15.2 are used for the mission space diagram inspection phase.

Task flow diagram inspection phase: The purpose of this phase is to verify the diagrams derived from UML activity diagrams, i.e., KAMA task flow diagrams (Tanrıöver and Bilgen 2007a, b). The activities in Table 15.3 are defined for the task flow diagram inspection phase.

15.4.3.2 Interdiagram Inspection

In this phase the interdiagram properties are verified (Tanrıöver and Bilgen 2007a, 2007b). For checking the interdiagram properties, we defined inspection tasks presented in Table 15.4. Note that tasks are not exhaustive; the lists may be augmented with newly identified properties pertaining to a specific modeling context.

TABLE 15.3

Task Flow Diagram Inspection Phase Tasks

1. Check for syntactic errors such as dangling nodes, initial nodes with more than one outgoing transitions.
2. Identify decision nodes.
   - 2.1 Check if all flows outgoing from the decision nodes have guards.
   - 2.2 Check the constraints on the guards to make sure that they do not overlap (overlapping such as constraint on one guard is x > = 0 and on the other x = < 0).
   - 2.3 Check if the guards define a complete set (such as x = > 0 and x < 0).
     - 2.3.1 Identify overlapping and incomplete conditions.
3. Identify fork nodes.
   - 3.1 Check if the fork node has only one entrance; if not, make sure that a task flow is not missed before the flow is joined.
   - 3.2 Check if all the flows from the fork node are joined by a (same) join node (nonstructurally joined nodes or fork nodes may indicate concurrency problems).
     - 3.2.1. If not, run the flows coming out of the fork node with UML's activity diagram (Petri Nets–like) control flow semantics.
     - 3.2.2. Identify livelocks and their causes.
4. Identify join nodes.
   - 4.1 Check if join nodes have only one exit transitions.
   - 4.2 If not, it is possible that the join node is placed too early; there is possibility that there is still a need for a parallel flow.
   - 4.3 Trace incoming transitions of the join nodes to make sure that all may eventually be activated.
   - 4.4 If not identify causes of deadlock.
5. If the task flow is complex (includes more than one fork node or composite decision nodes) trace each flow from the start to end.
   - 5.1 Make sure that every task may execute.
   - 5.2 Identify dead tasks.
6. Trace the flows reaching the final nodes.
   - 6.1 Make sure that they do not originate from a fork node.
   - 6.2 If they do, there is a possibility that some activities will terminate abruptly, try to identify such activities.
7. Identify loops by tracing through transitions.
   - 7.1 Run the localized loop with UML's activity diagram (Petri Nets–like) control flow semantics.
   - 7.2 Identify possible livelocks and their causes.
8. Identify activities with <input> and <output> entities.
   - 8.1 Make sure that if tasks use outputs of one another, they also follow the implied sequence in the control flow because a produced entity may be an input for another task, causing the task to never start or to prevent parallel flow.
   - 8.2 Identify deadlocks or redundancy.

TABLE 15.4

Interdiagram Inspection Tasks

1. Trace missions and check if they are modeled in task flow diagrams and vice a versa.
2. Compare ontology diagrams with corresponding subontology diagrams and make sure that there is only one subontology diagram for an entity in the upper ontology diagram.
3. Identify further decomposed tasks in task flow diagrams, make sure there is only one subtask flow diagram for a super task flow node.
4. Identify <inputs>, <outputs>in nonleaf task flow diagrams.
   - 4.1 Trace <inputs>, <outputs> in the next lower task flow diagram.
   - 4.2 Ensure that there is at least one <input> and/or <output> attached to the next lower task flow and identify missing <inputs> and/or <outputs> for the next lower task flow diagram.
5. Identify <input>, <outputs> entities in leaf task flow diagrams.
   - 5.1 Trace <inputs>, <outputs> entities in the task flow in the upper task flow diagram.
   - 5.2 Check if there is at least one <input> and/or <output> attached to the upper task flow and identify missing <inputs> and/or <outputs> in the leaf task flow.
6. Identify extended missions.
   - 6.1 Compare task flow diagrams of the mission with task flow diagram of the extended mission: the extended task flow diagram should be reachable by only extracting model elements from extending diagram.
7. Check each <input> and <output> entity in task flow diagrams, a corresponding entity has to exist in ontology diagrams.
8. Check all the actors in mission space diagrams are defined in organization diagrams.
9. Check if variables used in state chart diagrams are defined attributes of corresponding entity.
10. Check if operations used as transitions in entity state diagrams are defined in the corresponding entity diagram.

15.5 Case Studies

This section describes two case studies conducted to explore the applicability and effectiveness of the inspection approach presented above.

15.5.1 Case Study 1

The first case study was an exploratory study. We had developed an initial inspection process definition. The study aimed to test the applicability and to identify improvement possibilities for the initial version of the process.

15.5.1.1 The Setting

Three roles were identified for actors participating in the case study:

Modelers: Responsible for developing the conceptual model using the KAMA notation.

Inspector: Responsible for performing the inspection of the conceptual model developed.

Software Engineering Experts: Responsible for the defect approval and resolution in the inspection meeting.

Two modelers both experienced in UML modeling and KAMA notation had developed a conceptual model for a typical mission scenario. The conceptual model consisted of one mission space diagram, one command hierarchy diagram, five ontology diagrams, and 46 task flow diagrams at varying levels of structural decomposition with different levels of complexity and included a total of 179 model elements. The model was in its early stage of the CM development process (at the first iteration of three review stages) and was developed in a sketchy manner. For example, the entities did not include operations defined and any entity state diagrams. Hence during the inspection, only a set of the inspection tasks could be performed. Semantic checks with cardinalities for any of the structure diagrams were not necessary because cardinalities were not used. Similarly, the consideration of entity state diagram related properties were also left out of the scope of the inspection.

The conceptual model inspection was conducted in two main phases. Review of the conceptual model had been already performed informally during conceptual model development phases by the two modelers. Our inspection process was performed after this review. The defect detection and reporting was conducted by an inspector. This phase took 20 person hours. After the defect detection phase, an inspection meeting for validating the defects detected was planned. The inspector, modeler and two software engineering experts participated in this six-hour meeting. The outputs of this process were the corrected conceptual model and the verification report. Main sources of evidence and data of inspection were defect detection documentation and minutes of the inspection meeting.

15.5.1.2 Conduct of the Case Study 1

Applying the inspection tasks in Table 15.1, we identified only seven issues because the allowed relationship types in structural diagrams are limited in KAMA notation and the model belonged to an early modeling phase. As an example, a redundancy on the command hierarchy diagram in Figure 15.7 was identified. In command hierarchy diagrams sub/superior relation is a transitive relation derived from generalization metaclass of UML. When we

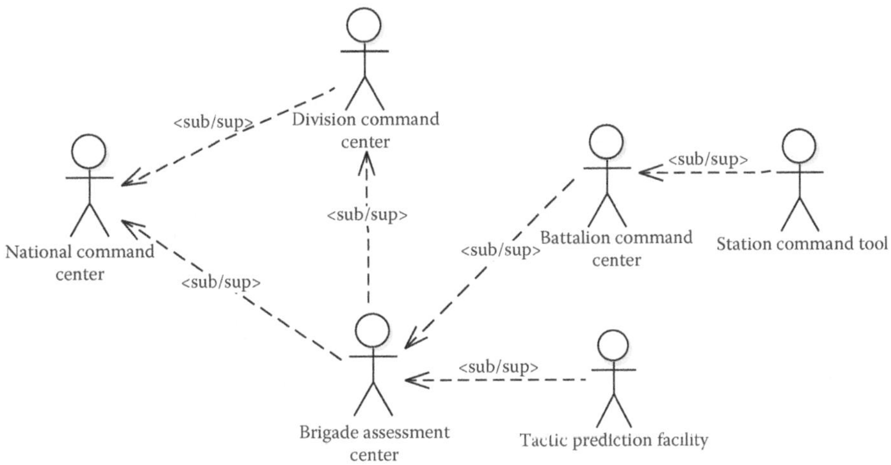

FIGURE 15.7
KAMA command hierarchy diagram with redundancy.

consider the "Brigade Assessment Center," "Division Command Center," and "National Command Center," the sub/super relation between "National Command Center" and "Brigade Assessment Center" forms a semantic redundancy, by "redundancy by transitivity" pattern.

Performing the tasks in Table 15.2, we identified 10 issues, such as missing extended and included missions, etc.

Performing the task flow review activity in Table 15.3, 23 issues were identified. In the "Watch Mission Region" task flow in Figure 15.8, since the entity "Identification/recognition data" can be an input to the task "Locate Allied Forces" only after being produced by "Search the Region" task, the fork node at the start has no effect on the flow. This has been identified by task 8 of Table 15.3. On the other hand, in "Develop Pointer Information" task flow, since the flows coming out of the fork node terminate with a decision node without a merge node, either of the tasks may terminate abruptly leaving dangling flows. This is an issue identified by task 6 of Table 15.3.

Furthermore, the interdiagram inspection tasks in Table 15.4 were performed and 29 issues were identified. For instance, by task 1 of Table 15.4, we identified nine entities used in task flow diagrams but not defined in ontology diagrams. According to refinement consistency, a subtask flow should show main task flow entities in higher or at least equivalent level of detail. As an example, consider the models in Figure 15.8. "Develop Communication Information" task flow is a sub-task flow of "Develop Pointer Information." However, although the output entity "Communication Intelligence Data" exists in "Develop Pointer Information" main task flow, associated or refining entities are not shown at all in "Compose Communication Intelligence" sub-task flow. By task 4 of Table 15.4, this has been identified as incompleteness.

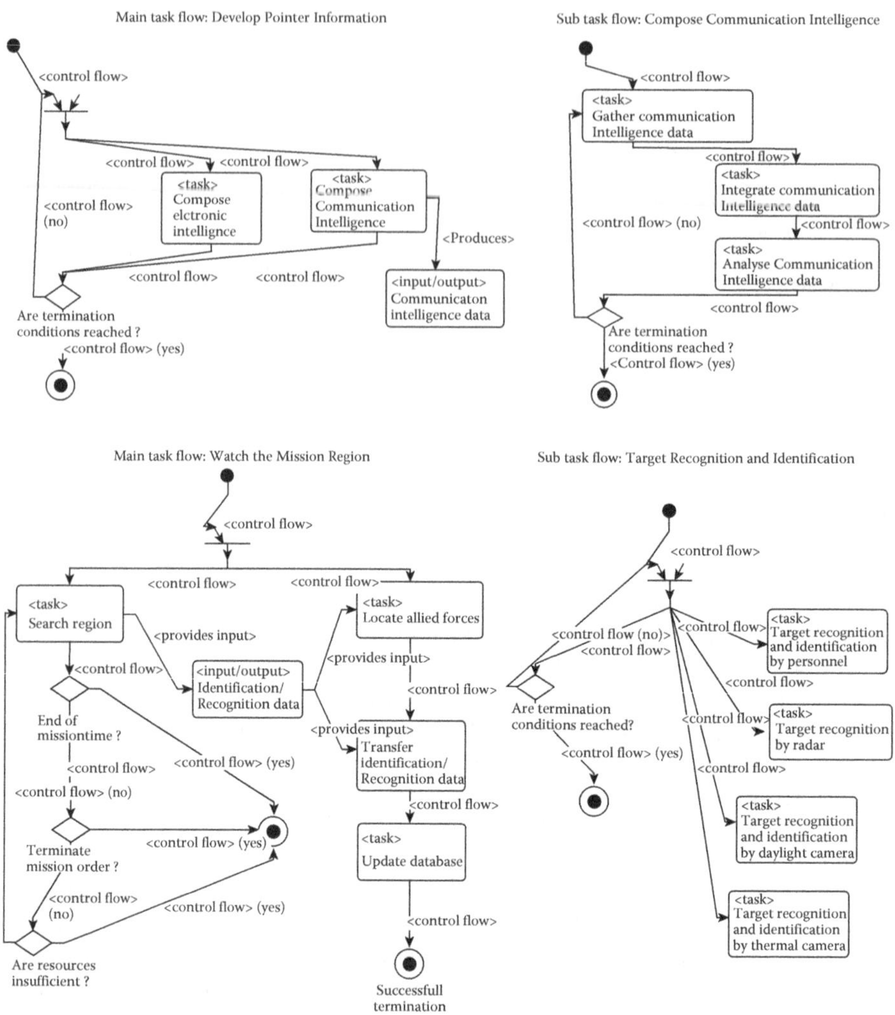

FIGURE 15.8
KAMA task flow diagram examples.

15.5.1.3 Discussion and Findings of Case Study 1

The defect detection phase of the inspection performed resulted in eighty-five identified issues. Ten of them were categorized as belonging to the major, seven of them to the moderate, and 68 of them to the minor levels of severity. An example of major issues was semantic deadlocks in the task flow diagrams and expert opinion was used for validation of the issues identified in the inspection. For this purpose, an inspection (six hours) meeting was held after the defect detection phase. The inspector, modeler, and two software engineering experts participated in the meeting. They agreed that, although some of the 85 issues signaled minor problems and some of them were not definitive defects, 39 of the identified issues included behavioral defects and were qualified as subtle and not easily detectable in ad hoc reviews. Seventeen of these issues were agreed to be definitive defects and 22 issues were identified as definitive incompleteness. They also agreed that these types of redundancies and contradictions are not easy to detect and deficiency patterns could help the inspectors to detect requirements related issues.

15.5.2 Case Study 2

The purpose of the second case study was to explore and evaluate the effectiveness of the improved inspection process.

15.5.2.1 The Setting

A similar inspection organization as in the first case study was used. However, the conceptual model used in this study was developed in a real life development environment. The model in case study 2 had already been subject to ad hoc review for two days by one UML expert. Also, a review meeting with the participation of six members of the development team was held. Later on, the conceptual model was subjected to a walkthrough that took five days. Four engineers from the conceptual model development team and three from the acquirer organization joined the meetings in this third phase. There were 150 issues identified. The issues identified were related mostly with validation. These were issues with task flow diagrams, incompleteness regarding entities, additional attributes and capabilities to the entities, definition of roles and actors. Our inspection-based verification was applied after all these three review activities were realized.

15.5.2.2 Conduct of the Case Study 2

Before each intradiagram inspection, the validation function of Enterprise Architect v6.5 (2006) (EA v6.5) was executed on each diagram. The tool's standard validation function, which included syntactic, wff, and other checks, signaled no errors. Then, the inspection tasks were performed. During the inspection the model tree browser was used and helped the inspector to manage the browsing (which may sometimes be rather complex) needed for interdiagram verification tasks. The facility of the EA 6.5 tool to view the class hierarchy tree was used to obtain all the lower-level entities transitively based on both aggregation and generalization relations. During the inspection of refinement relations, starting from the highest level, only the first sub level was checked when inspecting a given task flow diagram.

There were a few defects that were detected based on structural deficiency patterns. As an example, Figure 15.9 shows an occurrence of

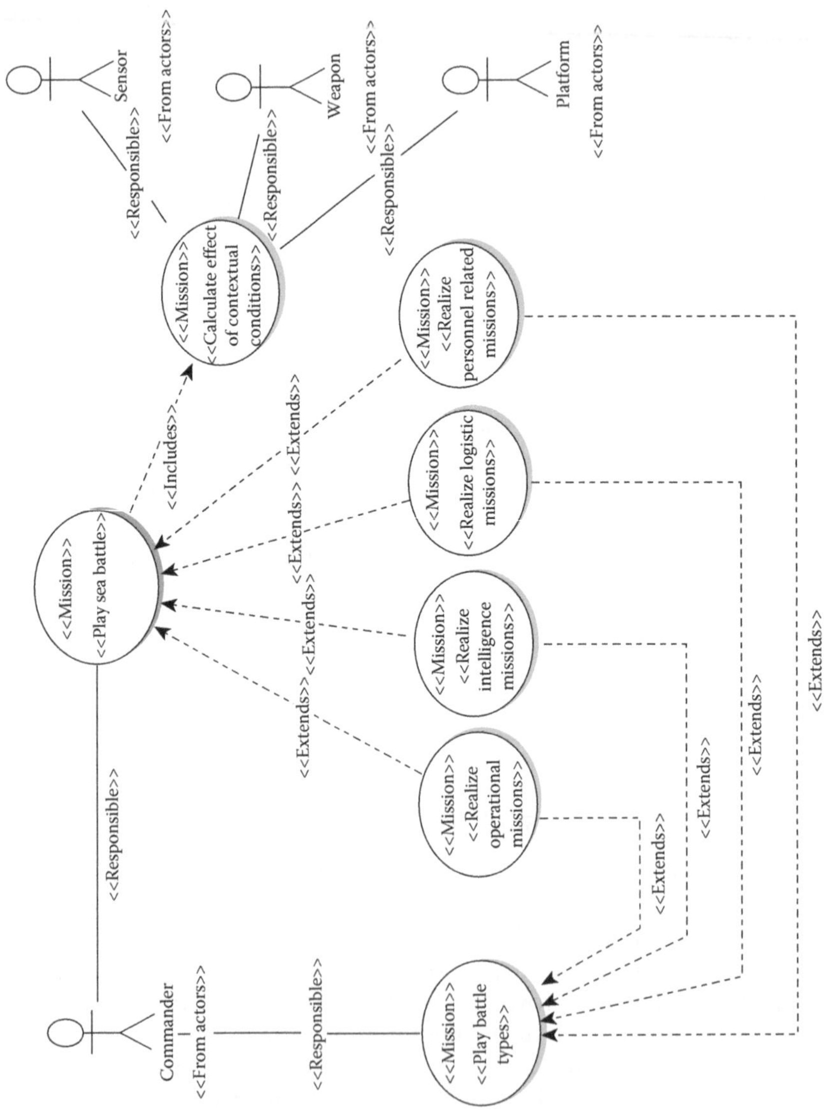

FIGURE 15.9
An occurrence of the multiple inheritance pattern in a mission space diagram.

the multiple inheritance pattern, as "extends" is a generalization type of relation. Note that, although the definition of the patterns is not formally given and use case patterns are not explicitly defined, this pattern could be identified as a structural issue by the inspector. This example shows that with the guidance of patterns initially presented, the inspectors easily identify similar deficiencies that were not previously formally specified.

15.5.2.3 Findings of the Case Study 2

As mentioned earlier, the conceptual model had gone through ad hoc review and walkthroughs conducted by inspectors. Inspectors were experienced in both UML and the domain. During this process 150 semantic and syntactic issues had already been identified and corrected. Even though the conceptual model was corrected and accepted to be valid, by applying our approach later, 58 additional semantic issues were identified. Thirty of these were acknowledged to be nontrivial and to have important implications by the inspectors. It was determined that spending the extra effort on applying the described inspection process after an ad hoc review was clearly worthwhile.

One important observation in the case study was that the model tree browser of the EA 6.5 tool proved to be very helpful for interdiagram verification tasks. The interdiagram inspection tasks were performed not as standalone activities but rather just after finishing the intradiagram inspection for that diagram. Thus, once a perspective is inspected, all the tasks related to that perspective were performed. This slight adaptation of the process has improved the inspection effectiveness, because in this way the inspector does not have to consider the same diagram twice for inter-view and intraview tasks.

15.6 Conclusions and Further Research

In this chapter, we have reviewed the literature on conceptual model validation and verification based on UML and presented an inspection process for conceptual models developed with a UML-based notation. The literature review first showed that, for UML-based models, there is not an agreed set of desirable properties, which may be identified depending on the purpose, development methodology and domain of modeling. Second, formal approaches and existing tools are either partial or cannot be practically used to reveal semantic issues during conceptual model verification.

We have presented an inspection process with tasks for different types of diagrams together with interdiagram properties. The set of properties and tasks can be augmented, depending on the intended purpose of the conceptual model, hence the AC. The set of desirable properties considered in this work have been determined specifically for high-level conceptual models. The inspection tasks for semantic properties are detailed, whereas many of the syntactic errors and simple interdiagram consistency checks can be eliminated through CASE tools. Through case studies, the process was shown to be effective in identifying semantic issues that may not be detected by the contemporary UML CASE tools and other inspection methods.

Tool support for the inspection process has not been considered. A single tool will not be enough to support the verification tasks but rather a set of tools should be identified. In general, environments such as Meta Edit, Open Architecture-ware and GME can be used to check properties related to syntax and simple consistency rules of the domain-specific notation. For task flow inspections, Petri Net analysis tools may be helpful if the view is too complex and critical. We also foresee that for structural view verification tasks ontology analysis tools such as extended entity relationship (EER)-conceptual (Compatangelo and Meise 2002) may be very helpful. However, in conceptual modeling, human comprehension is the essential aim, and tractable abstraction levels and sizes must be goals. Hence, it may be more appropriate and also more cost effective to integrate the verification tasks with the validation tasks that require human interpretation.

Another possible criticism that can be directed at the presented inspection approach is the lack of a risk perspective. To make sure that a conceptual model is fit for purpose in a cost effective way, V&V activities have to be focused on the most important aspects of the conceptual model.

Specifically, the modeling and simulation community has long acknowledged the need for a risk-based V&V process (e.g., Brade 2004, REVVAI/II and GMVVA [Generic Methodology for VV&A]). Work toward assessing and possibly enhancing the presented approach from a risk-driven perspective is outstanding.

We recognize that not all the semantic issues may be revealed, since only a set of common defects patterns are provided. On the other hand, for behavioral diagrams, we only guide the inspector by means of inspection tasks to facilitate identifying defects due to desirable properties. Especially when the number and complexity of diagrams participating in refinement or dependency relations increase, inspection of interdiagram properties becomes difficult. Hence, to what extent the approach is applicable to large scale complex models still needs further investigation. However, as conceptual models are incrementally developed, applying the proposed inspection process in each iteration will definitely help remove defects and result in increased model quality.

References

Amalio, N., and Polack, F. 2003. Comparison of formalization approaches of UML class constructs in Z and Object-Z. In *International Conference of Z and B Users (ZB 2003), LNCS 2561*. Springer-Verlag.

Ambler, W. S. 2005. *The Elements of UML 2.0 Style*. Cambridge, MA: Cambridge University Press.

Andre, P., A. Romanczuk, J. C. Royer, and I. Vasconcelos. 2000. Checking the consistency of UML class diagrams using Larch Prover. In *Proceedings of the Third Rigorous Object-Oriented Methods Workshop*. York, UK: BCS.

Argo. 2002. An open source UML case tool. Retrieved January 1996 from http://argouml.tigris.org/.

Balcı, O. 1997. Verification, validation and accreditation of simulation models. In *Proceedings of the 1997 Winter Simulation Conference*, ed. S. Andrad ttir, K. J. Healy, D. H. Withers, and B. L. Nelson. Atlanta, GA: IEEE.

Balci, O. 1998. Verification, validation, and accreditation. In *Proceedings of 1998 Winter Simulation Conference*, ed. E. F. Watson, J. S. Carson II, and D. J. Medeiros. Washington, DC: ACM.

Basili, V. R., S. Green, O. Laitenberger, F. Lanubile, F. Shull, S. Sorumgard, and M. V. Zelkowitz. 1996. The empirical investigation of perspective-based reading. *Empirical software engineering journal* 2(1):133–164.

Berardi, D., D. Calvanese, and G. De Giacomo. 2005. Reasoning on UML class diagrams. *Artificial intelligence* 168:70–118.

Berenbach, B. 2004. Evaluation of large, complex, UML analysis and design models. In *Proceedings of 26th International Conference on Software Engineering*, ICSE. Washington, DC: IEEE.

Boehm, B. W. 1984. Verifying and validating software requirements and design specifications. *IEEE Software* 1(1):75–88.

BOM. 2006. Base object model. Retrieved October 2007 from http://www.boms.info/.

Brade D. A. 2004. Generalized process for the verification and validation of models and simulation results. Dissertation, Fakultät für Informatik, Universität der Bundeswehr München, Neubiberg.

Briand, L., Y. Labiche, and L. O'Sullivan, L. 2003. Impact analysis and change management of UML models. Technical Report SCE-03-01. In *Proceedings of International Conference on Software Maintenance (ICSM)*. Washington, DC: Carleton University: IEEE.

Caplat, G. 2006. Sherlock environment. Retrieved April 2006 from http://servif5.insa-lyon.fr/chercheurs/gcaplat/.

Chang, L. P., D. Jong-Li, P. Lin-Yi, and J. Muder. 2005. Management and control of information flow in CIM systems using UML and Petri Nets. *International journal of computer integrated manufacturing* 18:2–3.

Compatangelo, E. and H. Meisel, H. 2002. Intelligent support to knowledge management: Conceptual analysis of EER schemas and ontologies. In *Internal Report, Dept. of Computing Science*, University of Aberdeen, Aberdeen, UK. Retrieved October 2007 from http://www.csd.abdn.ac.uk/research/conceptool/.

Csertan, G., I. Huszerl, Z. P. Majzik, and A. Patar. 2002. VIATRA: Visual Automated Transformations for formal verification and validation of UML models. In

Proceedings of the 17th International Conference on Automated Software Engineering. Edinburgh, UK: IEEE.

DMSO (Defense Modeling and Simulation Office) 2000a. Verification, validation and accreditation (VV&A) recommended practices guide. Retrieved December 2007 from http://vva.dmso.mil/.

DMSO (Defense Modeling and Simulation Office) 2000b. Conceptual model development and validation. Retrieved December 2008 from www.msiac.dmso.mil/vva/Special_Topics/Conceptual/conceptual-pr.PDF.

De Lara, J., and H. Vangheluwe. 2002. ATOM3: A tool for multi-formalism modelling and meta-modelling. *In Proceedings of FASE'02, LNCS,* 2306. Berlin, Germany: Springer.

DoD VV&A RPG. 2000. Special topic: Conceptual model development and validation. *DoD VV&A Recommended Practices Guide,* 11/30/2000. Retrieved December 2007 from http://vva.dmso.mil/.

DoD VV&A RPG. 2001. RPG reference document: V&V techniques, 15/08/2001. *DoD VV&A Recommended Practices Guide,* 11/30/2000. Retrieved December 2007 from http://vva.dmso.mil/.

Dupey, S., Y. Ledru, and M. Chabre-Peccoud. 2000. An overview of RoZ: A tool for integrating UML and Z specifications. In *12th Conference on Advanced information Systems Engineering, CAiSE'2000, LNSC,* 1789. Stockholm, Sweden: Springer.

Egyed, A. 2006. Instant consistency checking for UML. In *Proceedings of the 28th International Conference on Software Engineering (ICSE).* Shangai, China: IEEE.

Eishuis, R. and R. Weringua. 2004. Tool support for verifying activity diagrams. *IEEE transactions on software engineering* 30 (7):437–447.

Engels, G., J. Küster, R. Heckel, and L. Groenewegen. 2001. A methodology for specifying and analyzing consistency of object-oriented behavioral models. *Software engineering notes* 26(5). ACM.

Enterprise Architect 6.5. 2006. UML case tool. Retrieved October 2007 from http://www.sparxsystems.com.au/.

Fagan, M. E. 1976. Design and code inspections to reduce errors in program development. *IBM systems journal* 15(3):182–211.

FEDEP. 2000. Recommended practice for High Level Architecture (HLA), Federation Development and Execution Process (FEDEP), IEEE 1516.3.

Gagnon, P., F. Mokhati, and M. Mourad. 2008. Applying model checking to concurrent UML models. *Journal of object technology* 7 (1):59–84. Retrieved February 2008 from http://www.jot.fm/issues/issue_2008_01/article1/.

Garth, M., and G. Schulmeyer. 2002. Verification technology potential with different M&S development and implementation paradigms. *Foundations for V&V in the 21st Century Workshop (Foundations '02).* MD: SISO.

Gilb, T., and D. Graham. 1993. *Software inspections.* MA: Addison-Wesley Publishing Company.

GME (Generic Modeling Environment). 2006. Retrieved November 2006 from http://www.isis.vanderbilt.edu/projects/gme/.

ICOM. 2000. A prototype design tool for Intelligent Conceptual Modeling. Retrieved December 2007 from http://www.cs.man.ac.uk/òfranconi/icom/.

ITOP (International Test Operations Procedure). 2004. General procedure for modeling and simulation verification and validation information exchange, ITOP 1-1-002, WGE 7.2. Retrieved December 2007 from http://ftp.rta.nato.int/Public/Documents/MSG/.

Hue, A., Y. San, and Z. Wang. 2001. Verifying and validating a simulation model. In *Proceedings of the 2001 Winter Simulation Conference*. Arlington, VA: ACM.

Johnson, P., and D. Tjahjono. 1998. Does every inspection really need a meeting. *Journal of empirical software engineering* 3(1):9–35.

Karagöz, A., and O. Demirörs. 2007. Developing Conceptual Models of the Mission Space (CMMS): A meta-model based approach. *In Proceedings of Simulation Interoperability Workshop (SIW)*. Orlando, FL: SISO.

Karagöz, A., and O. Demirörs. 2008. *A Conceptual Modeling Notation*. Unpublished doctoral dissertation, Middle East Technical University, Ankara, Turkey.

Killand, T., and J. Borretzen. 2001. UML consistency checking. *Research Report SIF8094*. Institute for Datateknikk OG Informasjonsvitenskap, Trondheim, Norway.

Kim, S., and D. Carrington. 2000. A formal mapping between UML models and Object-Z specification and B. *Lecture Notes in Computer Science*, 1878, Berlin: Springer.

Laitenberger, O., and J. M. DeBaud. 2000. An encompassing life-cycle centric survey of software inspection. *Journal of systems and software* 50(1):5–31.

Law, A. M., and W. D. Kelton. 1999. *Simulation Modeling and Analysis*, 3rd ed. New York: McGraw-Hill.

Lemmers, A., and M. Jokipii. 2003. SEST: SE Specifications Tool-set. In *Proceedings of Fall Simulation Interoperability Workshop*. SISO.

Lilius, J., and I. P. Paltor. 1999. vUML: A tool for verifying UML models. *Technical report 272*, Turku, Finland: Turku Centre for Computer Science (TUCS).

Lindland, O. I., G. Sindre, and A. Sølvberg. 1994. Understanding quality in conceptual modeling. *IEEE software* 11(2): 42–49.

Litvak, B., S. Tyszberowicz, and A. Yehudia. 2003. Behavioral consistency validation of UML diagrams. In *Proceedings of the 1st International Conference on Software Engineering and Formal Methods*. Brisbane, Australia: IEEE.

Marcano, R., and N. Levy. 2002. Using B formal specifications for analysis and verification of UML/OCL models. In *Workshop on Consistency Problems in UML-Based Software Development, 5th International Conference on the UML*. Dresden, Germany: IEEE.

Meta Edit. 2007. A case tool for domain specific software development. Retrieved January 2007 from http://www.metacase.com.

Minas, M. 2002. Specifying graph-like diagrams with DiaGen, *Electronic notes in theoretical computer science 72*(2), 102–111, Amsterdam, The Netherlands: Elsevier.

MOF 2.0. 2004. *Meta Object Facility core specification*. Retrieved December 2005 from http://www.omg.org.

Mota, E., M. Clarke, A. Groce, W. Oliveira, M. Falcão, and J. Kanda. 2004. Veri Agent: An approach to integrating UML and formal verification tools. *Electronic notes in theoretical computer science* 95:111–129.

Murata, T. 1989. Petri Nets: Properties, analysis and applications. *Proceedings of the IEEE* (77).

NATO. 2007. Verification, validation, and accreditation of federations. Retrieved November 2007 from http://www.rta.nato.int/search.asp#MSG-019.

Ober, I. 2004. Harmonizing design languages with object-oriented extensions and an executable semantics. Unpublished doctoral dissertation, Institute National Polytechnique de Toulouse, Toulouse, France.

OCLE. 2005. OCL Environment. Computer Science Research Laboratory, Babes Boyls University, Romania. Retrieved December 2006 from http://lci.cs.ubbcluj.ro/ocle/index.htm.

Ohnishi, A. 2002. Management and verification of the consistency among UML models. In *Proceedings of Workshop on Knowledge-Based Object-Oriented Software Engineering* (KBOOSE), LNCS. Malaga, Spain: Springer.

Open Architecture-ware. 2007. A platform for model driven development. Retrieved October 2007 from http://www.openarchitectureware.org/.

Queralt, A., and E. Teniente. 2006. *Reasoning on UML Class Diagrams with OCL Constraints, Conceptual Modeling: ER , LNCS.* Berlin: Springer.

Pace, D. K. 2000. Simulation conceptual model development. In *Proceedings of the Spring Simulation Interoperability Workshop*. Retrieved November 2005 from www.sisostds.org.

Porter, A. A., L. G. Votta, and V. R. Basili. 1995. Comparing detection methods for software requirements inspections: A replicated experiment. *IEEE transactions on software engineering* 21(6), 563–575.

Poseidon. 2006. UML Case Tool. Retrieved October 2006 from http://www.gentleware.com/.

Rational. 2004. Rational case tool. Retrieved October 2006 from http://www-306.ibm.com/software/rational/.

REVVA 2. 2005. VV&A process specification (PROSPEC) user's manual, v1.3. Retrieved October 2007 from http://www.revva.eu/.

Sargent, R. G. 2001. Some approaches and paradigms for verifying and validating simulation models. In *Proceedings of the 2001 Winter Simulation Conference.* Arlington, VA: ACM.

Schinz,I., T. Toben, C. Mrugalla, and B. Westphal. 2004. The rhapsody UML verification environment. In *Proceedings of Second International Conference on Software Engineering and Formal Methods (SEFM).* Beijing, China: IEEE.

SD Metrics. 2007. List of object oriented design rules. Retrieved December 2007 from http://www.sdmetrics.com/LoR.html#LoR.

SEDEP. 2007. Synthetic Environment Development and Exploitation Process: Euclid RTP 11.13. Retrieved December 2007 from http://www.euclid1113.com/.

Sourrouille, J. L., and G. Caplat. 2002. Constraint checking in UML modeling. In *Proceedings International Conference Software Engineering and Knowledge Engineering (SEKE 2002).* ACM.

Sourrouille, J. L., and G. Caplat. 2003. A pragmatic view on consistency checking of UML models. In *Workshop on Consistency Problems in UML-Based Software Development II, Workshop Materials,* ed. L. Kuzniarz, Z. Huzar, G. Reggio, J. L. Sourrouille, and M. Staron. IEEE.

Statemate-Magnum. 2007. A case tool for UML verification. Retrieved April 2007 from http://www.ilogix.com/products/magnum/index.cfm.

Tabu. 2004. Tool for the active behavior of UML. Retrieved April 2007 from http://www.cs.iastate.edu/~leavens/SAVCBS/2004/posters/Beato-Solorzano-Cuesta.pdf.

Taentzer, G. 2003. AGG: A graph transformation environment for modeling and validation of software. In *Proceedings of Application of Graph Transformations with Industrial Relevance (AGTIVE'03).* Springer.

Tanrıöver, Ö. 2008. An inspection approach for conceptual models of the mission space in a domain specific notation. Unpublished PhD Thesis, Middle East Technical University, Ankara, Turkey.

Tanrıöver, Ö., and S. Bilgen. 2007a. An inspection approach for conceptual models for the mission space developed in domain specific Notations of UML. In *Software Interoperability Workshop Papers*. Orlando, FL: SISO.

Tanrıöver, Ö., and S. Bilgen. 2007b. An inspection approach for conceptual models in notations derived from UML: A case study. In *Proceedings of Symposium on Computer and Information Sciences*. Ankara, Turkey: IEEE.

Travassos, G. H., F. Shull, J. Carver, and V. R. Basili. 2002. Reading techniques for OO design inspections. University of Maryland Technical Report, April(OORT V.3). Retrieved December 2007 from http://www.cs.umd.edu/Library/CS-TR-4353/CS-TR-4353.pdf.

UML Superstructure. 2005. Unified Modeling Language 2.0 superstructure specification. Object Management Group, retrieved December 2005 from http://www.omg.org/uml/.

Unhelkar, B. 2005. *Verification and Validation for Quality of UML 2.0 Models*. Hoboken, NJ: Addison Wesley.

Van der Straten, R., T. Mens, J. Simmons, and V. Jenkers. 2003. Using description logic to maintain consistency between UML models. In *Proceedings of UML 2003: Model Languages and Applications. LNCS:2863*. Springer.

Zhao,Y., X. Fan, Y. Bai, H. C. Vang, and W. Ding. 2004. Towards formal verification of UML diagrams based on graph transformation. In *Proceedings of the International Conference on E-Commerce Technology for Dynamic E-Business*. Beijing: IEEE.

Part V

Domain-Specific Conceptual Modeling

16

Conceptual Modeling Evolution within US Defense Communities: The View from the Simulation Interoperability Workshop

Dale K. Pace

CONTENTS

16.1 Introduction

Simulation-related conceptual modeling is a challenging and complex topic. Insights can be gained about factors influencing development of conceptual modeling ideas by examining the continuing evolution of simulation-related conceptual modeling and the approaches used in various communities. Why different simulation communities have addressed this topic in such varied ways often becomes much clearer from a consideration of conceptual modeling history.

This chapter provides historical perspective on the evolution of simulation-related conceptual modeling within the US military and defense simulation communities.* Simulation-related conceptual modeling in the US military and Department of Defense (DoD) has been influenced strongly by efforts initiated or directed by the Defense Modeling and Simulation Office (DMSO), an organization that began in 1991 and evolved into the Modeling and Simulation Coordination Office (M&S CO) about 2006. Much, but not all, of DMSO's conceptual modeling influence manifested itself within the Simulation Interoperability Standards Organization (SISO).

SISO originated with a small conference held April 26 and 27, 1989, called Interactive Networked Simulation for Training. The group was concerned that activity occurring in networked simulation was occurring in isolation. They believed a means to exchange information between companies and groups would enable networked simulation technology to advance more rapidly. Once the technology began to stabilize and mature, there would be a need for standardization to capture technology and community consensus. The conferences soon developed into the Distributed Interactive Simulation (DIS) Workshops. They focused on creating standards based on the major project SIMNET, which was established as the baseline standard in the early 1990s. In late 1996, in light of the development of the high level architecture (HLA), the DIS organization transformed itself into a more functional organization called SISO.

The SISO Simulation Interoperability Workshop (SIW), in its forums at the semi-annual workshops and by interactions among participants throughout the year, is where many of the ideas, concepts, and interoperability processes for HLA simulations were thrashed out. SIW is also where many conceptual modeling ideas were presented and refined by the varied perspectives of the different simulation communities represented. Some of DMSO's conceptual modeling ideas were discussed within the semi-annual DIS Workshops before SIW began in 1996. Such conceptual modeling discussions, which continued in SIW as well as with others outside SIW, helped to clarify and vet conceptual modeling ideas being developed by DMSO endeavors.

For the past two decades, the main SISO conferences (first DIS and then SIW) have met twice a year. Typically at each of these meetings, more than a hundred papers will be presented in addition to the work performed within the groups and forums of the conference to draft, review, and revise guidance being developed through SISO. These main conferences have been supported by active email interchanges among participants and by a large number of ancillary meetings of groups and forums as they grapple with technical issues being addressed. Since 2001 there has also been an annual

* A glossary at the end of the chapter lists acronyms used in the chapter.

European SIW (with 25–100 presentations) whose papers and discussions are fully integrated into SISO.

Three main streams of conceptual modeling stimulated by DMSO/M&S CO have interacted with one another, both within SIW and elsewhere, sometimes in competitive ways and sometimes broadening and honing ideas and concepts for all. One stream is the Conceptual Model of the Mission Space (CMMS), later renamed Functional Description of the Mission Space (FDMS). A second stream is simulation conceptual modeling (SCM) as expressed in the DoD *Recommended Practices Guide* (*RPG*) for modeling and simulation (M&S) verification, validation, and accreditation (VV&A). The third stream is development of the Federation Conceptual Model (FCM), a conceptual model for a collection of simulation applications working in concert as embodied in the DMSO HLA Federation Development and Execution Process (FEDEP).*

This chapter focuses (1) on simulation-related conceptual modeling ideas reflected by these three streams, most of which were discussed extensively within SIW, and (2) on where simulation-related conceptual modeling ideas are within the simulation communities of SIW in early 2009. Even though it causes a bit of repetition in the chapter, development of conceptual modeling ideas in each of these three streams is treated individually. Then how they merge is discussed. Some of the material mentioned in the following background section presage points that are addressed more fully later in the chapter. Material is presented in this way for reader convenience. It allows each of the chapter sections to be coherent without dependence upon material in the other sections of the chapter.

The next section of this chapter presents historical background about SCM evolution within US Defense communities. The section after that addresses conceptual model implications of the application context. Then a section will examine the parallel approaches to conceptual modeling by the *RPG* and the FEDEP. That is followed by a section that considers what has been done in the SIW Simulation Conceptual Modeling Study Group (SCM SG) and its evolution into a SIW Standing Study Group (SSG). A number of persistent problems related to conceptual modeling are identified and discussed before conclusions and final comments are presented.

16.2 Historical Background

Prior to initiation in the mid-1990s of the three streams mentioned above, the classical approach to problem solving used in science, engineering, and

* In HLA parlance, a *federate* is a single simulation and a *federation* is a collection of simulations (federates) working together. Various terms have been applied to such collections of simulations working together: networked simulations, distributed simulations, and advanced distributed simulations, as well as federations.

business since long before the days of computer simulation was the basis for development of most simulations within the US military and Defense community. This approach involved five basic steps:

1. Identify/specify what is to be accomplished. This is done through objectives, requirements, goals, problem definition, etc.
2. Plan the approach to accomplish the objectives. This is done through conceptual analysis, conceptual modeling, design of the approach, etc.
3. Implement the planned approach by coding the program, building the device, etc.
4. Test, then correct/modify/improve, and demonstrate the implemented approach accomplishes the objectives. This involves the disciplines of verification and validation (V&V), test and evaluation (T&E), etc.
5. Use the implemented approach to accomplish objectives.

Formal processes were not used widely for simulation-related conceptual modeling within the US military or Defense community before the mid-1990s, even though the idea of the conceptual model as the connecting link between the reality to be simulated and the computerized model had been noted in various simulation paradigms. One of the best known of such paradigms is the "Sargent Circle," which was developed in the 1970s by Dr. Robert Sargent of Syracuse University. The paradigm showed where V&V fit in the simulation development and use process (Schlesinger 1979). This paradigm is still cited by various V&V guides for simulation today (e.g., AIAA 1998, ASME 2006).

It was understood that a conceptual model (whether named as such or not) underlay and led to simulation design and implementation; however, only rarely before the mid-1990s was a simulation-related conceptual model explicitly and completely defined or documented. In addition, at that time there was no general agreement about what items and processes were involved in a simulation-related conceptual model. Consequently, it was difficult in most cases to perform either conceptual validation of simulation designs or validation assessments of simulations for conditions not tested specifically. This severely limited simulation credibility and utility. Modification and evolution of a simulation often ran into unnecessary problems because of lack of information about assumptions, processes, limitations, and algorithms of the conceptual model upon which the simulation had been built. As a result, much of the motivation and discussion of simulation-related conceptual modeling has come from simulation VV&A practitioners since conceptual modeling has such a major impact on simulation VV&A. In SIW, prior to establishment of the Conceptual Model Study Group in 2003 (see section 16.5), many of the conceptual modeling papers and presentations were in the VV&A Forum.

During the late-1980s, a number of significant criticisms were made of M&S in the US military and DoD by the Defense Science Board (DSB), Government Accounting Office (GAO), and other responsible parties. This coupled with growing appreciation for the potential value of M&S in Defense led Assistant Secretary for Force Management and Personnel Christopher Jehn and Defense Director of Research and Engineering Charles Herzfeld to sponsor a Simulation Policy Study in 1990, which was led by retired Army General Paul Gorman. The study recommended that DoD establish a M&S organization to look across the military services with the objective of reducing duplication of effort and facilitating interoperability. In June of the following year (1991), DMSO* was established in response to the Defense Simulation Policy Study recommendations and Army COL Ed Fitzsimmons, who had been Executive Secretary of the Defense Simulation Policy Study, was appointed as the first DMSO Director.

DMSO's early efforts focused on creating a DoD M&S strategy that would reduce duplication of M&S effort within DoD and facilitate M&S interoperability. The strategy developed built upon a three part common technical framework. The parts of the common technical framework were (1) HLA to enable more substantive M&S interoperability, especially for military and Defense simulation, (2) CMMS† to provide a common world view, and (3) data standards (DoD M&S Master Plan 1995). In addition, the DoD M&S Master Plan also emphasized M&S VV&A as part of the M&S infrastructure needed to improve M&S credibility.

As part of DMSO's common technical framework, CMMS was given a great deal of attention. "Conceptual Models of the Mission Space (CMMS) are a first abstraction of the real world activities associated with a particular mission area. Such conceptual models provide an entities, actions, tasks, and interactions (EATI) representation of the military mission space" (Hollenbach and Alexander 1997). By its emphasis upon more extensive employment of knowledge engineering techniques than was normally used in Defense simulation development, CMMS sought to provide (Sheehan et al. 1998) the following:

- A disciplined procedure to systematically acquire knowledge
- A set of information standards

* In 2006, in conjunction with realignment of various responsibilities among DoD organizations, DMSO evolved into the Modeling and Simulation Coordination Office (M&S CO) which has similar but not identical responsibilities to those which DMSO had previously.

† About 2000, the term CMMS was replaced by the term Functional Description of the Mission Space (FDMS), which had the same meaning as CMMS, in order to emphasize the functional (vice "conceptual") nature of the simulations (as desired by the operational community). This also reduced confusion between FDMS and "conceptual model," whether *conceptual model* was applied to a single simulation (federate) or to a collection of simulations functioning together as a federation in a distributed simulation.

- A decomposition of real-world, military operations
- A singular means for establishing reuse opportunities
- A library of reusable conceptual mission space models

By the late 1990s, there was considerable confusion about conceptual models as they pertain to simulation. There were four main reasons for the confusion. First was use of the words *conceptual model* in CMMS, which was concerned with abstraction of a military mission space from authoritative sources in EATI terms. Second was the idea of the conceptual model as the link between simulation requirements and design, as it was being developed for the Web-based *Recommended Practices Guide* (*RPG*) for M&S VV&A by a team under DMSO direction. Third was the conceptual model idea being developed for the FEDEP to standardize development processes for HLA federations because it was realized that although the Federation Object Model (FOM) could ensure communication compatibility within the federation, it did not ensure representational compatibility among the federates. The FCM became the mechanism to ensure representational compatibility within a federation. All three of these simulation-related conceptual model ideas were percolating within the DMSO and SIW* communities. Fourth, in addition, other ideas about conceptual modeling, such as the database-oriented ones associated with the *Journal of Conceptual Modeling*, were also being communicated and discussed. This confusing variety of connotations for the term conceptual model continues to this day (Druid et al. 2006). Resolution of the differences in connotations for simulation-related conceptual modeling becomes easier when application context is brought into the picture, as will be done later in this chapter.

Timelines can provide perspective on conceptual modeling related to simulation, at least within the US military arena. All work under DMSO direction was oriented primarily toward simulation by or for US Defense communities. HLA had been directed by senior DoD leadership to be the architecture for distributed simulation within DoD (Kaminski 1996). Work on CMMS began in the mid-1990s, and as noted earlier the name changed to FDMS about 2000 but the concept stayed the same. The CMMS/FDMS idea was migrating to the Knowledge Integration Resource Center (KIRC) about 2002, which seems to have been where FDMS went after trying to integrate FDMS with the Defense Modeling and Simulation Resource Center. Personnel changes and funding decreases caused the DMSO-sponsored CMMS/FDMS effort to atrophy. The most substantial continuation of the FDMS idea directly seems to have been done at the Swedish Defense Research Agency (FOI). Their evolution of the FDMS idea is called the Defence Conceptual Modeling Framework (DCMF) (Kabilan and Mojtahed 2006).

* SIW is the primary venue where various parts of IEEE Standard 1516 relative to HLA implementations were developed prior to their balloting for acceptance as an IEEE standard.

The basic idea that CMMS/FDMS developed of providing an authoritative description of the military application domain for a simulation with identification and description of the entities, processes, and interactions (the EATI construct) has been generalized in what is called *domain modeling*. In software engineering, domain modeling has been very helpful in facilitating software reuse. "By systematically representing (or modeling) the functions, objects, data, and relationships of applications in the domain, domain modeling is used to define what the applications are, what the applications do, and how the applications work" (Krut and Zahman 1996).

The FCM idea that had appeared in Version 1.5 of the DMSO HLA FEDEP in late-1999 (DMSO 1999) became codified in 2003 with adoption of IEEE Standard 1516.3-2003 *Recommended Practice for High Level Architecture Federation Development and Execution Process* (*FEDEP*).* It is expected that the systems engineering approach embodied in the FEDEP may be expanded in scope beyond HLA implementations and given a new name, Distributed Simulation Engineering and Execution Process (DSEEP), with a new IEEE standard number (1730) if it should progress to the status of an IEEE standard.

The first edition of the DoD *Recommended Practice Guide* (*RPG*) for M&S VV&A was published in 1996. It was well received, and shortly thereafter plans were made to upgrade the *RPG* and make it available in a Web-based format. In 1998, Simone Youngblood,† the DMSO Technical Director for VV&A, assembled a team to develop the Web-based Millennium Edition of the *RPG*. Many of the ideas for materials that would be included in the *RPG* were presented and discussed at SIW‡ and Society for Computer Simulation International (SCS) conferences. The special topic on simulation conceptual models in the Millennium Edition (Build 1) of the *RPG* (2000) focused on new simulation developments. The next revision (Build 2) to the *RPG* (2001) expanded conceptual modeling discussion to include legacy simulations. The current (Build 3) version of the *RPG* (2006) addresses conceptual modeling both for new simulation developments or modifications, and for legacy simulations developed without explicit conceptual models.

In the spring of 2003, another element entered the conceptual modeling fray at SIW. At the Spring SIW, a SCM SG led by Jake Borah was formed "to conduct preliminary investigation on the best practices of SCM and to establish recommendations for pursuit of the topic within the scope of the SISO, if appropriate."

---

* The FEDEP extends beyond US military simulation since it is the subject of a NATO standardization agreement, STANAG 4603: *Modeling and Simulation Architecture Standards for Technical Interoperability: High Level Architecture* (*HLA*).

† Simone Youngblood also has chaired the SIW VV&A Forum since SIW began. Previously she had been an active participant in DIS VV&A activities.

‡ From 2001 on, "SIW" is used for both the SIW sessions in the US and the Euro SIW sessions.

16.3 Conceptual Model Characteristics and Application Context

What is a conceptual model? As noted earlier, CMMS is the "first abstraction of the real world activities associated with a particular mission area" (Hollenbach and Alexander 1997). Mojtahed et al. (2005) in their elaboration of CMMS/FDMS into the Defence Conceptual Modeling Framework (DCMF) for the Swedish Defence Research Agency (FOI) "summarise a definition of conceptual models in the context of computer information systems as: *Conceptual models are abstractions of a real world domain of discourse. They are intended to capture the semantics, pragmatics and to an extent the syntactic of the domain being modelled.*" The CMMS, FDMS, and DCMF concept for a conceptual model is relatively independent of the simulation application. This characteristic of CMMS was recognized early (Lewis and Coe 1997). In this regard, CMMS/FDMS/DCMF is akin to ideals in the *Journal of Conceptual Modeling*, which are concerned with data, modeling, design, and implementation issues related to databases.

For other definitions of simulation-related conceptual models, simulation application context has a significant impact on conceptual model connotation. This will be illustrated for DIS, for simulation conceptual models, and for FEDEP FCMs. Application context also impacts simulation implementation independence. Implementation independence is generally considered to be a positive attribute for a conceptual model since it enhances reuse potential of the conceptual model or its parts. The closer to simulation design one comes, the less implementation independence the conceptual model can have.

In DIS communities of the early 1990s, conceptual model referred to the agreement between the simulation developer and user about what the simulation was to do (Pace 2000). Later the DIS glossary definition for *conceptual model* became, "A statement of the content and internal representations which are the developer's concept of the model. It includes logic and algorithms and explicitly recognizes assumptions and limitations" (DIS 1995). The current description of simulation conceptual model in the *RPG* follows this approach.

The *VV&A Recommended Practices Guide* (*RPG* Build 3 2006) says, "A simulation conceptual model is frequently described as the bridge between the Developer and the User. It serves as a primary mechanism for clear communication among simulation development personnel (e.g., software designers, code developers, system engineers, system analysts) and members of the user community (e.g., Users, functional area subject matter experts [SMEs], testers, V&V Agents, Accreditation Agents)." For new simulations or simulation modifications, the simulation conceptual model is driven by M&S requirements. The simulation conceptual model encompasses *all* M&S requirements so that specifications (i.e., the detailed guidance upon which M&S design is based) for the M&S development or modification may be developed from the conceptual model. This kind of conceptual model permits a simulation design that fully captures the M&S requirements so that the simulation will have the capability to satisfy simulation objectives in its intended applications.

For a legacy simulation that was constructed without an explicit conceptual model, the simulation conceptual model has a different function. It uses a full description of the M&S implementation to create the simulation conceptual model so that assessments of the appropriateness (or limitations) of M&S applications may be determined from the simulation conceptual model for situations that are not tested directly. For such a legacy simulation, the simulation conceptual model provides a solid basis for decisions about modifications to the simulation.

Figure 16.1 from the *RPG* illustrates these two perspectives on simulation conceptual models, one for new simulations or simulation modifications, and the other for legacy simulations developed without an explicit conceptual model. As shown in Figure 16.1, the simulation conceptual model has three primary components: the simulation context, the mission space, and the simulation space. "The simulation context provides authoritative information about the user and problem domains to be addressed in the simulation based on the M&S requirements of the intended application" (*RPG* Build 3 2006). Thus, most of the CMMS/FDMS information pertinent to a simulation as the "first abstraction of the real world" becomes part of the simulation context in the *RPG* simulation conceptual model. This information establishes

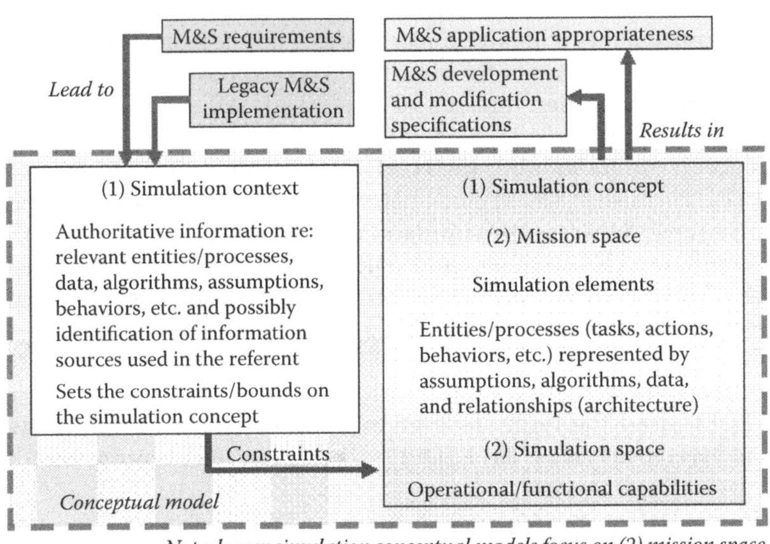

FIGURE 16.1
Simulation conceptual model components.

constraints and boundary conditions for the simulation concept that guides articulation of the detailed specifications that focus the simulation design.

In early discussions of the *RPG* simulation conceptual model, questions were raised about the need for both *mission space* and *simulation space* parts of the conceptual model. It was decided that it would be very helpful to separate the representational aspects of the conceptual model (i.e., the *mission space*) from those aspects of the conceptual model concerned with simulation implementation (i.e., the *simulation space*). This separation of representational aspects of the conceptual model from implementation aspects (such as being required to run on particular hardware or operating systems, run in real-time or some multiple of real-time, support particular kinds of display systems, etc.) has proved very helpful, especially when representational compatibility of various simulations (federates) to be used in a distributed simulation (federation) has to be assessed. It permits focus on representational issues: what level of resolution and accuracy is expected from the representation, what elements are treated explicitly within the conceptual model, what are the assumptions and pedigrees of the algorithms selected, etc.

In the spirit of maximizing implementation independence of the conceptual model, the *simulation space* is restricted to implementation implications from the requirements that the simulation must satisfy. Other implementation decisions about the simulation are left to simulation design and are not part of the simulation conceptual model; although in practice simulation conceptual models often have design elements included in them beyond what is essential to satisfy M&S requirements fully. Obviously the kind of simulation (live, virtual, or constructive)* as well as the amount of implementation related requirements will have major impacts on the extent of material in the simulation space portion of the simulation conceptual model.

In diagrams of earlier versions of the FEDEP (HLA versions 1.x prior to publication in IEEE Std 1516.3) (Lutz 2003), there seemed to be a significant difference in connotation for the FCM and the simulation conceptual model of the *RPG* because the FCM came before federation requirements and drove them in a sense, whereas the simulation conceptual model of the *RPG* comes after simulation requirements and is driven by them. However, the difference was only apparent and not a real difference in substance. The apparent difference arose because the FEDEP used the term "federation objectives" (which preceded the FCM and drove it as input to it) in the way that the *RPG* used "M&S requirements" (which drive the simulation conceptual model). The FEDEP used the term "federation requirements" (which followed the FCM and were shown in "old FEDEP" diagrams as an output from the FCM)

* In the early mid 1990s, US defense simulation communities began to use the terms *live, virtual,* and *constructive* to indicate aspects of simulation implementation. *Live* meant actual military forces and systems such as tanks, aircraft, ships, and personnel were involved in the simulation. *Virtual* meant simulators such as the simulators used to train aircraft or tank crews were involved in the simulation. *Constructive* meant the simulation was contained completely in computer code and did not involve either real systems or simulators.

in the way that the *RPG* used "M&S specifications" (the detailed information that enables a simulation design to be developed that *fully* satisfies M&S requirements).

The current version of the FEDEP (in IEEE Standard 1516.3-2003) removes the "apparent" difference by having conceptual analysis that is driven by federation objectives lead to the FCM, which in turn leads to federation design. "Federation requirements" come out of conceptual analysis in parallel to the conceptual model and drive assessment of federation results. This makes the conceptual model connotations for the simulation conceptual model (oriented toward federates) compatible with the connotation for the FCM, which addresses a collection of federates used together in a federation. The conceptual model in the FEDEP is a description of "what the [simulation or federation] will represent, the assumptions limiting those representations, and other capabilities needed to satisfy the user's requirements" (IEEE Std 1516.3-2003).

The basic function of the simulation-related conceptual model has been clearly identified as (1) the link between objectives/requirements and simulation specifications/design and (2) a vehicle for effective communication among the various simulation stakeholders and other interested parties. This is true both for individual simulations (federates) and for combinations of simulations (federations).

Conceptual modeling issues identified below, which need consideration once simulation-related conceptual modeling functions are established, will be addressed later in this chapter:

- What information should a conceptual model contain?
- How should a conceptual model be described and documented?
- How can a conceptual model be developed effectively and efficiently?

16.4 Parallel Paths: *RPG* Simulation Conceptual Model and FEDEP Federation Conceptual Model (FCM)

This section of the chapter explains why conceptual model guidance developed under DMSO leadership lacked the specific details many desired. The section then addresses conceptual model functions, content, and documentation format.

16.4.1 Unmet Desire for a Prescriptive Approach

From the mid-1990s when DMSO-sponsored conceptual modeling work began to the present, discussions within SIW (and elsewhere) of ideas leading to the *RPG* description of a conceptual model and the FCM criticized the lack of detailed specific guidance about conceptual model development, content, and format. A more prescriptive approach than the descriptive approach of the *RPG* and FEDEP was desired.

How does a *prescriptive* approach differ from *descriptive* approaches used in the *RPG* and the FEDEP? A *prescriptive* approach provides a specific list of things a conceptual model must do and items it must document; a *descriptive* approach provides general guidance about what a conceptual model should do and an example list of items it should document. A *prescriptive* approach specifies the content, format, and detailed structure of the conceptual model and its documentation; a *descriptive* approach permits leeway regarding such.

Numerous suggestions were made about how simulation-related conceptual modeling guidance could become more prescriptive. Some proposed use of Zeigler's Discrete Event System Specification (DEVS) (Zeigler 1999). A number of people suggested use of Unified Modeling Language (UML) constructs. There were other ideas also, many of which were based upon knowledge engineering precepts.

The importance and potential value of a prescriptive approach were recognized and not disputed. A prescriptive approach is more likely to cause a conceptual model to be developed that conforms fully to the guidance for it. Thus, a prescriptive approach tends to improve the quality of items developed using it as well as improving the efficiency with which an item can be developed (i.e., reducing resources required to develop a conceptual model and facilitating reuse). However, the team developing simulation conceptual model material for the *RPG* concluded that the variety of simulations and their applications that the *RPG* was intended to support were too great for a prescriptive approach that could be applied broadly enough to be viable, and therefore pursued the descriptive approach found in the *RPG*.

US military and Defense simulations run the gamut from detailed science and engineering simulations used for trade studies in weapon system design and exploration of fundamental physical, chemical, and biological phenomena to highly aggregated simulations supporting Defense policy analyses and senior commander war games. The Defense community also uses a full spectrum of business simulations to support DoD financial and personnel planning. Simulations address logistical systems for office and business supplies as well as for movement of troops and materiel. Defense M&S support medical resource requirements estimations in plans for various military operations. Defense simulations might be live involving actual personnel and systems, virtual using training simulators, or constructive simply within a computer. Defense simulation-related conceptual modeling guidance is intended to support the full spectrum of Defense simulation, and therefore cannot be prescriptive as might be possible for simulations within a narrow application domain.

In recent years, much of the continuing desire for more prescriptive approaches to simulation-related conceptual models seems to be driven by hope of automating aspects of conceptual model development. Significant progress has been made in automating some aspects of conceptual model development since early visions of automating simulation model generation, such as put forth by Mayer and Young (1984). Much of the progress in automating aspects of conceptual model development come from using tools developed mainly for other simulation or software activities, such as applying process modeling tools to conceptual modeling (Heavey and Ryan 2006). The greatest progress has been made for restricted application domains, such as illustrated by US Air Force and NASA applications of the Adaptive Modeling Language (AML) (Blair and Love 2002, NASA Tech Briefs 2006), and in describing conceptual models in simulation development paradigms such as UML (Tanriover and Bilgen 2007), It is hoped that such progress will continue, and the spectrum of applications to which such automation may be applied will increase. However, the scope of US military and Defense simulation still prevents a prescriptive approach to simulation-related conceptual model development that can apply to *all* varieties of Defense simulation. Table 16.1 indicates some of the simulation varieties and diversities that prevent use of a prescriptive approach to conceptual model development.

TABLE 16.1

Selected Defense Simulation Varieties and Diversities

Characteristic	One Extreme	Another Extreme
Time/Progress Method	Continuous Simulation	Discrete-Event Simulation
Facility/Run-time Constraints	"Live" with military systems & people in the loop that must be run in real-time	Computer-code only simulation with no run-time constraints
Level of Detail/ Aggregation	First principle physics code	Simulation with aggregate representation of large (theater level) military forces
Simulation Application Domain	Simulation of policy, financial, or personnel matters	Simulation of product flow on a manufacturing floor
Mechanisms Represented	Simulation of atmospheric transport and diffusion for chemical or biological agent	Simulation of disease progression in humans at various levels of resolution
Capacity for Human Interaction with the Simulation While It Runs	Training simulator for helicopter pilots with interactive visualization	Constructive simulation with no interactive capabilities

Note: (The spectrum of simulations shown here illustrates why a prescriptive approach to conceptual model development is beyond current capabilities if the approach is to apply to all simulation varieties. Potential simulation differences indicated would impact significantly a conceptual model for the simulation. These differences affect how the conceptual model should be developed, what it should contain, and how it should be documented. For example, the conceptual model for a missile seeker simulation with a facility, such as an anechoic chamber, has to address control of facility temperature and humidity so that reliable results can be obtained. Such capabilities are not required for constructive simulations that are comprised only of computer software.)

16.4.2 Functions of Federate and Federation Conceptual Models

For a single simulation (i.e., a *federate* in HLA terminology) modification or new development, the function of the simulation conceptual model is to connect M&S requirements to the detailed specifications for a simulation design that can satisfies M&S requirements fully. According to the *RPG,* this is the function of the simulation conceptual model even if the simulation is complex and involves people, hardware, software, or systems in the loop. Figure 16.1 illustrates these functions for the simulation conceptual model. For a legacy simulation, the simulation conceptual model identifies and describes capabilities and limitations of the simulation so that appropriate assessment can be made of potential simulation applications.

During the early development of the FEDEP, it was thought by some that the FOM would be sufficient to assure federate compatibility in a federation. However, early experiments with HLA federations showed that something more than the FOM and HLA rules were needed to ensure federation objectives could be satisfied. The FOM addressed communications among federates, but it did not fully address the logical issues of representational compatibility since federates in a federation may have different levels of resolution in their representations, different assumptions, etc. So the FCM became part of the FEDEP. The FCM describes what the simulated world of the federation will look like and how it will function. The FCM is the key for determining representational compatibility among federates in a federation and for identifying federate characteristics that may be needed (if not present) in order for the federation to satisfy its objectives. This identification of gaps in capability by the FCM can lead to identification of modifications needed by federates or of new federates that will be required for the federation to be able to satisfy its objectives.

In a combination of simulations used together in a federation, FCM helps to organize use of individual federates so that federation objectives can be achieved, which leads to the federation design. Often the conceptual analysis involved in federation development is more concerned with determining what can be achieved with the collection of federates and communication capabilities available for the federation than it is in determining federate characteristics needed to satisfy federation objectives. This is because schedule and resource constraints may restrict the federation to use of available federates. Federation capability (relative to objectives for the federation) may be constrained by capabilities and limitations of federates available to it, including the real systems and personnel available for use in the federation.

Both the *RPG* and the FEDEP FCM made a significant contribution to simulation professionalism by making the conceptual model an explicit and distinct artifact of simulation development, whether federate or federation. Prior to the *RPG* and FEDEP FCM, few simulations had explicit conceptual models as artifacts of simulation development. The information that should be found in a conceptual model might be scattered among design papers, users and analysts' manuals, etc., if the information were documented at all. Existence of an explicit conceptual model creates a particular item where anyone involved with simulation development or use can find the information needed for assessment of simulation appropriateness for a particular application, whether used as an individual simulation by itself or in a federation with other simulations. Determination of compatibility among federates in a federation is always an important consideration.

16.4.3 Conceptual Model Content

What information should a simulation-related conceptual model contain? That question has engendered much debate and discussion. There have been many suggestions about what information should be contained in the conceptual model and how it should be documented. In 1999, a nine-element description was suggested for simulation conceptual models (Pace 1999):

1. Model version or portion identification
2. Identification of the simulation developer and pertinent points of contact (POCs)
3. Simulation purpose and requirements
4. Overview of a simulation based upon the conceptual model
5. General assumptions of the conceptual model
6. Identification of possible states, tasks, actions, behaviors, relationships, interactions, events, parameters, and factors for entities and processes represented in the conceptual model
7. Identification of algorithms (pedigrees and assumptions)
8. Simulation development plans
9. Summary and synopsis

RPG guidance about what information a conceptual model should contain has evolved to the example list of information included in a simulation conceptual model shown in Table 16.2.

FCM describes the entities and actions that need to be included in the federation to achieve all federation objectives. As far as possible, the FCM should be an implementation-independent representation that serves as a vehicle for transforming federation objectives into functional and behavioral capabilities. Thus, the FCM, just as the simulation (federate) conceptual model, provides traceability from federation objectives to federation design implementation. As with the federate conceptual model, there is not a prescribed list of information items for the FCM. The VV&A Overlay for the FEDEP adopted in December 2007 as a recommended practice (IEEE 1516.4) focuses on ensuring that the FCM achieves objectives specified for the FCM.

TABLE 16.2

Example List of Information Included in a Simulation Conceptual Model

1) Simulation descriptive information
   - model identification (e.g., version and date)
   - POCs
   - model change history
2) Simulation context (per intended application)
   - purpose and intended use statements
   - pointer to M&S requirements documentation
   - overview of intended application
   - pointer to FDMS and/or other sources of application domain information
   - constraints, limitations, assumptions
   - pointer to referent(s) and referent information
3) Simulation concept (per intended application)
   - mission space representation (simulation elements & simulation development description)
   - simulation space functionality
4) Simulation elements, including
   - entity definitions (entity description, states, behaviors, interactions, events, factors, assumptions, constraints, etc.)
   - process definitions (process description, parameters, algorithms, data needs, assumptions, constraints, etc.)
5) Validation history, including
   - M&S requirements and objectives addressed in V&V effort(s)
   - pointer to validation report(s)
   - pointer to simulation conceptual model assessment(s)
6) Summary
   - existing conceptual model limitations (for intended application)
   - list of existing conceptual model capabilities
   - conceptual model development plans

Source: Based on Department of Defense Modeling and Simulation Coordination Office, *VV&A Recommended Practices Guide, RPG Build 3.0,* September 2006.

The exact content of a simulation-related conceptual model will vary with the kind of simulation and its application. For example, processing speed for elements of the simulation becomes a matter of concern when the simulation interacts with real systems that function in real-time. This requires the conceptual model to give consideration to computational resources required by the algorithms identified in the conceptual model, and at times some of the algorithms will need to be replaced with faster-running approximations. On the other hand, normally the conceptual model for a constructive simulation would not need to address such factors.

Well done and well documented simulation (federate) conceptual models facilitate development of a FCM that might use one or more of the federates with such conceptual models (Pace 2001). Information in the federate conceptual models enables conceptual analysis of the FEDEP to determine compatibility and appropriateness of the various federates relative federation objectives. Not so well-done conceptual models for federates in a federation make it both much more difficult to develop an appropriate FCM, and more difficult to perform V&V on the federation. This is because information needed to support assessment of representational capability of federates and to support V&V of the federation may be difficult or impossible to discover for federates without well-done conceptual models for those federates.

16.4.4 Conceptual Model Documentation Format

The preceding section addressed conceptual model content. Now it is appropriate to ask, How should a conceptual model be documented? Conceptual model documentation format should accomplish two objectives: (1) ensure that the simulation design team fully understands what the simulation must do so that an appropriate simulation design can be developed, and (2) facilitate communication with all simulation stakeholders so that all fully understand simulation capabilities, limitations, and assumptions. It should be remembered that the stakeholders include the simulation development team and simulation users, those involved in assessing the simulation (such as V&V personnel), SMEs used in simulation development and/or assessment, those impacted by results from the simulation, simulation sponsors, and perhaps others.

A variety of formats have been used to document simulation-related conceptual models (Pace 1999). Originally the most common one encountered was the *ad hoc method*. In this approach, information items of the sort that one might like to have in a conceptual model were scattered among design papers, user and analyst manuals, code comments, etc. Often conceptual models described by the ad hoc method were incomplete, inconsistent, and not updated for continued development and evolution of the simulation.

Another approach to documenting simulation-related conceptual models is the *design accommodation method*. In this, the simulation developer uses the descriptive format, such as UML, that has been chosen to support simulation design to describe and document the conceptual model. There are advantages to such an approach:

- It minimizes opportunity for misunderstanding and error as the simulation developer transforms the conceptual model into the simulation design.
- It facilitates keeping the conceptual model current with evolution of the simulation.

However, there are drawbacks to this approach.

- Most simulation development formats such as UML do not have convenient mechanisms for capturing assumptions, algorithm pedigrees, POCs, simulation development plans, and other such information that is part of a well-done simulation-related conceptual model.

- Use of a particular simulation development format can limit capability of the conceptual model as a communication vehicle among the variety of parties with interests in the simulation (sponsors, users, assessment personnel, etc., as well as simulation development personnel) because some of them may not be adequately familiar with the descriptive format (such as UML) used for the simulation design.

A third approach to describing simulation-related conceptual models employs knowledge engineering techniques. This method was emphasized in the CMMS/FDMS paradigm. It has also been continued by a variety of others (e.g., Firat 2001, Kabilan and Mojtahed 2006). This approach has the benefit of forcing the conceptual model to have more formality and logical consistency than it might otherwise have. This approach also has substantial potential for efficiency and reuse of conceptual elements as they employ "standard" compositions and formats. However, it is not always possible to describe the complexity of some conceptual model easily in such formats.

Efforts to automate aspects of simulation-related conceptual model development normally use the design accommodation approach or a knowledge engineering approach, or a combination of the two. Such automation endeavors include development of tools that support conceptual model development. Within particular application domains, encouraging progress is being made.

A fourth approach considered during development of *RPG* simulation conceptual model ideas is the *scientific paper method*. This approach employs the normal way of developing a scientific paper (or report). This material tends to identify assumptions more completely, be more explicit about algorithms and their development, uses standard mathematical and technical conventions, and is more rigorous in its specifications of limitations associated with the simulation conceptual model. This method of conceptual model description also is the most amenable to robust support for conceptual validation reviews and most accommodating to simulation reuse and modification. It was the approach to describing simulation-related conceptual models preferred by the team developing conceptual model materials for the *RPG*.

The FEDEP and the VV&A overlay for it both identify a number activities related to development, use, and evaluation of the FCM, but IEEE Standards 1516.3 (FEDEP) and 1516.4 (VV&A Overlay) do not prescribe particular documentation formats for these activities.

16.5 SIW Conceptual Model Study Group

By late 2001, the simulation (federate) conceptual model described in the *RPG* and the FCM of the FEDEP had basically stabilized. Their approaches for conceptual model development and documentation have changed little since 2001.

Unfortunately, although the importance of an explicit and distinct simulation conceptual model was frequently emphasized within the SIW VV&A community and elsewhere, development of a simulation conceptual model of the sort described by the *RPG* has not been done in many simulation developments. This should not imply that explicit conceptual models have not been developed. There are a number of program-specific examples within the US Defense community, such as the maritime component of the Synthetic Theater Operations Research Model (STORM), where explicit and distinct simulation conceptual models of the sort described by the *RPG* have been developed. However, failure to develop an explicit and distinct conceptual model continues to be a problem in many simulation developments. Assessments have been published showing that absence of a *RPG*-like simulation conceptual model can make simulation conceptual validation more difficult and can increase resource requirements since it can be more difficult to reuse simulation components with the information provided by a well-done simulation conceptual model (e.g., Metz 2000).

During 2002, Jack ("Jake") Borah began to emphasize the need to go beyond where things were with simulation-related conceptual modeling. He thought a conceptual modeling study group within SIW could begin to identify conceptual modeling best practices and possibly develop a conceptual modeling overlay for the FEDEP (Borah 2002). In the spring of 2003, SIW established the SCM SG with Borah as its leader.

During the next 2 years, in addition to interim reports at the Euro-SIWs and Fall SIWs, the Study Group published an article in *Simulation Technology* (Borah 2003), created a SCM bibliography (in 2008, the bibliography reached its sixth version, which is available in the Standing Study section of the SISO Web site, http://www.sisostds.org), and developed a set of foundational documents for a simulation conceptual modeling Product Development Group (PDG) within SIW. "This set of documents consisted of the SCM Introductory Statement, the SCM Topics Listing, the SCM Terminology for Definition, and the SCM Taxonomy of Concepts" (Borah 2005).

In July 2007, the SIW SCM SG transitioned into a SSG.* The focus of the SIW Simulation Conceptual Modeling (SCM) group has evolved from developing a possible conceptual modeling overlay to the FEDEP to a focus on expansion of FEDEP Step 2 (Conceptual Analysis Activity), Activity 2.2 (Develop Federation Conceptual Model) (SIW SCM SSG, Spring 2007 meeting).

In 2007, the SIW SCM SSG also began to interact significantly with the NATO Conceptual Modeling Study Group (MSG-058) and plans to include

* A SISO study group typically is established for a limited time and is expected to evolve into a product oriented group or into a Standing Study Group (SSG). A SSG within SISO is "established to represent a specific community or national group, to mature a potential standard ... SSGs may have an indefinite life span." (SISO Web site, http://www.sisostds.org).

material from the NATO Conceptual Modeling Study Group in development of conceptual modeling best practices as a balloted community product (personal e-mail communication from Borah, 13 January 2009).

In 2008, Simulation Conceptual Model SSG focus evolved to work with the DSEEP model development, helping with the conceptual model definition and conceptual model development process. The FEDEP is a generalized systems engineering process for building and executing HLA federations. In 2007 as part of the SISO for periodic review of approved products (such as IEEE Std 15.16.3-2003) to ensure their continued relevance, it was decided to expand the systems engineering process embodied within the FEDEP to "all users of distributed simulations" in a product now called DSEEP. It "is intended as a high-level process framework into which lower-level systems engineering practices native to any distributed simulation user can be easily integrated."

16.6 Persistent Problems

This section identifies a number of persistent problems impacting simulation-related conceptual modeling theory and practice. Overcoming some of these problems will require advances in conceptual modeling theory (refinement of conceptual modeling definitions, conceptual model content, documentation formats, processes for conceptual model development and use, etc.) and other problems will require changes in the way simulation community members behave if progress is to occur.

Conceptual modeling is bigger than simulation-related conceptual modeling. This was noted earlier, with an example of nonsimulation conceptual modeling being the *Journal of Conceptual Modeling* with its database orientation. Insights about conceptual modeling from the nonsimulation communities have been used by simulation-related conceptual modeling where pertinent. Often these insights have come to simulation-related conceptual modeling via knowledge engineering approaches. The comments of this section are restricted to simulation-related conceptual modeling for simulations mainly used within the US military and Defense community. It should be noted that such simulations include those related to NATO and other US allies; so this perspective is not as restrictive as it might sound at first.

16.6.1 Failure to Develop Explicit and Distinct Simulation-Related Conceptual Models

The most serious persistent problem in simulation-related conceptual modeling in this author's perspective is the frequent failure of a simulation development or modification to produce an explicit and distinct simulation conceptual model as an artifact. Reasons for the persistence of this problem include the following:

- Few simulation development or modification contracts specify a conceptual model as a product required by the contract.
- Lack of a widely accepted standard paradigm for simulation development and modification that specifies artifacts normally produced in quality simulation developments (with the conceptual model as one of the artifacts)—this comment mainly applies to single simulations (federates) since a standard paradigm (the FEDEP) exists and is used for HLA combinations of simulations (federations) in accordance with IEEE Std 1516.3-2003.
- Lack of professionalism by simulation developers.

There are enough simulation development horror stories of avoidable problems that were encountered because of lack of an explicit conceptual model for knowledgeable simulation developers to know the importance of producing an explicit simulation conceptual model. Normally such stories are heard privately since simulation developers are reluctant to air such problems publicly, or to document them in publications.

This author considers failure of simulation communities to insist that all simulations within their sphere have explicit conceptual models to be a lack of professionalism. M&S practitioners have been striving to advance M&S as a profession. Their efforts include work on development of an M&S body of knowledge (e.g., Oren 2005) and development of various standards for professional certification in M&S.* Perhaps as progress occurs in M&S professional standards, failure to develop simulations without explicit conceptual models will become unacceptable to all M&S professionals.

What must happen for this problem of simulations being developed without explicit conceptual models to be overcome? It took a DoD edict that only distributed simulations using HLA would be funded to make the US Defense community use HLA as extensively as it now does. It probably will take a similar level of financial incentive for all (or at least most) simulation developers to always produce an explicit and distinct conceptual model during simulation development. As system designs and operational policies become based more heavily on simulation results, perhaps the threat of litigation questioning the appropriateness of such results will stimulate more emphasis on explicit conceptual models for simulations to provide the solid rational basis for conceptual validation

* For example, the Modeling and Simulation Professional Certification Commission (M&SPCC) under the auspices of the National Training and Simulation Association (NTSA) is involved in such.

that a simulation conceptual model can provide (especially for those areas in which data are limited or lacking) will create the financial incentive for always having an explicit conceptual model as an artifact of simulation development.

Successful simulation developments and modifications with explicit and distinct conceptual models demonstrate viability of including such conceptual models in simulation development and modification. It is time for simulation developers to move beyond the simulation development equivalent of spaghetti code, which is an appropriate analogy for simulations developed without explicit conceptual models.

16.6.2 Diversity of Applications

As noted earlier, the diversity of simulations for which guidance in the DoD *RPG* is intended precluded the possibility of prescriptive guidance about conceptual model; hence, conceptual modeling guidance in the *RPG* is descriptive. A sound military principle is "Divide and conquer." It applies here. Prescriptive conceptual modeling guidance is viable for restricted application domains. For that application domain, a precise definition of conceptual modeling for simulations in that application domain can be developed. The exact information content of such conceptual models and their documentation format can be specified. Development and use processes can be defined explicitly. Other aspects of conceptual modeling can also be addressed in detail for that application domain. Perhaps after conceptual modeling successes that can be captured and summarized as best practices for a few dozen application domains have been demonstrated, more general prescriptive conceptual modeling guidance can be developed based upon common elements of such best practices.

16.6.3 Excessive Expectations for Simulation-Related Conceptual Modeling

The potential of formal approaches from knowledge engineering, from use of UML, and from other simulation-design methodologies to improve conceptual modeling, especially in automating conceptual model development and facilitating reuse of conceptual model components is great, but thus far such approaches have been most successful in particular application domains. Sometimes discussion of such formal approaches to simulation-related conceptual modeling leaves the impression that these formal approaches to simulation-related conceptual modeling will be able to address all varieties of simulation applications, including what presently can only be done with less formal approaches to conceptual modeling, and that as a result of extensive use of the formal approaches, many of the simulation problems stemming from inadequate conceptual models will be avoided in the future. This author believes that kind of impression creates excessive expectations that are unlikely to be achieved in the near term.

Perspectives presented at the Spring 2008 meeting of the SIW SCM SSG provide a counterbalance to the kind of excessive expectations mentioned above. That "no single, monolithic conceptual model can satisfy the needs of all stakeholders" was presented as a premise in this meeting. This comment is in the context of a conceptual model related to a single simulation. Acceptance of such a premise will help to reduce excessive expectations for simulation-related conceptual modeling, and make expectations more realistic. It was also noted in that meeting that not all future ways simulation-related conceptual models may be useful are currently known and understood. A conclusion suggested at the meeting was that composable approaches to conceptual model construction should be pursued to help ensure greatest future utility for simulation-related conceptual models.

16.6.4 Resource Limitations

Overcoming difficult problems requires substantial resources. Borah noted that lack of resources for those directly involved in the SIW conceptual modeling study group was the biggest obstacle preventing the study group from making as much progress as desired (personal e-mail communication from Borah, 13 January 2009).

There are a number of substantial development efforts currently related to simulation-related conceptual models and tools for them, a few of which are identified here. A number of program-specific simulations within the US Defense community provided adequate resources and direction for explicit conceptual models of the sort described in the *RPG* to be part of simulation development. Activities by the Swedish Defence Research Agency (FOI) were mentioned earlier. A substantial effort addressing conceptual model V&V as well as tool support for conceptual modeling is going on in Turkey under the leadership of Semih Bilgen (e.g., Tanriover and Bilgen 2007). Development of conceptual modeling material for the evolving DSEEP supplement to the HLA FEDEP and the NATO Conceptual Modeling Study Group (MSG-058) were mentioned earlier. Some of these efforts (and others) are addressed elsewhere within this book.

In 2006, Stewart Robinson provided a broader perspective on conceptual modeling than is usually presented by identifying a number of research areas that need to be addressed for substantial progress in simulation-related conceptual modeling (Robinson 2006a, 2006b). The research areas identified included conceptual modeling definition and requirements, methods for conceptual model development and representation, conceptual model assessment, and effective means of teaching conceptual modeling. Whether or not adequate resources come forth to support such research fully remains to be seen.

16.7 Final Comments and Conclusions

This chapter has provided historical context for several major conceptual modeling developments within the US military and Defense simulation communities. This context should provide increased understanding about continuing issues in simulation-related conceptual modeling.

For new simulation developments and for simulation modifications, ample guidance is provided in the DoD *RPG* about the simulation conceptual model for an explicit and distinct conceptual model to be produced as an artifact of simulation development or modification even though that guidance is descriptive instead of prescriptive. For HLA federations, the FEDEP of IEEE Std 1516.3 provides guidance about the FCM and the VV&A overlay to the FEDEP of IEEE Std 1516.4 provides guidance that will help ensure the FCM accomplishes its functions acceptably. In the future, DSEEP materials are expected to provide comparable guidance for all users of distributed simulations.

Simulation developers, whether of a single simulation (federate) or a combination of simulations working in concert as a distributed simulation (federation), should follow the available guidance about simulation-related conceptual models so they can avoid the kinds of problems that arise from not having the kind of conceptual model needed.

Much room for improvement in simulation-related conceptual modeling exists. Various efforts are underway to advance conceptual modeling capabilities within US Defense simulation communities and elsewhere. These efforts are expected to improve efficiency in conceptual model development, both by increased automation and by employment of proven methods in conceptual model development. These efforts also may increase potential reuse of conceptual model components, especially with an organization or particular simulation development.

Appendix: Glossary

AIAA	American Institute of Aeronautics and Astronautics
ASME	American Society for Mechanical Engineers
CMMS	Conceptual Model of the Mission Space
DCFM	Defence Conceptual Modeling Framework
DEVS	Discrete Event System Specification
DIS	Distributed Interaction Simulation
DMSO	Defense Modeling and Simulation Office
DoD	United States Department of Defense
DSB	Defense Science Board

DSEEP	Distributed Simulation Engineering and Execution Process
EATI	Entities, Actions, Tasks, and Interactions
FCM	Federation Conceptual Model
FDMS	Functional Description of the Mission Space
FEDEP	Federation Development and Execution Process
FOI	Swedish Defense Research Agency
FOM	Federation Object Model
GAO	Government Accounting Office
HLA	High Level Architecture
IEEE	Institute of Electrical and Electronics Engineers
KIRC	Knowledge Integration Resource Center
M&S	Modeling (or sometimes "Model" or "Models") and Simulation(s)
M&S CO	Modeling and Simulation Coordination Office
M&SPCC	Modeling and Simulation Processional Certification Commission
NTSA	National Training and Simulation Association
POC	Point of Contact
RPG	*Recommended Practices Guide*
SCS	Society for Computer Simulation International
SISO	Simulation Interoperability Standards Organization
SIW	Simulation Interoperability Workshop
SME	Subject Matter Expert
STORM	Synthetic Theater Operations Research Model
T&E	Test and Evaluation
UML	Unified Modeling Language
US	United States
V&V	Verification and Validation
VV&A	Verification, Validation, and Accreditation

Acknowledgments

The author appreciates helpful and constructive reviews of draft material for this chapter by Simone Youngblood and Jack ("Jake") Borah.

References

About SISO. 2009. Simulation Interoperability Standards Organization (SISO) Web site, http://www.sisostds.org, accessed on March 18, 2010.

AIAA. 1998. American Institute for Aeronautics and Astronautics AIAA G-077, Guide for Verification and Validation of Computational Fluid Dynamics Simulation.

ASME. 2006. American Society for Mechanical Engineers Guide V&V 10—2006, *Guide for Verification and Validation in Computational Solid Mechanics.*

Blair, M., and M. Love. 2002. Scenario based affordability assessment tool. NATO RTO AVT Symposium on *Reduction of Military Vehicle Acquisition Time and Cost through Advanced Modelling and Virtual Simulation*, 22–23 April 2002, Paris, France (RTO-MP-089).

Borah, J. 2005. Simulation Conceptual Modeling (SCM) final report, submitted to SISO Standards Activities Committee. SISO Web site, http://www.sisostds.org, accessed on March 18, 2010.

Borah, J. 2003. Simulation conceptual modeling study group gets rolling. *Simulation technology magazine* 6(3).

Borah, J. 2002. Conceptual modeling—How do we move forward?—The next step. Paper 054 in *Fall 2002 Simulation Interoperability Workshop.*

Department of Defense Modeling and Simulation (M&S) Master Plan, DoDD 5000.59-9P, October 1995.

Department of Defense Modeling and Simulation Coordination Office. 2006. *VV&A Recommended Practices Guide, RPG Build 3.0.* http://vva.msco.mil/ (accessed January 26, 2009).

DMSO. 1999. Defense Modeling and Simulation Office, *High Level Architecture Federation Development and Execution Process (FEDEP) Model*, Version 1.5, December 8, 1999.

Distributed Interactive Simulation (DIS). 1995. *A glossary of modeling and simulation terms for distributed interactive simulations.* Institute for Simulation and Training, University of Central Florida.

Druid, L., K. Johansson, P. Stahl, and P. Asplund. 2006. Methods and tools for simulation conceptual modeling, Paper 06E-SIW-029 in *2006 European Simulation Interoperability Workshop.*

Firat, C. 2001. A knowledge based look at Federation Conceptual Model development. Paper 024 in *European 2001 Simulation Interoperability Workshop.*

Heavey, C., and J. Ryan. 2006. Process modelling support for the conceptual modelling phase of a simulation project. In *Proceedings of the 2006 Winter Simulation Conference*, 801–808.

Hollenbach, J. W. and W. L. Alexander. 1997. Executing the modeling and simulation strategy, making simulation systems of systems a reality. *Winter Simulation Conference 1997 Proceedings.*: 948–954.

IEEE Std 1516.3-2003, Recommended Practice for High Level Architecture (HLA) Federation Development and Execution Process (FEDEP).

IEEE Std 1516.4-2007. 2007. IEEE Recommended Practice for Verification, Validation, and Accreditation of a Federation: An Overlay to the High Level Architecture Federation Development and Execution Process.

Kabilan, V., and V. Majtahed. 2006. Introducing DCMF-O: Ontology suite for defence conceptual modelling. Paper 028 in *European 2006 Simulation Interoperability Workshop.*

Kaminski, P. 1996. DoD USD (AT&L) Memorandum, subject: DoD High-Level Architecture (HLA) for Simulations, September 10, 1996.

Krut, T., and N. Zahman. 1996. *Domain Analysis Workshop Report for the Automated Prompt Response System Domain*, Special Report CMU/SEI-96-SR-001, May 1996.

Lewis, R. O., and G. Q. Coe. 1997. A comparison between the CMMS and the conceptual model of the federation. Paper97F-SIW-001 in *Fall 1997 Simulation Interoperability Workshop.*

Lutz, B. 2003. IEEE 1516.3: The HLA Federation Development and Execution Process (FEDEP). Paper 03E-SIW-022 in *2003 European Simulation Interoperability Workshop.*

Mayer, R. J. and R. E. Young. 1984. Automation of simulation model generation from system specifications. In *Proceedings of the 1984 Winter Simulation Conference,* 570–574.

Metz, M. 2000. Comparing the Joint Warfare System (JWARS) conceptual model to a conceptual model standard. Paper 129 in *Fall 2000 Simulation Interoperability Workshop.*

Mojtahed, V., M. Lozano, P. Svan, et al. 2005. DCMF: Defence Conceptual Modeling Framework. Swedish Defence Research Agency (FOI) Report FOI-R--1754–SE, November 2005.

NASA Tech Briefs. 2006. *Adaptive Modeling Language and Its Derivatives.* June 1, 2006. http://www.techbriefs.com/content/view/116/34/ (accessed August 2009).

Oren, T. I. 2005. Toward the body of knowledge of modeling and simulation. Paper 2025, *Interservice/Industry Training, Simulation, and Education Conference (I/ITSEC) 2005.*

Pace, D. 2001. Impact of federate conceptual model quality and documentation on assessing HLA federation validity. Paper 014 in *European 2001 Simulation Interoperability Workshop.*

Pace, D. 2000. Conceptual model development for C4ISR simulations. *In 5th International Command and Control Research and Technology Symposium (ICCRTS),* 24–26 October 2000, Canberra, Australia.

Pace, D. 1999. Development and documentation of a simulation conceptual model. Paper 017 in *Fall 1999 Simulation Interoperability Workshop.*

Robinson, S. 2006a. Issues in conceptual modelling: Setting a research agenda. In *Proceedings of the Third Operational Research Society Simulation Workshop (SW06),* 165–174. Birmingham, UK: Operational Research Society.

Robinson, S. 2006b. Conceptual modeling for simulation: Issues and research requirements. In *Proceedings of the 2006 Winter Simulation Conference,* 792–800.

Schlesinger, S. 1979. Terminology for model credibility. *Simulation* 32(3): 103–104.

Sheehan, J., T. Prosser, H. Conlay, et al. 1998. Conceptual Models of the Mission Space (CMMS): Basic concepts, advanced techniques, and pragmatic examples. Paper 127 in *Spring 1998 Simulation Interoperability Workshop.*

Tanriover, R., and S. Bilgen. 2007. An inspection approach for conceptual models in notations derived from UML: A case study. In *22nd International Symposium on Computer and Information Sciences, 2007, International Symposium on Computer and Information Sciences (ISCIS) 2007.*

Zeigler, B. P. 1999. A theory-based conceptual terminology for M&S VV&A. Paper 064 in *Spring 1999 Simulation Interoperability Workshop.*

17

On the Simplification of Semiconductor Wafer Factory Simulation Models

Ralf Sprenger and Oliver Rose

CONTENTS

17.1 Introduction

In the globalized world with competition and the focus on productivity, the main goal in semiconductor wafer manufacturing is to maximize the output of the fabrication facilities (in short, wafer fabs) under due date constraints. In such an optimized environment, it is essential to have good prediction of fab behavior after a breakdown of critical machines and how to meet the due dates in such a situation. To that end, simulation is a very important tool. Simulation can also be used to test various parameters of machines to find better settings (e.g., as in Rose 2003).

Modeling manufacturing environments in all of its details result in complex models, which have the big disadvantage that simulation studies can become very time consuming. Therefore, we developed several approaches to build simple models that lead to shorter simulation runtimes.

It is abundantly clear that a high degree of simplification of a complex fab model leads to relatively high deviations of the interesting fab characteristics but also to a very short simulation time (e.g., Hung and Leachman 1999). In contrast to this, a low degree of simplification leads to low deviations in the characteristics but a higher simulation time.

When designing a simulation conceptual model, the model should reproduce the behavior of the real world related to the model objectives (Robinson 2008). Objectives are, for example, to maximize the throughput or to minimize the inventory level in a fab. Consequently, the simulation model must mimic the behavior of the real world concerning these characteristics to a sufficient degree. In addition, the model should be as simple as possible to reduce the development and simulation time. This chapter focuses on models with a very high degree of simplification and we show different methods to reach a sufficient degree of accuracy concerning different model objectives.

In the next section, we give an overview about the already existing approaches, both with a high and a low degree of simplification. In the third section, we describe the complex wafer fab model, which is used to estimate the parameters of the simple models, and the procedure to calibrate the simple models. Then we present two variants of a simple model. The first one focuses on a good prediction of the average lot cycle times whereas the second focuses on predicting lot cycle time distributions. We also provide some results and discuss the limitations of the approaches. In the last section, we give a conclusion.

17.2 Related Work

Robinson (2008) provides an overview on model simplification ideas for conceptual models. In most cases, simple models are not built from scratch but are simplifications of existing models. To that end the modeler removes scope and detail from a given model or represents its components in a simplified manner. Most authors work with the one or more of the following approaches: removing unimportant components of the model, using random variables to replace parts of the model, considering less detail for the range of variables in the model, and combining components of the model into new and simpler components.

The main idea of our approach to reduce the complexity of the model for a semiconductor wafer fab is to replace machines in the fab with delay elements, i.e., with random variables representing large parts of the production plan of a lot (e.g., Hung and Leachman 1999). This means to model a few machines in detail and to replace the processing steps a lot would take at deleted machines by dummy machines. These dummy machines mimic the delay the lots would need at the replaced steps. The main issue is to delete as many machines as possible with the objective to reduce the complexity of the fab and to minimize the number of operations during simulation while the simple model behaves as much as possible as the complex one. Highly utilized machines have a big influence on the fab and the behavior of the lots, and should not be considered for deletion. In Jain and Lim (1999b), different levels of detail have been tested. One idea is to model only the bottleneck machine group in detail. Alternatively more highly utilized machines can be modeled if the accuracy concerning the behavior of the simple and the complex model are inadequate. Our approaches focus on the high degree of simplification bottleneck only modeling approach to maximize the savings in simulation time.

There are different ways to set the processing times of the dummy machines. Rose (1998) uses exponential distributions, whereas Hung and Leachman (1999) try static values and quartile-uniform distributions. The quartile-uniform variant assumes the distribution is uniform between quartile points. Calculating delay values using this method is faster than the distribution method of Rose. Our approaches in this chapter use Rose's variant because the exponential distributions match the distributions in the complex model at best.

The next problem is the dynamic adaptation to different utilizations of the fab. Peikert, Thoma, and Brown (1998) try to adapt the dummy delays by calculating the raw processing time of the dummy steps and multiplying it with the flow factor (cycle time divided by the raw processing time) of the lots at a specific utilization. This method may be acceptable, but in scenarios where the different parts of the fab have differently utilized machines, this approach may influence the results to a high degree. Further dynamic adaptation approaches are not available in the literature so far. In this chapter, we extend Rose's approach with dynamic distributions to adapt to different release rates of lots into the fab, or in a more general sense to different fab workloads.

The intention of the application of simple models is to match some or ideally all important fab performance characteristics of the complex model. In the related work, different types of simple models are tested with regard to some of these characteristics. Jain, Lim, and Low (1999a) split a fab into a few independent parts and use the bottleneck only approach for each part. They focus on the average cycle time and compare a detailed fab model with their reduced model. The simple models fail to approximate the average cycle time. Peikert, Thoma, and Brown (1998) also use a bottleneck-only approach. Their interest is to assist the design of a wafer fab with the aim to understand fab's behavior and to optimize operator deployment and the usage of dispatch rules. Hung and Leachman (1999) use different levels of model reduction and predict the cycle time of lots with an acceptable accuracy.

However, none of these approaches provides a sufficient adaptation to a change in lot release rates. In this chapter, we extend the previous approaches of one of the authors to overcome this weakness. This approach is as simple as possible and provides a high simulation speed-up. The related work can be found in Rose (1998, 1999, 2000a, 2000b, and 2002) and will be described in the following sections.

17.2.1 Basis of the New Approach

Rose's simple model approach is shown in Figure 17.1. As mentioned, we chose the bottleneck only approach and modeled the bottleneck machine group in all its details in our simple model. The bottleneck machine is usually not the first machine that a lot enters when entering a fab. The machines between the production start of a lot and the first arrival at the bottleneck queue are modeled with a delay distribution. Another delay distribution is used to model the machines at the end of the production process that lots will pass after their last departure from the bottleneck tool group. Another important property of wafer fabs is the repeated processing of the lots at the bottleneck tool group. A wafer contains up to 50 layers, which require similar processing steps. This repeated processing of lots at the bottleneck tool group makes it necessary to extend the model with a loop back to the bottleneck. In the previous approaches of Rose a static delay distribution is used to model the delay of the lots between single bottleneck processing steps. Static distributions mean that all lots of a product will be delayed according to the same distribution. We have three dummy machines in the simple model, in total. For each of these dummy machines one distribution for each product in the fab was assigned.

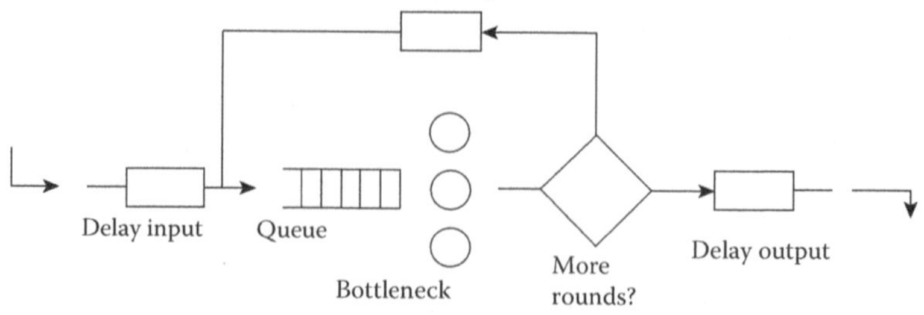

FIGURE 17.1
Simple model with delay distribution in the loop.

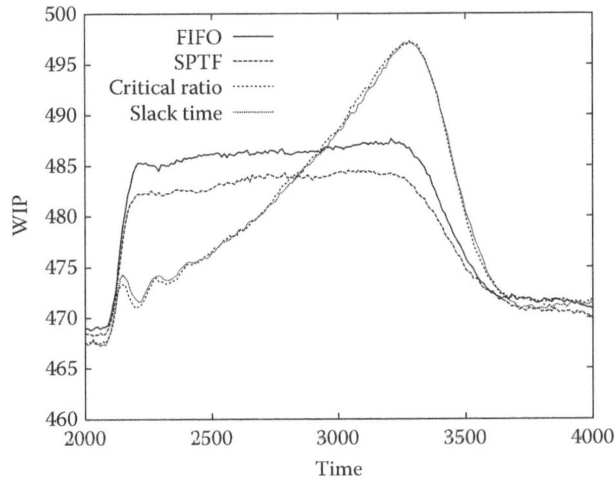

FIGURE 17.2
WIP evolution.

17.2.2 Predicting Fab Behavior Over Time

Figure 17.2 shows the inventory (WIP = Work In Progress) over time after a breakdown of all bottleneck machines. Dozens to hundreds of replications are necessary to generate these curves so that our simple model approach has to be used. Typical research questions for bottleneck breakdowns are to find appropriate dispatch rules to meet due dates or to avoid the dramatic increase of the inventory/WIP level after a bottleneck breakdown. It is possible to test this within an acceptable time horizon only with a simple model. In the figure, first-in first-out (FIFO), shortest processing time fist (SPTF), critical ratio (CR), and slack time are compared. A detailed description of this problem and the different dispatch rules can be found in Rose (1998).

17.2.3 Predicting Cycle Times

Predicting cycle times works well with the simple model if the workload and product mix that is released into the simple fab model is the same as it was in the complex model during calibrating the simple model. This leads to one of the weaknesses of the old approach. In Figure 17.3, the characteristic curves of the complex model and the simple model are compared. The simple model was adjusted with one simulation run of the complex model at a release rate of 70%. Consequently, the mean cycle time of the lots matches the value of the complex model only for this workload. At high workloads (beyond 90%) the mean cycle time increases. The reason for this evolution of the curve is the nonexisting workload limitation of the simple model. Only the bottleneck workcenter has a limiting effect, which appears only at higher workloads. A detailed description of this problem can be found in Rose (2000a).

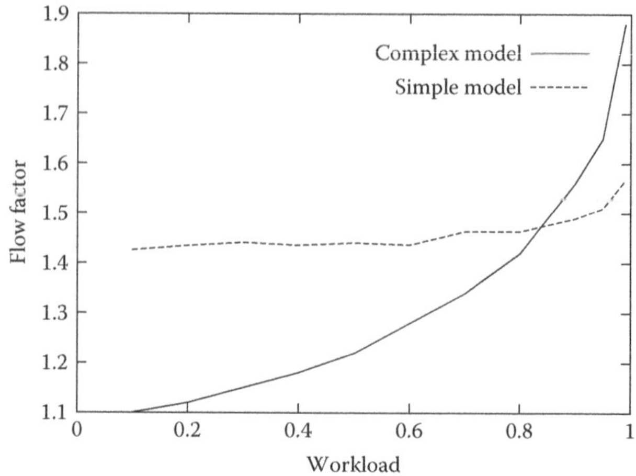

FIGURE 17.3
Characteristic curve.

17.3 New Approaches: An Introduction

Our new approaches are based on the approach discussed in the last section. So far static distributions have been used in the loop to model the complex fab behavior. In our new approaches, we make the distributions in the loop dependent on the number of lots in this loop, which is an indirect indicator of the fab workload. This means that at the time a value is measured (e.g., a delay value of a lot in the loop) in the complex model simulation run (which is used to adjust the simple model), the number of lots in the marked section with the label "lots in loop" of the fab (Figure 17.4) is used to assign this value to one of an array of distributions. At the time a lot needs to be delayed in the loop of the simple model the number of lots in the loop will be counted and the respective distribution will be used to sample a delay value for this lot.

The difference between our two approaches is the following: The first approach uses a delay distribution modeling the delay of a lot in the loop, whereas the second approach intends to generate the interarrival time distributions of lot arrivals at the bottleneck queue.

17.3.1 Complex Model

We chose the Measurement and Improvement of MAnufacturing Capacities (MIMAC) model 6 from the SEMATECH test bed (Fowler and Robinson 1995) as our complex wafer fab model test case. It consists of 228 workstations. Nine products are built in this fab. Every product has about 300 processing steps. Therefore, this fab is complex enough to be comparable to a

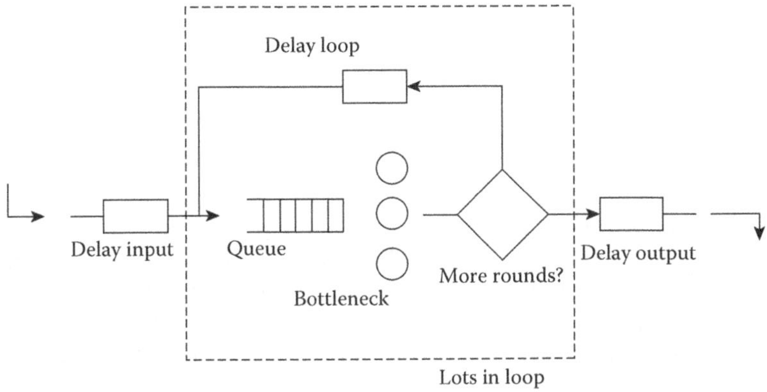

FIGURE 17.4
Simple model with number of lots in the loop.

real wafer fab. There are some limitations concerning the operators of the bottleneck machine group. If this group is not only responsible for the bottleneck machines, the operators have to be deleted due to the nonavailability of other machines, other than these bottleneck machines in the simple model. This procedure is necessary in our case. One solution to avoid this problem could be to restrict a particular group of operators in the complex model to the bottleneck machine group. If this is the case, they can be modeled with all characteristics in the simple model, too. Alternatively, it may be possible to compute availability times of operators at the bottleneck (depending on the utilization of the fab) and use these values to adjust operator characteristics in the simple model. However, we did not consider these alternatives. The second limitation is that products that are not passing through the bottleneck must be deleted. In our fab this applies to three products so that six products remain in our fab model. Deleting products in a fab, results in different characteristics of the fab, e.g., in different cycle times of all lots due to the lower utilization. Therefore, we have deleted these three products in our complex fab to solve this problem. A possible alternative could be that additional workcenters have to be included into the simple model.

17.3.2 Required Characteristics for Calibrating the Simple Model

The following characteristics are necessary for calibrating the simple model:

- Delay distribution of every product at the first part of the production process (may depend on the number of lots in the loop).
- Delay distribution of every product at the last part of the production process (may depend on the number of lots in the loop).

- Distributions in the loop:
  - Delay approach: Delay distribution of every product in the loop depending on the lots in loop value.
  - Interarrival time approach: Interarrival distribution of lots from the loop at the queue of the bottleneck machine group depending on the lots in loop value.
- Number of rounds in the loop for every product (we use the average number of rounds of a product in the loop and use this value to route a lot into the loop or to release it through the last distribution of our simple model. If the number of loops varies due to rework for example, a distribution for the number of loops may be required).
- Parameters of the bottleneck machine group:
  - Number of machines.
  - Meantime between failure (MTBF) distribution, meantime to repair (MTTR) distribution.
  - Processing times.
  - Set-up times, set-up rules.

17.3.3 Computing Distributions

A long simulation run of the complex model with a wide range of workloads is necessary to compute enough values for the array of dynamic (discrete empirical) distributions. We tested different run lengths to initialize the model and found that a 23-year-long run of the complex model is sufficient to compute the distributions with an acceptable accuracy. Later, we show a method to shorten this long pilot run. We started the complex model run with a workload of 10% and increased the workload every simulated year by 5%. The run finishes 3 years at 99% workload. Based on this very long run, we obtain delay data on a wide range (between 5 and 160) of lots in loop. The lots-in-loop evolution over time is depicted in Figure 17.5. This method is used for both, the delay and the interarrival time approach described in this chapter.

17.4 Predicting the Characteristic Curve Using the Delay Approach

In this approach, we use delay distributions for each product in the loop and choose a lots in loop interval width of five units. According to Figure 17.5, approximately a maximum of 160 lots can be in the loop at the same time. As a consequence, we have 32 distributions for six products and therefore

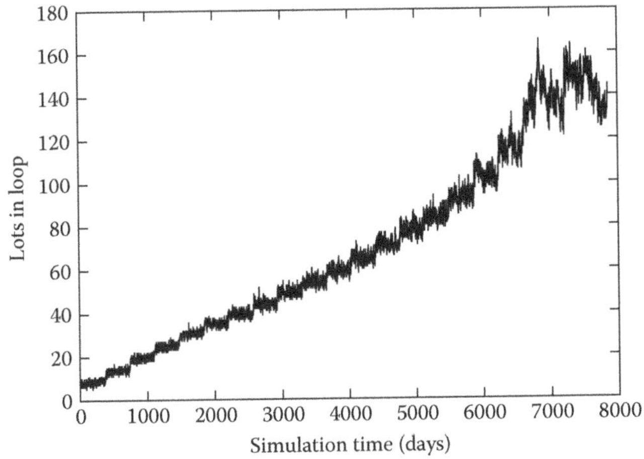

FIGURE 17.5
Single run to generate the delay distributions.

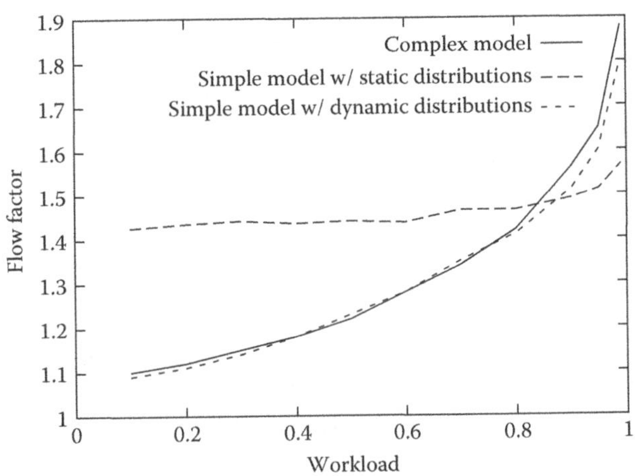

FIGURE 17.6
Characteristic curve.

192 loop delay distributions being dynamically chosen depending on the lots in loop value (indirectly modeling the workload of the fab).

For all of the experiments, we made 25 replications of 10-year runs with a warm-up period of 1 year.

17.4.1 Characteristic Curve

The characteristic curve is shown in Figure 17.6. Concerning the mean cycle time, the new approach matches the complex fab model to a high degree. Only at high workloads the deviation slightly increases.

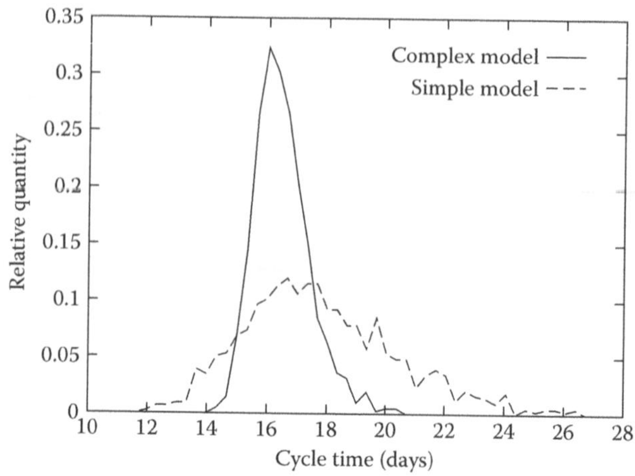

FIGURE 17.7
Cycle time distribution.

17.4.2 Cycle Time Distributions

Concerning the cycle time distribution the variance of the simple model is too high (Figure 17.7). This is due to the overtaking of lots as described in Rose (2000a). In the complex model, only a small percentage of the lots in the loop overtake one another. Only if reworking or batch processing occurs or dispatch rules other than FIFO are used, does the overtaking of lots of the same product take place. Overtaking between lots of different products may occur and depends on the different processing step sequences. In contrast to this, in the loop of the simple model every lot will be delayed independently from the previous one. Therefore, lot overtaking happens rather often. In every round of the lots in the loop of the simple model, the variance of the cycle time distribution increases. Caused by the independent delays of the lots, some lots will have statistically low delay times in every loop and others statistically high ones. In contrast to that, if overtaking happens only in a very limited way (as in the complex model), all lots show a similar cycle time and the variance is lower.

17.5 Predicting the Cycle Time Distribution with the Interarrival Time Approach

To achieve a better cycle time distribution estimate, we want to avoid overtaking of lots in the loop. Hence, we use a dummy machine with a FIFO queue to meet this requirement. A lot arriving at the loop, will be stored in the dummy queue and the machine after this queue releases the lots depending on the interarrival time characteristics at the bottleneck queue. Instead of delay distributions, we now have to determine the appropriate array of interarrival time distributions of lots from the loop at the queue of the bottleneck machine group. This approach still uses dynamic distributions. This is necessary because the cycle time estimates of an interarrival time model without dynamic adaptation to the workload will not match the expected ones. The model will run full or low on lots. This is caused by the indirect modeling of the delay in the loop using interarrival times. The interarrival times change significantly with a different workload of the fab whereas the delay in the loop changes only slightly. We have chosen an interval width of one lot in loop for our experiments because this approach is much more sensitive to deviations in the distributions than the delay approach.

17.5.1 Modeling Interarrival Times

We tested three possibilities to model interarrival times in the simple model. All of them have advantages and disadvantages and will be described in the following.

17.5.1.1 Interarrival between All Lots

The first version (version 1) uses one distribution for all products. The model parameters are determined by measuring each interarrival time independently of the kind of product of the lots (Figure 17.8). There is only a single distribution for all lots-in-loop numbers. This works well but as soon as the product mix changes the cycle time results become bad. For the MIMAC model about 160 distributions have to be generated (one for each lots in loop value, cf. section 17.4). The following versions require a higher number of distributions.

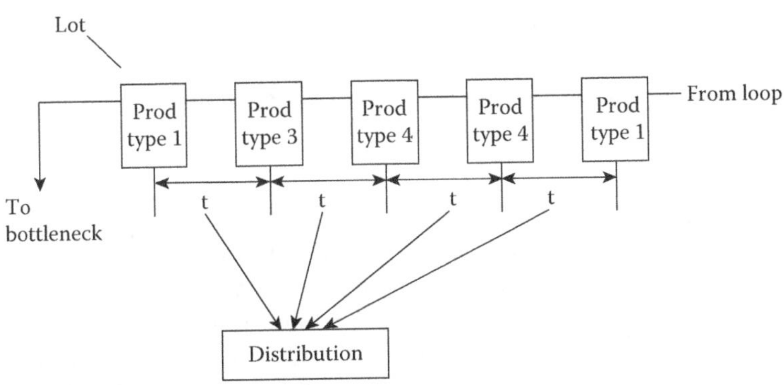

FIGURE 17.8
Interarrival between all lots.

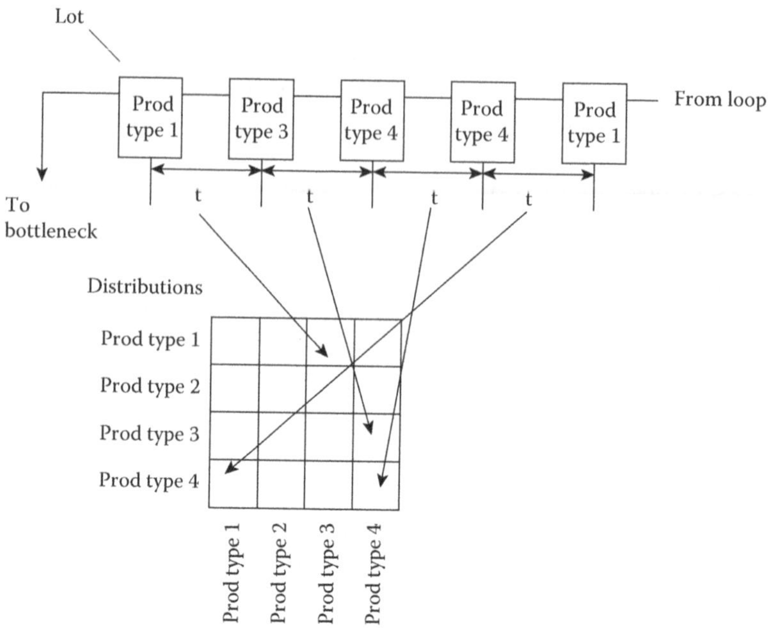

FIGURE 17.9
Interarrival between lots with all combinations of products.

17.5.1.2 Interarrival between Lots with All Combinations of Products

The second version (version 2) uses the interarrival times between all combinations of products (Figure 17.9). In this case, the number of distributions is equal to the square of the number of products because we consider each combination of proceeding lots. In addition, we need distributions for each lots in loop value. This means values for 5760 (6 products squared, yielding 160 distributions each) distributions are required. Therefore, we had to find a method to shorten the time for the simulation run of the complex model, which is discussed later in this chapter.

17.5.1.3 Interarrival between Lots of the Same Product

In the first two versions no overtaking of lots in the loop is possible. But overtaking happens, even if it is relatively limited. If the delay in a loop of one product is much shorter than the one of another product, overtaking will happen more often. Therefore we developed a third model version (version 3).

In the complex model, we measure the interarrival time of lots of the same product (Figure 17.10). Consequently, we have one distribution for every product. That means 960 (6 times 160) distributions in the case of the MIMAC model. In the simple model there is one dummy machine per product. This machine releases the lots of its product according to the distribution of this product. In the first two versions, the dummy machine releases the lots with

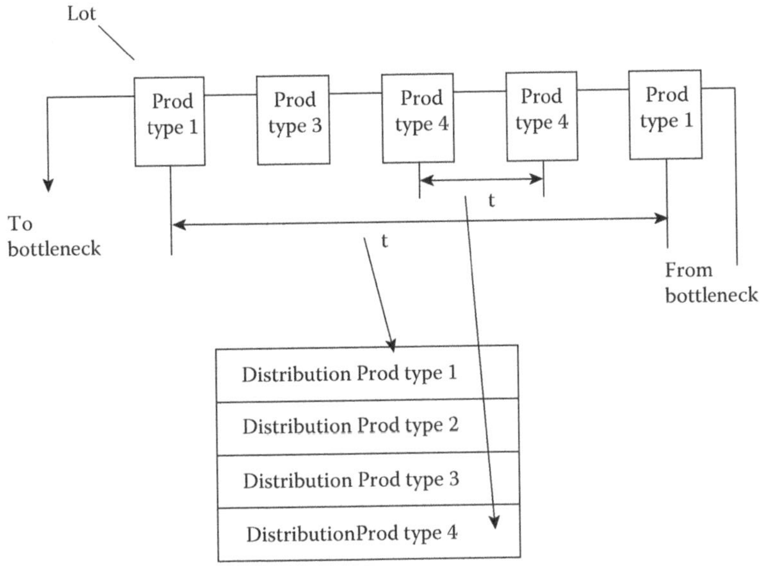

FIGURE 17.10
Interarrival between lots of the same product.

an interarrival time sampled from the distribution that is chosen according to the corresponding total number of lots in loop. For this third version every dummy machine has to use the number of lots of the respective product in the loop to compute and select the corresponding distribution from the array. This is due to the independent release of lots of different products in the loop of this version.

It is necessary that the number of lots of a specific product in the loop of the simple model matches approximately the corresponding number in the complex one to achieve a good mimic of the complex model. In the first two versions, this can be done by the calibration with the total lots in loop value because all products are in the same queue. But if the total lots in loop value is used in the third version the total lots in loop value matches approximately the value of the complex model but the products lots in loop value differs significantly from the respective complex model value after some time and the behavior of the simple model does not match the behavior of the complex model.

The model version that should be used depends on the complex fab model. If overtaking of lots of different products occurs to a considerable degree, the third version might be better. If no product mix changes are planned, the first version should be used. If the product mix changes, the third version seems to be relatively bad. The second version is very good but needs a long simulation run to compute the distributions. The first version is in almost all cases just as suitable and has the advantage that a lower number of distributions is necessary.

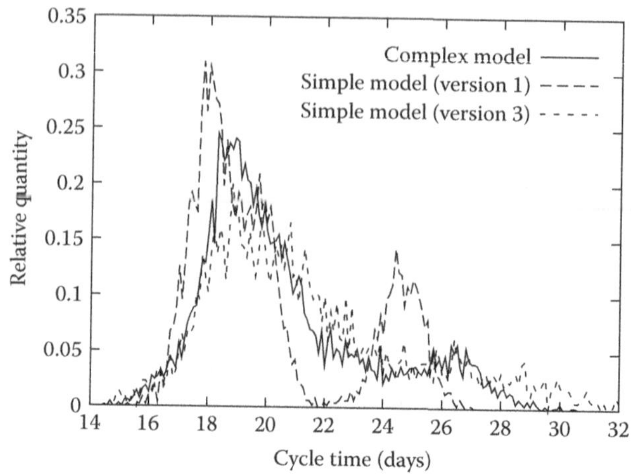

FIGURE 17.11
Absolute cycle time distribution.

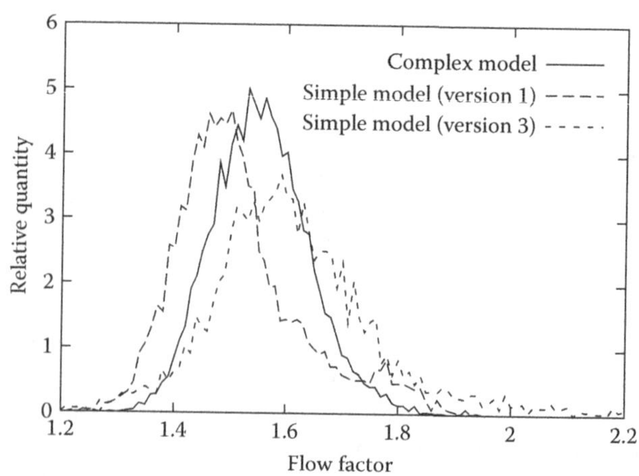

FIGURE 17.12
Relative cycle time distribution.

17.5.2 Cycle Time Distribution

Figure 17.11 shows the absolute cycle time distributions for two of the three versions (version 2 turns out like version 1). Figure 17.12 depicts the cycle times divided by the raw process times (the flow factor of the products). In Table 17.1, we compare the deviations of some cycle time moments for all products. Version 1 leads to a shift in the expected values whereas version 3 leads to a broader variance than in the complex model.

TABLE 17.1

Deviation of the Expectation Value, the Variance and the Relative Divergence of the Expectation Values

Product	Moment	Version 1	Version 3
All Products	Expectation	–0.58	0.78
	Variance	–0.60	3.73
	Relative deviation	–2.8%	3.8%
Product 1	Expectation	2.04	1.61
	Variance	–0.11	6.56
	Relative deviation	11.9%	9.4%
Product 2	Expectation	–1.07	0.23
	Variance	–0.08	0.88
	Relative deviation	–5.7%	1.2%
Product 3	Expectation	–0.76	2.21
	Variance	–0.09	6.45
	Relative deviation	–4.0%	11.7%
Product 4	Expectation	–0.87	0.02
	Variance	–0.15	1.06
	Relative deviation	–4.2%	0.1%
Product 5	Expectation	1.24	1.14
	Variance	0.16	1.75
	Relative deviation	5.3%	4.9%
Product 6	Expectation	–1.92	1.51
	Variance	0.04	1.80
	Relative deviation	–7.3%	5.7%

17.5.3 Characteristic Curve

The characteristic curve chart is shown in Figure 17.13. Version 1 is similar to the one of the previous approach. Version 3 is slightly higher but lies within an acceptable range. Version 2 matches version 1 and is not shown.

17.5.4 Minimizing Adjustment Time

To be able to compute the high number of distributions for some of the simple models we developed a method to shorten the simulation time.

17.5.4.1 Nonlinear Regression

The values have shifted exponentially distributions with a density function of

$$f(x) = -ae^{-bx} + 1,\ a,\ b,\ x \in \Re^{+}$$

with the parameters a and b.

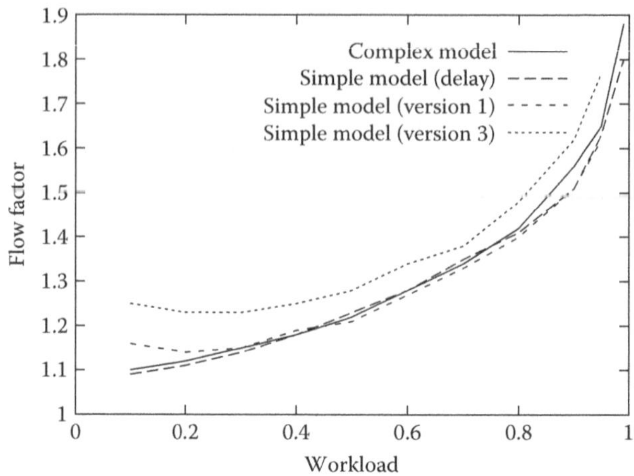

FIGURE 17.13
Characteristic curve.

17.5.4.2 Regression over Regression

Regression can be applied for the parameters *a* and *b*, too. Parameter *a* is described by a first-degree polynomial, whereas *b* is described by a shifted exponential function:

$$a(y) = cy - d \text{ and } b(y) = -he{-}iy + j,\ c, d, h, i, j \in \Re^{+},\ y \in \aleph,$$

where *y* is the number of lots in the loop.

17.5.4.3 Application of the Methods

At first, the above-mentioned two methods must be used. If not enough values are generated (e.g., the simulation run was too short), the missing distributions can be approximated with the function described above. The regression over regression method could be applied for intervals where enough values exist to generate a distribution. The approach requires a good accuracy concerning the distributions. This method prevents bad distribution parameters and enables us to shorten the adjustment time.

With the first version, we were able to shorten the simulation time from 23 to three and a half years without measurable influence on the results of the simple model.

17.5.5 Modeling Overtaking

As already mentioned above, no overtaking occurs in the simple model by applying the first two versions. The deviation in contrast to the overtaking in the complex model shifts the mean value of the products' cycle time

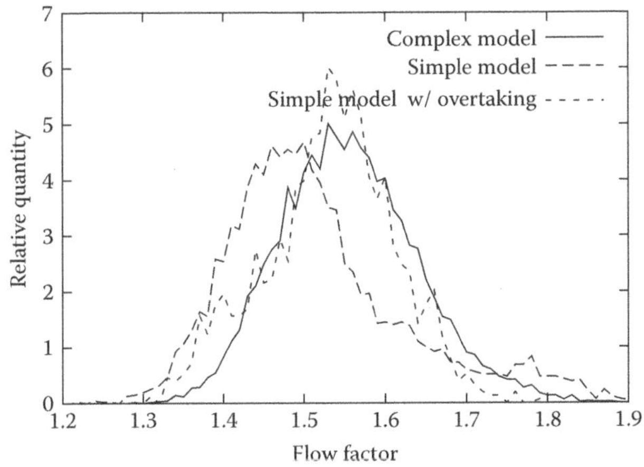

FIGURE 17.14
Simple model with modeled overtaking.

distributions. We achieved a better matching of the curves to the complex model results by modeling overtaking in the first version.

At first, we measure the amount of overtaking in the complex model as the number of lots a product overtakes in the loop on an average. The result is a list that contains an overtaking value that specifies the number of overtakings for every product. Based on this list, we change the order of the lots in the queue of the loop in the simple model. Every new lot arriving at the queue will overtake exactly the number of lots as described in the list. This allows a better match of the curves. The results are depicted in Figure 17.14. In Table 17.2 the deviations of some distributions' moments for all products are compared.

17.5.6 Disadvantages

The main problem of the interarrival time approach is that the delay of the lots in the loop are indirectly modeled with interarrival times. The average delay of a lot in the loop can be estimated by multiplying the number of lots in the loop with the average interarrival time. There are scenarios where the loop runs low on lots (e.g., if a big bottleneck breakdown occurs). Consequently, the delay in the loop is low. In this case the lots accumulate at the queue of the bottleneck. In general, this is not a problem but in the case of a non-FIFO order of lots at the bottleneck queue some lots may speed-up and leave the fab soon. This leads to a number of finished products that is considerably higher than in the complex model. One example is shown in Figure 17.15 for the CR dispatch rule for different target flow factors (the mean cycle time is 1.66). This is a scenario where the interarrival time approach did not work. For such investigations the delay-based approach should be used.

TABLE 17.2

Deviation of the Expectation Value, the Variance and the Relative Divergence of the Expectation Values: Overtaking Included in Modeling

Product	Moment	w/o Overtaking	w/Overtaking
All Products	Expectation	–0.58	0.20
	Variance	–0.60	–1.34
	Relative deviation	–2.8%	–1.0%
Product 1	Expectation	2.04	0.42
	Variance	–0.11	–0.14
	Relative deviation	11.9%	2.4%
Product 2	Expectation	–1.07	–0.26
	Variance	–0.08	0.12
	Relative deviation	–5.7%	–1.4%
Product 3	Expectation	–0.76	0.26
	Variance	–0.09	–0.18
	Relative deviation	–4.0%	–1.4%
Product 4	Expectation	–0.87	0.01
	Variance	–0.15	–0.02
	Relative deviation	–4.2%	0.0%
Product 5	Expectation	1.24	0.77
	Variance	0.16	–0.02
	Relative deviation	5.3%	3.3%
Product 6	Expectation	–1.92	–1.62
	Variance	0.04	0.32
	Relative deviation	–7.3%	–6.2%

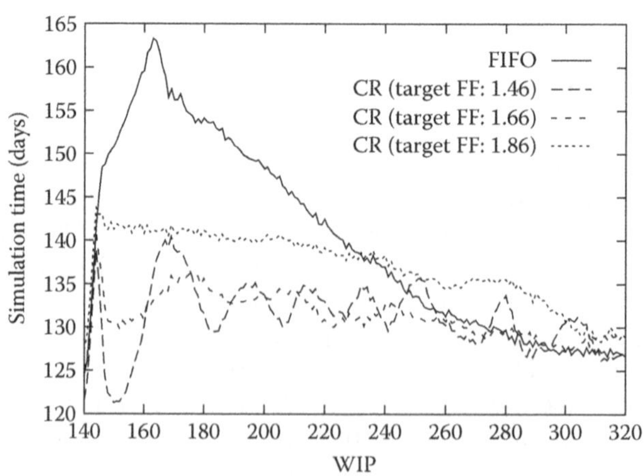

FIGURE 17.15
Different dispatch rules during a big breakdown.

17.6 Conclusions

In this chapter, we presented a simple model for semiconductor wafer fabs or similar fab types where a reentrant lot flow occurs. These approaches can help to reduce the complexity of conceptual models. We introduced two new approaches solving several problems of previous simple model approaches. In particular, concerning the characteristic curve (depicting the flow factor over utilization of the fab and the products' cycle time distributions, the new approaches were considerably better. In addition, we developed a method to shorten the model calibration time, which facilitates the application of a large number of distributions that might be necessary for some versions of the simple models.

The two approaches can also be used to model parts of a fab. Perhaps if the behavior of the complete fab is too complex to be modeled with only one simple model, the discussed techniques can be combined to model the complete fab. For example if the delay of the first loops defer to a high degree from later loops, the fab can be modeled with two or more simple models that are connected in series. Alternatively, each loop could be modeled separately. The discussed overtaking techniques can also be modified to achieve a better mimic of the complex fab. For example the overtaking behavior between different loops can be investigated. However, this makes the simple model more complex and the adjustment and simulation time will increase.

Despite the successful improvements in modeling several fab characteristics, there are still issues left; especially the behavior of the interarrival time approach. For example the delay approach leads to very good performance predictions in the case of a changed product mix, whereas the interarrival approach might fail. It is also important which of the interarrival versions is used. The first version turns out to be very good in most of the scenarios. The second one might be better but the long adjustment time is a big disadvantage. The interarrival approach needs distributions with a high degree of accuracy whereas this is not very important for the delay approach. Furthermore, the application of a non-FIFO dispatch rule is currently not possible with the interarrival approach. In contrast to this the delay approach can be used with other dispatch rules.

References

Fowler, J., and J. Robinson. 1995. *Measurement and Improvement of Manufacturing Capacity (MIMAC) Final Report*. Sematech Inc.

Hung, Y.-F., and R. Leachman. 1999. Reduced simulation models of wafer fabrication facilities. *International journal of production research* 37:2685–2701.

Jain, S., C.-C. Lim, and Y.-H. Low. 1999. Bottleneck based modeling of semiconductor supply chains. *International Conference on Modeling and Analysis of Semiconductor Manufacturing*, 340–345.

Jain, S., and C.-C Lim. 1999. Criticality of detailed modeling in semiconductor supply chain simulation. In *Proceedings of the 1999 Winter Simulation Conference*, 888–896.

Peikert, A., J. Thoma, and S. Brown. 1998. A rapid modeling technique for measurable improvements in factory performance. In *Proceedings of the 1998 Winter Simulation Conference*, 1011–1015.

Robinson, S. 2008. Conceptual modeling for simulation Part I: Definition and requirements. *Journal of the operational research society* 59:278–290.

Rose, O. 1998. WIP evolution of a semiconductor factory after a bottleneck workcenter breakdown. In *Proceedings of the 1998 Winter Simulation Conference*, 997–1003.

Rose, O. 1999. Conload: A new lot release rule for semiconductor wafer fabs. In *Proceedings of the 1999 Winter Simulation Conference*, 850–855.

Rose, O. 2000. Estimation of the cycle time distribution of a wafer fab by a simple simulation model. In *Proceedings of the SMOMS '99e*, 133–138.

Rose, O. 2000. Why do simple wafer fab models fail in certain scenarios? In *Proceedings of the 2000 Winter Simulation Conference*, 1481–1490.

Rose, O. 2002. Some issues of the critical ratio dispatch rule in semiconductor manufacturing. In *Proceedings of the 2002 Winter Simulation Conference*, 1401–1405.

Rose, O. 2003. Accelerating products under due-date oriented dispatching rules in semiconductor manufacturing. In *Proceedings of the 2003 Winter Simulation Conference*, 1346–1350.

Rose, O. 2007. Improving the accuracy of simple simulation models for complex production systems. In *Proceedings of the 2007 INFORMS Simulation Society Research workshop*.

Part VI

Conclusion

18

Conceptual Modeling: Past, Present, and Future

Durk-Jouke van der Zee, Roger J. Brooks, Stewart Robinson, and Kathy Kotiadis

CONTENTS

18.1 Introduction

Conceptual modeling is probably the most difficult part of the process of developing and using simulation models (Law 1991). Despite this fact, conceptual modeling is largely ignored at conferences and in the literature. This book addresses this issue by considering the body of research for the field as it is beginning to develop. In this way it aims to create a point of reference, highlight current research and identify avenues for future research.

The objective of this chapter is to provide an overall summary and assessment of the current state of research in conceptual modeling as set out in the previous chapters, and to highlight the opportunities for future research. As a starting point for this chapter we will use the research agenda set out as a list of research themes (shown in Table 18.1) in the editorial of a recent special issue on conceptual modeling (Robinson 2007). It builds on earlier literature reviews by Robinson (2006, 2008), and the outcomes of a themed day on conceptual modeling following SW06 (The Operational Research Society Simulation Workshop 2006).

TABLE 18.1

Research Themes for Conceptual Modeling

The Problem/Modeling Objectives Domain (P)	The Model Domain (M)
1. Use of "soft OR" as a basis for determining a simulation conceptual model [ch. 4, 9, 10].	1. Identifying dimensions for determining the performance of a conceptual model [ch.1, 2, 4, 5, 7].
2. How best to work with subject matter experts in forming a conceptual model [ch. 4, 5, 7–10].	2. Comparing different models in the same problem domain [ch. 2].
3. How to organize and structure the knowledge gained during conceptual modeling [ch. 11–15].	3. Studying expert modelers to understand how they form conceptual models [ch. 3].
4. Alternative sources of contextual data/information for conceptual modeling, including paper, interview and electronic sources.	4. How software engineering techniques might aid simulation conceptual modeling [ch. 8, 11–16].
5. Developing curricula to include conceptual modeling in university and industry courses on simulation [ch. 3, 16].	5. Adopting/developing appropriate model representation methods [ch. 5, 6, 11–16].
	6. Exploring methods of model simplification [ch. 1, 2, 17].
	7. Identifying, adapting and developing conceptual modeling frameworks [ch. 4–8, 16].
	8. Refining models through agreement between the modeler and stakeholders—"convergent design" [ch. 4, 5, 16].
	9. Exploring the creative aspects of modeling [ch. 10].
	10. Understanding the organizational diffusion and acceptance of models [Ch. 4, 5, 8–10, 12, 13].
	11. Investigating the impact of other modeling tasks on the conceptual model (iteration in the simulation life cycle).
	12. Understanding the effect of throw-away models versus models with longevity—for example, the time spent on conceptual modeling, documentation and organizational diffusion.

Source: Robinson, S., *Journal of simulation*, 1(3), 2007.

Taking each of Parts I–V of the book in turn (Sections 18.2–18.6), the chapters in the book are reviewed and cross-referenced to the research themes in Table 18.1. The themes are numbered 1–5 for the problem domain and 1–12 for the model domain and are referenced by a letter representing the domain (P or M) and their number in the table. For example, M6 refers to the model domain theme "Exploring methods of model simplification." Finally, in Section 18.7, Table 18.1 is reconsidered in the light of the research reported in this book.

18.2 Foundations of Conceptual Modeling

The foundations of conceptual modeling include the definition and aims of conceptual modeling. Research on conceptual modeling is still at the early stages with no agreed definition of exactly which parts of the modeling process are included. Robinson, in Chapter 1, reviews a range of definitions and, indeed, the definitions vary in the different chapters of this book (see also the preface for comments on this). However, as Chapter 1 explains, it is generally agreed that conceptual modeling is concerned with the earliest steps of the modeling process from identifying the problem at the start of the project up to (but not including) building the model. Steps here refer to a logical sequence (e.g., the need to understand the system in order to decide what to include in the model, in order to then build the model) rather than timing. The issues addressed may be reconsidered later in the project and tasks in the project may take place in parallel or with lots of iteration (for example, decisions on the content for one part of the model may be taken while other parts are being built). The initial part of a project involves a number of tasks and the differences in definitions concern exactly which of these should be included under the heading "conceptual modeling." The core of conceptual modeling would seem to be the decisions taken as to which aspects of the system being studied to include and exclude from the model (or as Chapter 1 puts it, "what is going to be modeled and how"). The conceptual model is the combination of these decisions (whether represented explicitly or not) and facets of the conceptual model identified in Chapter 1 are that it is a "simplified representation of the real system" and "independent of the model code or software." The development and general acceptance of a clear definition of conceptual modeling in simulation would be very helpful in facilitating dialogue and avoiding misunderstanding, as well as helping to establish the profile of this important research area.

However, of more practical importance than the definition is to improve the way that the modeling tasks involved in conceptual modeling (or the initial steps in a modeling project) are carried out. A starting point for this is considering the aims of conceptual modeling. In Chapter 2, Brooks relates conceptual modeling to the other tasks in the modeling project and states that the aim in conceptual modeling should be to "select the conceptual model that will lead to the best overall project outcome." Consequently, it is important to consider the interrelationships between conceptual modeling and the other modeling tasks. At the conceptual modeling stage the assessment of alternative conceptual models (i.e., the decisions being considered for what is going to be modeled) should therefore involve predicting the impact on the rest of the project. In particular, this includes how the resulting model, once it is built, will perform and Chapter 2 sets out 11 aspects of model performance (M1). A small-scale experiment is then described (Section 2.8) to investigate the relationship between model characteristics and model performance, in which master's students built different models and aspects of performance were measured (M2). The indications were that size in terms of the number of model elements and connections affects understanding, whereas complex logic affects build time. Ideal characteristics for models are sometimes proposed as part of guidelines for the choice of model and, for example, Chapter 1 sets out four requirements for conceptual models (validity, credibility, utility, and feasibility) (M1). A better understanding of relationships between characteristics and performance should help in the choice of conceptual model and therefore more experiments in the future could provide valuable information.

Also discussed in Chapters 1 and 2 is model simplification, since conceptual modeling can be viewed as deciding which assumptions and simplifications to make regarding the system being studied. Alternative conceptual models can be developed by simplifying existing ones. Although there are a number of papers on simplification methods, more research in this area, particularly on the circumstances in which the different methods are effective, could be informative for conceptual modeling (M6).

Another foundation of conceptual modeling research is to understand, evaluate and learn from current practice. This can include finding out what experts do and comparing experts with novices. Wang and Brooks in Chapter 3 collected data on the time spent on different topics by an expert and nine groups of students (M3). The expert project was a real consultancy project and the student projects were also studies of real systems as part of a university course. This enabled comparison of the expert against the students (novice simulation modelers). There were, however, limitations in some of the data collection and as a result of the projects all being different. Carrying out this sort of research and getting good data is difficult. However, the results are potentially very valuable in learning lessons from experts, identifying differences between experts and novices, and comparing approaches in different domains. A comparison of experts and novices can provide a basis for conceptual modeling teaching (P5). Therefore further research could be carried out following modelers during projects and collecting data on what they do. An experiment where experts and novices tackle the same problem would provide the best comparison of the effect of expertise and experience. Other ways of obtaining data on current practice includes questionnaires or interviews, and some of the results from a survey of simulation experts regarding conceptual modeling and the modeling process are reported in Wang and Brooks (2007).

18.3 Conceptual Modeling Frameworks

A modeling framework provides a specific set of steps that guide a modeler through development of a conceptual model. This characteristic distinguishes modeling frameworks from principles advocating the creation of simple models by an evolutionary model development and/or model pruning (Robinson 2008).

Modeling frameworks are not new, cf. Chapter 1. Shannon (1975) proposes a stepwise approach including (i) specification of the model's purpose, (ii) specification of the model's components, (iii) specification of the parameters and variables associated with the components, and (iv) specification of the relationships between the components, parameters, and variables. Furthermore, Nance and Pace devised frameworks that relate primarily to the development of large-scale models in the military domain (Nance 1994; Pace 1999, 2000). For simulation for modeling of operations systems, work on conceptual modeling frameworks is limited. Recent papers by Guru and Savory (2004) and van der Zee and Van der Vorst (2005) propose conceptual modeling frameworks in some more detail, addressing physical security systems and supply chains, respectively.

In Table 18.2 we summarize the chapters' contributions to the development of conceptual modeling frameworks (M7). The table typifies framework construction by the chapter title, elementary starting points underpinning the research, the domain being addressed, research focus, conceptual modeling activities addressed, actors involved, and resources offered. Conceptual modeling activities are related to the activities, as described by Robinson (Chapter 4). Actors refer to parties involved in conceptual modeling activities, as they are explicitly mentioned and considered by authors. Resources refer to tools, methods, and principles offering support to the actors.

Analysis of Table 18.2 allows us to set some guidelines for future research on the development of conceptual modeling frameworks. As a first step we make some observations on the construction of current modeling frameworks, by comparing their essential characteristics.

- *Research starting points and focus*: frameworks show various alternative choices. The respective choices seem to find their origin in specific actors' needs (P2; M8,10), i.e.:
  - *Modeler*: conceptual modeling activities should be explicitly specified, and ordered. Execution of conceptual modeling activities should be supported by principles, methods, and tools.
  - *Stakeholders (domain experts, staff, etc.)*: conceptual models should serve as a means of efficient and effective communication.
  - *Programmer*: conceptual models should allow for an unambiguous specification of model code.
- *Domain*: most frameworks start from a wide field of application. Furthermore, research with respect to the business and military domain seem to develop along separate lines. The choice of domain is reflected in case examples for illustrating the frameworks' applications. However, little empirical test results are provided.

TABLE 18.2

Conceptual Modeling Frameworks: A Classification

Chapter	Chapter title	Starting points	Domain	Focus	Activities	Actors	Resources
4	A framework for simulation conceptual modeling	• Little discipline in conceptual modeling • Lack of a framework for developing simulation conceptual models of operations systems	Operations systems in business	• Identify a sequence of activities • Associate guidelines and methods with activities	All	• Modelers (novice, experts) • Clients • Domain experts	• Ordering and detailing of key activities • Guidelines and methods supporting activities • Simple means for documenting outcomes of activities
5	Developing participative simulation models—framing decomposition principles for joint understanding	• Need for participative engineering & modeling following from new business configurations and management concepts • Implicit modeling of key logistic elements	Manufacturing, supply chains	• Highlight foundations of insightful modeling, i.e., decomposition principles relevant for a field of interest. • Guide and illustrate construction and application of domain related reference architectures for specifying simulation models	Specify model contents	• Modelers • Stakeholders, especially staff	• Decomposition principles for simulation modeling • Reference architecture for manufacturing and supply chain simulation • Object-oriented notation for model specification

6	The ABCmod conceptual modeling framework	• No language for adequately characterizing behavior of Discrete Event Dynamic Systems because of their diversity and complexity • Practice of directly leaping into the intricacies of computer software instead of the model.	Discrete Event Dynamic Systems	• Defining non-software-specific language for specifying Discrete Event Dynamic Systems, clearly separating structural and behavioral aspects. • Specifying conceptual models at low and high levels of abstraction	Specify model inputs, contents, and outputs.		• Specification language addressing entities and their behavior, underlying data, model inputs and outputs at both a high and low level of detail.
7	Conceptual modeling notations and techniques	• No systematic guidance on how to develop a conceptual model.	Military, business	• Review of approaches, frameworks and methods. • Identify future work opportunities	All	• Modelers • Domain experts	• Description of approaches. • Comparison of approaches for the availability of a stepwise method, notation and tool support.
8	Conceptual modeling in practice: a systematic approach	• No adequate methodology for specification of conceptual models addressing practical needs	Business	• Employing and extending software engineering principles for conceptual modeling starting from the "waterfall software life cycle"	All	• Modelers • Stakeholders	• Specification of requirements document—what the model should do • Specification of the design document—how the model should do it.

- *Activities*: frameworks distinguish among two notions of conceptual modeling activities, i.e., either including or excluding problem understanding and modeling objectives (P1,2).
- *Actors*: approaches differ widely with their notion of actors, i.e., no distinct actors, a single actor, or multiple categories of actors (M7,8).
- *Resources*: there are many opportunities to make good use of software engineering techniques in both detailing activities and supportive tools.

Our interpretation of the above observations suggests a number of implications for future research. Firstly, there is a need for researchers to clarify the demands underlying the construction of their frameworks. Specific issues concern the researchers' choice of domain, modeling requirements (cf. Section 18.2), and their viewpoints on (potential) actors' roles, interests and meaning for the type of projects considered (P2; M1,7,8,10). Consequently, more-focused frameworks may (have) to be developed—and empirically tested—that serve respective actors better.

Frameworks for the military and business domains seem to evolve along separate lines. This suggests a potential for exploiting and combining efforts on constructing conceptual modeling frameworks for applications in military and business, also see Section 18.6. The same is true for the use of software engineering techniques in supporting simulation studies (M4).

Three views on conceptual modeling activities seem to emerge. A first area—not mentioned in Table 18.2—addresses a deepening of the notion of problem understanding and modeling objectives; also see soft systems approaches (Section 18.4, P1). A second area studies conceptual model specification "in the small," i.e., essentially model inputs, contents, and outputs (M5). Finally, a third area combines both types of activities. It starts from a somewhat more procedural focus, by identifying a series of key modeling activities, addressing a wide(r) field of application. In turn, good practices and tools are associated with the respective activities. The approach taken allows for adaptation and extension of frameworks for new insights, domain specifics and alternative uses (M7,8). In principle, all three views offer promising avenues for future research. In part, avenues will concern straightforward extensions of current frameworks—still very much in their infancy. Other avenues follow from a more grounded focus on the domain and actors to be addressed by the frameworks, also see above (P1,2; M1,4,5,7,8).

18.4 Soft Systems Methodology for Conceptual Modeling

Soft systems methodology (SSM) is a method that is applied to problem structuring, and could be used for any type of problem—it is not specifically related to simulation. In a simulation study, problem structuring is usually the first step including identifying the problem, finding out about the system and setting objectives. This leads on to conceptual modeling and some of these tasks may even be regarded as part of conceptual modeling (depending on the definition used) (P1).

In Chapter 9, Pidd provides an overview of problem structuring approaches with a particular focus on SSM. He argues that these can be very important in making sure that the right problem is tackled, and in identifying different stakeholders and their point of view. This can facilitate effective communication with and involvement of the stakeholders, which can be critical for the ultimate acceptance of the modeling results by the stakeholders (M10).

Kotiadis describes the use of SSM in a health care simulation project in Chapter 10. Working with stakeholders, SSM was used to understand the problem situation and, in addition, certain aspects of the SSM process were adapted to produce the simulation objectives as an outcome. The role of problem structuring in helping to develop a good conceptual model is likely to be particularly important in complex projects. Kotiadis argues that the use of SSM can develop a level of trust with the client (M10) and can encourage creativity in the early stages of the modeling process (M9). She also states that there are only a few simulation papers that include the use of SSM, and these are mainly health care applications.

Given the limited literature on SSM in simulation, future research in this area could include reporting and evaluating the use of SSM and other soft operational research (OR) tools in simulation studies to provide greater knowledge of how to use them in the most effective way. It may be that some of the tools need to be adapted in the simulation context and that they can be linked directly with developing aspects of the conceptual model, as in the example described in Chapter 10. Much more research is also needed on the role of creativity in modeling.

18.5 Software Engineering for Conceptual Modeling

There are many tools and approaches that have been developed to assist the process of modeling a system, and a substantial section of the book is devoted to software engineering for conceptual modeling, which might be described as "conceptual engineering." The five chapters (Chapters 11–15) mention many such methods and, for example, Ryan and Heavey in Chapter 12 refer to a survey that listed over a hundred tools (Kettinger et al. 1997). Many of these tools are general or intended for other modeling approaches rather than being designed specifically for discrete-event simulation modeling.

There is some overlap with part II of the book (conceptual modeling frameworks) with, for example, the ABCmod framework of Chapter 6 being an environment (with a software tool that is under construction) as well as a framework. A framework may provide a set of steps or guidelines for conceptual modeling, whereas software engineering is concerned with tools to assist these steps.

In Chapter 11, Liston et al. term the use of tools or techniques for process modeling as a "modelcentric" (as opposed to a documentcentric) approach. They highlight the lack of a standard approach in simulation. Since the Unified Modeling Language (UML) attempts to provide a standard for software development they evaluate the potential of the Systems Modeling Language (SysML), which is based on UML, for conceptual modeling. SysML is a graphical modeling language that enables nine types of diagram to be produced to describe a system. Standardization by using common diagram formats could have benefits in enabling modelers to understand models built by others more easily, which would facilitate reuse, teamwork, and model interoperability (Willard 2007). Some SysML tools allow stepping through diagrams to see the sequence of activities, which can help with validation. However, general limitations and weaknesses of SysML discussed by Liston et al. in Chapter 11 include too much scope for freedom and interpretation by the user and a substantial learning curve to master the language.

Liston et al. describe the retrospective use of SysML for a simulation project. Based on this case study, they found advantages in using SysML for simulation in enabling a modular approach with cross-referencing of information, and in providing good knowledge management by holding the logic of the system. SysML also enables alternative system representations (such as simplifications) to be produced by copying and editing the original, which can then be compared with the original. Limitations identified from use in the case study included not being able to include a sketch of the production facility as one of the diagrams, and the lack of certain elements that are in common use in other diagram formats. Another current limitation is that tools enabling the use of SysML online do not appear to be available as yet.

During conceptual modeling, the modeler needs to understand the system and then decide how to model it. This potentially requires two representations—one of the system and one of the model. Liston et al. suggest that SysML may be suitable for including both representations using its viewpoint and views features.

SysML therefore has potential for assisting with several of the conceptual modeling tasks. However, it is a fairly new tool and so more research is required to assess and evaluate its usefulness for conceptual modeling across a range of applications. This could include the identification of additional requirements and Liston et al. suggest that model libraries may need to be developed for specific domains. There is also scope for research on improving software to make SysML easier to learn and use, and to enable online collaboration.

Ryan and Heavey, in Chapter 12, present their Simulation Activity Diagrams (SAD) for process modeling. They set out criteria for evaluating process modeling tools under the main categories of the ability to describe the different aspects of a discrete-event system, the ease of use and understanding, the ability to use concepts that would be understood by system personnel, and the visualization capability. There is a particular focus here on the tool being easy to use and, with the latter two criteria, on facilitating good communication with and involvement of staff in the organization (M10). From a review of existing process modeling tools (Ryan and Heavey 2006) they found that all the tools they looked at had weaknesses in at least some of these areas. They therefore developed the SAD technique with the aim that it would perform well under each of the criteria as well as supporting project teamwork. The technique produces a SAD diagram that includes entities, resources, states, actions (making up the activities and events), queues, information, and the relationships between all these elements. The technique also has an elaboration language for providing additional information. Ryan and Heavey have also developed a prototype software application, PMS (Process Modeling Software), in which SAD models can be built.

Chapter 12 also describes part of one of the five case studies so far carried out by Ryan and Heavey to evaluate SAD and compares the SAD model with an IDEF3 model for the same case study. They identify various aspects of SAD that require further development including multiple modeling views to enable alternative conceptual models to be built and compared, and a step through facility. It also requires further usage for additional validation and evaluation.

In Chapter 13, Onggo emphasizes the importance of representing the conceptual model in a way that it can be understood easily by the different stakeholders (M10). This matches with some of the criteria of Ryan and Heavey in Chapter 12. Starting from the methods for documenting a conceptual model reported in a survey (Wang and Brooks 2007), he categorizes the methods as textual, pictorial, and multifaceted. The different categories are discussed and evaluated with examples from a generic hospital simulation project.

Some advantages of textual representation are considered by Onggo to be speed and flexibility, but with the disadvantages of potential ambiguity, inability to use mathematical methods for verification, and possible communication problems depending partly upon how well it is written and tailored to the stakeholders. Pictorial representation of conceptual models is usually by diagrams, and activity cycle diagrams, process flow diagrams (using the business process diagram as an example) and event relationship graphs are discussed. In general, a picture can be very effective in helping to communicate complex information. The multifaceted representation consists of several elements and these may be a mixture of text and diagrams. Onggo discusses UML and SysML as examples of this.

He also describes his proposed unified conceptual model (Onggo 2009), which is a multifaceted approach. For describing the problem he proposes the use of an objective diagram with the possible addition of a purposeful activity model for the objectives, an influence diagram for inputs and outputs, a business process diagram with text for the system contents to be included in the model, and text and a data dictionary for data requirements. The model representation depends on the type of model with an activity cycle diagram or event relationship graph suggested for discrete-event simulation, a stock and flow diagram or causal loop diagram for system dynamics, and a flow chart, business process diagram, or UML activity diagram for agent-based simulation. However, much more testing of these methods is required for a more complete evaluation of their usefulness in different circumstances.

Tolk et al. take a different focus in Chapter 14, by concentrating on model interoperability and composition. They define interoperability for two systems as meaning "they are able to work together to support a common objective" and so relates to issues such as software and exchange of data. Composability refers to more abstract modeling issues regarding whether it makes sense to combine models together (e.g., they do not have contradictory assumptions). Based on these definitions they state that "interoperability of simulation systems requires composability of conceptual models." Therefore conceptual model composability is one of the requirements for it to be appropriate and feasible to combine simulation models or systems together.

Tolk et al. consider that the goal in their area of research is for conceptual models to be constructed in a format that enables them to be machine understandable for automatic reasoning about and combining of these models. They describe the Levels of Conceptual Interoperability Model (LCIM), where the addition of each of six successive levels (in this order: technical, syntactic, semantic, pragmatic, dynamic, conceptual) enables greater interoperation. They consider that data engineering addresses the first three of these levels and part of the pragmatic level. It consists of the administration, management, alignment, and transformation of data and they discuss each of these steps. Process engineering deals with the pragmatic and dynamic layers and, similar to data engineering, Tolk et al. suggest the four steps of administration, management, alignment, and transformation. Constraint engineering covers the conceptual level of identifying the assumptions, constraints, and simplifications. For each of the three engineering methods (data engineering, process engineering, constraint engineering) Tolk et al. discuss what is required for the information to be machine readable. However, they conclude that "the solutions provided by current standards as described in this chapter are not sufficient," indicating scope for plenty of further work in this area.

In Chapter 15, Tanriöver and Bilgen discuss the verification and validation of conceptual models built using UML. The general relationships are that the conceptual model can be compared with the real system to assess whether it is a suitable representation for the purpose of the study (validation), the built simulation model can be compared with the conceptual model to assess whether it has been built correctly (verification), and the conceptual model itself can be examined and tested for certain desirable properties such as consistency and lack of redundancy.

Tanriöver and Bilgen review the verification and validation literature for simulation conceptual modeling. They note that the general verification and validation principles and methods for the different simulation tasks can be applied, but that there is a lack of literature on the internal verification of conceptual models. They also review the literature on verification and validation of UML models where there are various lists of desirable properties, and some formal techniques and tools for different aspects of verification. They advocate a number of advantages of an informal inspection approach for verification of UML conceptual models rather than formal methods (such as the inspection approach being easier to apply and understand, and more suitable for the subjective nature of conceptual models where the assessment may require expert evaluation). They then develop a systematic inspection process including looking for specific "deficiency patterns" in the UML diagrams, which might indicate contradictions or redundancy, and carrying out a defined set of inspection tasks. They apply this to two case studies in the military domain. A considerable number of issues with the conceptual models are identified by the process in both cases, demonstrating the potential usefulness of the method. Continuing the research in this area could include identifying required or desirable internal properties for conceptual models in UML or other languages, further development and testing of the inspection process, adding a risk perspective to the process, and the development of software tools to assist the process.

In terms of the research themes of Table 18.1, these five chapters particularly focus on organizing and structuring the information about the conceptual model (P3), the use of software engineering techniques (M4), and model representation methods (M5).

Combining the comments in these chapters, there are various characteristics of software tools that are desirable in some or all circumstances:

- Quick to learn
- Easy to use
- Enable all aspects of the conceptual model to be captured
- Enable the conceptual model to be changed easily
- Enable alternative conceptual models to be compared
- Produce suitable documentation that is easy to understand
- Good visualization features
- Facilitate online collaboration
- Facilitate reuse, interoperability, and composition
- The data are machine readable

There is potential for considerable further research on investigating the relative importance of these characteristics (and possibly others), and testing, evaluating, and developing new and existing tools. Another area is to investigate whether it is feasible, or even desirable, to enable the automatic generation of model code using the software tools.

18.6 Domain-Specific Conceptual Modeling

Part V of the book discusses domain-specific conceptual modeling and considers two important fields of application: military and business. Chapter 16 reviews progress made on conceptual modeling for simulation within the military domain, whereas Chapter 17 addresses approaches for model simplification in semiconductor manufacturing.

In Chapter 16, Pace makes clear how much of the progress on conceptual modeling for simulation in the military domain may be related to discussions within the Simulation Interoperability Standards Organization (SISO). Starting from the respective discussions, he distinguishes between three (interacting) main streams of conceptual modeling, i.e.:

- Functional descriptions of the mission space, i.e., abstraction of the real-world activities associated with a particular mission.
- Simulation conceptual modeling as expressed in the DoD *Recommended Practices Guide (RPG)* for modeling and simulation (M&S) verification, validation, and accreditation (VV&A).
- Conceptual modeling for a collection of simulation applications working in concert.

Starting from a thorough historical review of developments on standards, methods, and tools supporting the three streams, Pace isolates a number of persistent problems impacting simulation related conceptual modeling theory and practice:

- Frequent failure in producing an explicit and distinct conceptual model as an artifact. Pace points out that there may be three reasons for this: (i) a lack of professionalism (P5), (ii) the existence of contracts being unclear on the need for a conceptual model, and (iii) the lack of a widely accepted standard paradigm for simulation development and modification, including the conceptual model as one of its artifacts (M7). This is found especially true for single simulations. For combinations of simulations such a paradigm exists.

- The diversity of simulations hinders guidance in being prescriptive (M7).
- Excessive expectations with respect to the descriptive power of formal approaches. Typically they only work well for specific domains (M4).
- Resource limitations.

Pace's suggestions for future research—derived from the above observations—include approaches being tailored toward more-specific domains and the need for a commonly accepted framework for simulation model development—which includes conceptual modeling as a productive phase (M4,7,8). The precision of a restricted domain may be helpful in specifying the information content of the conceptual model in greater detail. This would allow methods to become more prescriptive, and therefore more supportive in model specification. Composable formal approaches, i.e., approaches which may be tailored to a more narrow field of interest, may underlie such methods (M4,5).

Sprenger and Rose (Chapter 17) address model simplification for semiconductor manufacturing. Semiconductor manufacturing is considered one of the fastest growing industries in the world today, due to a strong growth on the use of integrated circuits for networking, storage components, telecommunications/wireless, consumer, computer, and storage systems. Semiconductor manufacturing systems are among the most complex production systems due to the intricate manufacturing processes involved as well as product variety, and short product life cycles (Mathirajan and Sivakumar 2006). In turn, system complexity sets high demands on simulation model simplification in order to guarantee model feasibility.

Sprenger and Rose make clear how a model of complex systems such as semiconductor fabs may be simplified in a systematic way by leaving out/simplifying shop floor elements (M6). They distinguish among three steps:

1. Redefining the shop configuration by focusing on the bottleneck and/or heavy loaded systems, while replacing the remainder of the shop by one or more dummy machines.
2. Modeling the dummy machines by delay functions expressed in terms of statistical distributions.
3. Calibrating the choice and parameters of distributions based on shop system characteristics such as utilization and product routing.

Note how an iterative pattern is foreseen for steps 1–3, as there may be no a priori fit of the simplified model.

In their study, Sprenger and Rose illustrate the need and means for model simplification in a specific context. Relevance of such approaches is increasing due to growing complexity of business configurations. This suggests room for more elaborate methods of simplification (M6). Typically, they will start from basic techniques for model simplification (cf. Chapters 1 and 2), which are combined, extended, and/or detailed for a specific domain of interest. As an essential feature methods should address procedures for testing the fit of the simplified model.

18.7 The Current State of Conceptual Modeling and Future Research

The work reported in this book addresses most of the research themes listed in Table 18.1 to some extent, as shown by the chapter references in the table. The only themes with no references are P4, M11, and M12.

Theme P4 refers to considering different sources of information about the problem and how they might be used. The chapters on SSM (Chapters 9 and 10) do include methods for problem structuring, and many of the chapters in the book are concerned with how to analyze and organize the conceptual model data. What seems to be missing is the identification of different sources that are or might be used and the benefits and issues with each. A starting point would be the analysis of the information currently used for understanding the problem and the system. There is some information on this in the survey results of Wang and Brooks (2007).

Theme M11 is concerned with how other modeling tasks have an impact on the conceptual model. Many chapters consider the impact of the conceptual model on subsequent tasks, but not the iteration that often takes place whereby the conceptual model is revised as a result of work on later tasks. Again, in fact, there are some results on this in the Wang and Brooks (2007) survey.

Theme M12 looks at comparing how the expected life of a model affects conceptual modeling. This is not explicitly addressed by the research described in this book, but it is implicit in the work being carried out in the different modeling domains—especially military and business. A related theme is comparing different models in the same problem domain (M2), but this is only addressed in the book to a limited extent by a small-scale experiment in Chapter 2.

Some of the other themes only have a small amount of coverage in the book. In particular M9 about exploring creativity in modeling is only cross-referenced to Chapter 9 where it is suggested that SSM can promote creative thinking. Among the other themes, M2 and M3 are also only cross-referenced to one chapter each. By contrast, there are seven themes cross-referenced to five or more chapters (P2, P3, M1, M4, M5, M7, M10).

Overall, the fact that this book includes some research taking place under 14 out of the 17 themes indicates promising progress in conceptual modeling research and reflects greater interest in this area over recent years.

Nevertheless, the discussions in this chapter show that there are many unanswered questions and opportunities for further research under all the themes in Table 18.1. Conceptual modeling is a vital step in any simulation project and research on it has the potential for making great contributions to the success of simulation studies. We hope that this book will help to inspire further research on the important topic of conceptual modeling for discrete-event simulation.

Acknowledgment

Table 18.1 is based on Table 1 in Robinson, S. 2007. The future's bright the future's ... Conceptual modelling for simulation!: Editorial. *Journal of Simulation* 1(3): 149–152. © 2008 Operational Research Society Ltd. Reproduced with permission of Palgrave Macmillan.

References

Guru, A., and P. Savory. 2004. A template-based conceptual modeling infrastructure for simulation of physical security systems. In *Proceedings of the 2004 Winter Simulation Conference,* ed. R.G. Ingalls, M.D. Rossetti, J.S. Smith, and B.A. Peters, 866–873. Piscataway, NJ: IEEE.

Kettinger, W.J., J.T.C. Teng, and S. Guha. 1997. Business process change: A study of methodologies, techniques, and tools. *MIS quarterly* 21(1): 55–80.

Law, A.M. 1991. Simulation model's level of detail determines effectiveness. *Industrial engineering* 23(10): 16–18.

Mathirajan, M., and A.I. Sivakumar. 2006. Literature review, classification and simple meta-analysis on scheduling of batch processors in semiconductor. *International journal of advanced manufacturing technology* 29: 990–1001.

Nance, R.E. 1994. The conical methodology and the evolution of simulation model development. *Annals of operations research* 53: 1–45.

Onggo, B.S.S. 2009. Towards a unified conceptual model representation: A case study in health care. *Journal of simulation* 3(1): 40–49.

Pace, D.K. 1999. Development and documentation of a simulation conceptual model. In *Proceedings of the 1999 Fall Simulation Interoperability Workshop*. http://www.sisostds.org, accessed March 19, 2010.

Pace, D.K. 2000. Simulation conceptual model development. In *Proceedings of the 2000 Spring Simulation Interoperability Workshop*. http://www.sisostds.org, accessed March 19, 2010.

Robinson, S. 2006. Issues in conceptual modelling for simulation: Setting a research agenda. In *Proceedings of the Third Operational Research Society Simulation Workshop (SW06),* ed. J. Garnett, S. Brailsford, S. Robinson, and S. Taylor, 165–174. Birmingham, UK: The Operational Research Society.

Robinson, S. 2007. The future's bright the future's … Conceptual modelling for simulation!: Editorial. *Journal of simulation* 1(3): 149–152.

Robinson, S. 2008. Conceptual modelling for simulation part I: Definition and requirements. *Journal of the operational research society* 59(3): 278–290.

Ryan, J., and Heavey, C. 2006. Process modelling for simulation. *Computers in industry* 57: 437–450.

Shannon, R.E. 1975. *Systems Simulation: The Art and Science*. Englewood Cliffs, NJ: Prentice-Hall.

Wang, W., and R.J. Brooks. 2007. Empirical investigations of conceptual modeling and the modeling process. In *Proceedings of the 2007 Winter Simulation Conference*, ed. S.G. Henderson, B. Biller, M.-H. Hsieh, J. Shortle, J.D. Tew, and R.R. Barton, 762–770. Piscataway, NJ: IEEE Computer Society Press.

Willard, B. 2007. UML for systems engineering. *Computer standards & interfaces* 29: 69–81.

Index

A

C

D

K

L

M

For Product Safety Concerns and Information please contact our
EU representative GPSR@taylorandfrancis.com Taylor & Francis
Verlag GmbH, Kaufingerstraße 24, 80331 München, Germany